The subject of elliptic curves is one of the jewels of nineteenth-century mathematics, originated by Abel, Gauss, Jacobi, and Legendre. This book presents an introductory account of the subject in the style of the original discoverers, with references to and comments about more recent and modern developments. The treatment combines three of the fundamental themes of mathematics: complex function theory, geometry, and arithmetic.

After an informal preparatory chapter on rational functions, Riemann surfaces, and the like, the book follows a historical path, beginning with practical examples of elliptic integrals and the discovery of Abel and Gauss that the inversion of such an integral yields an elliptic function. This is followed by chapters on Jacobi's theta functions, modular groups and modular functions, Abel's and Hermite's work on the quintic, Kronecker and Weber's imaginary quadratic field, and the Mordell–Weil theorem on the rational points of elliptic curves.

Requiring only a first acquaintance with complex function theory, this book is an ideal introduction to the subject for students of mathematics and physics. The many exercises with hints scattered throughout the text give the reader a glimpse of further developments.

ELLIPTIC CURVES

ELLIPTIC CURVES

Function Theory, Geometry, Arithmetic

HENRY McKEAN
New York University

VICTOR MOLL
Tulane University

PUBLISHED BY THE PRESS SYNDICATE OF THE UNIVERSITY OF CAMBRIDGE
The Pitt Building, Trumpington Street, Cambridge CB2 1RP, United Kingdom

CAMBRIDGE UNIVERSITY PRESS
The Edinburgh Building, Cambridge CB2 2RU, United Kingdom
40 West 20th Street, New York, NY 10011-4211, USA
10 Stamford Road, Oakleigh, Melbourne 3166, Australia

First published 1997

Printed in the United States of America

Typeset in Times

Library of Congress Cataloging-in-Publication Data

McKean, Henry P.
Elliptic curves : function theory, geometry, arithmetic / Henry
McKean, Victor Moll.
p. cm.
Includes bibliographical references.
ISBN 0-521-58228-8 (hardcover)
1. Curves, Elliptic. I. Moll, Victor. II. Title.
QA567.2.E44M38 1997 96-36605
516.3'52 – dc20 CIP

A catalog record for this book is available from
the British Library.

ISBN 0-521-58228-8 hardback

In Memory of GRETCHEN WARREN

To Lisa, Alexander, and Stefan

I have always an idea in my mind, a certain confused picture, which shows me, as in a dream, a better form than I have used; but I cannot grasp it and develop it. And even this idea is only on a middling plane. From this I conclude that the works of those rich and great minds of ancient days are very far beyond the utmost stretch of my hopes and imagination. Their writings not only satisfy me to the full, but astound me and strike me with wonder. I appreciate their beauty; I see, if not the whole, at least so much that I cannot possibly aspire to equal it. Whatever I undertake, I owe a sacrifice to the Graces, to obtain their favor.... But they always leave me in the lurch. With me everything is rough; there is a lack of grace, charm, and beauty. I am incapable of making things show for all that they are; my style adds nothing to the matter. That is why my subject must be a solid one, with plenty to grip, and one that shines with its own lustre.

Montaigne, *On Presumption*

Book 2, Chapter 17, *Essays*, trans. J. M. Cohen, Penguin Classics, New York, 1958.

Contents

Preface

The subject of **elliptic curves** such as $y^2 = (1-x^2)(1-k^2x^2)$, with $k^2 \neq 0, 1$, and of the corresponding **elliptic integrals** $\int_0^x y^{-1}dx$ and the **elliptic functions** which invert them, is one of the jewels of nineteenth-century mathematics. Abel, Gauss, Jacobi, and Legendre are the masters here. It is a subject that combines in the most attractive way three of the fundamental themes of mathematics: complex function theory, geometry, and arithmetic. Naturally, it is possible to emphasize just one of these, but that we think is to miss the point, and so we have tried to keep a fair balance among them.

Chapter 1 is preparatory: It deals with rational functions, fractional linear substitutions, projective curves, Riemann surfaces, coverings, and the like, emphasizing low genus 0 or 1, but with glimpses of the higher genera $g = 2$ or more. The presentation is often informal, but what is not actually proved is made plausible, and we thought it better to give a bird's-eye view of the fundamental facts than to get too much involved in the necessary technicalities.

Chapter 2 begins with practical examples of elliptic integrals and the discovery of Abel and Gauss that the inversion of such an integral yields an **elliptic function**, that is, a function of rational character on a complex torus produced by taking the complex plane $\mathbb{C}$ modulo a lattice $\mathbb{L} = \mathbb{Z}\omega_1 \oplus \mathbb{Z}\omega_2$. The remarkable geometric fact emerges that the class of elliptic curves $\mathbf{X}$, such as $y^2 = (1-x^2)(1-k^2x^2)$, and the class of complex tori are one and the same; in particular, $\mathbf{X}$ carries a law of addition of rational character. The general discussion is mostly in the style of Weierstrass, though Jacobi's viewpoint is also explained and that of Legendre is sometimes preferred. The development is illustrated by occasional mechanical and other applications.

Chapter 3 is occupied by Jacobi's **theta functions**. These are not functions on the complex torus $\mathbf{X} = \mathbb{C}/\mathbb{L}$ but rather on its universal cover $\mathbb{C}$, obeying simple transformation rules: $\vartheta(x+1) = \vartheta(x)$, $\vartheta(x+\omega) = f\vartheta(x)$ with a factor $f = e^{a+bx}$. The subject is developed only so far as to permit several striking geometric and arithmetic applications, such as the projective embedding of complex tori, counting the number of representations of a whole number as the sum of two squares, and the spectacular continued fractions of Ramanujan.

Chapter 4 is devoted to the **modular group** $\Gamma_1 = PSL(2, \mathbb{Z})$ and some of its simpler arithmetic subgroups, and to the fundamental cells and absolute invariants attached to them. Quick explanations: Two complex tori are conformally equivalent if and only if their **period ratios** $\omega = \omega_2/\omega_1$ are related by the action $\omega \mapsto (a\omega + b)(c\omega + d)^{-1}$ of a 2×2 integer matrix $[ab/cd]$ of determinant 1.These substitutions form the group Γ_1. The period ratio ω is taken in the upper half-plane, so the quotient of the latter by Γ_1 is a list of conformally inequivalent tori. Dedekind's **absolute invariant** of Γ_1 supplies numbers: It is a function j of rational character in the period ratio ω of $\mathbf{X}$, distinguished by the fact that two such curves $\mathbf{X}$ are conformally equivalent if and only if their absolute invariants match. The Jacobi modulus k^2, viewed as a function of the period ratio, fills the same office of absolute invariant for the modular group Γ_2 of second level comprised of modular substitutions congruent to the identity modulo 2. The highpoint is the **modular equation** relating the absolute invariants of two elliptic curves one of which covers the other. This part stems from Abel, Gauss, and Jacobi and has extraordinary arithmetic applications.

Abel proved that the general quintic is not solvable by radicals. Hermite discovered that the extra ingredient required is, so to speak, a single new "radical," namely, the eighth root of the Jacobi modulus k^2, in terms of which the roots of the modular equation of level 5 may be expressed. Chapter 5 explains this tour de force.

Chapter 6 tells the even more remarkable story, due to Kronecker and Weber, of the **class field** of the imaginary quadratic field $\mathbb{Q}(\sqrt{D})$ for a square-free whole number $D < 0$. The class field is the biggest (unramified) extension of $\mathbb{Q}(\sqrt{D})$ with commutative Galois group and is produced from $\mathbb{Q}(\sqrt{D})$ in the most surprising manner: by the adjunction of certain class invariants $j(\omega)$ attached to the integral ideals of $\mathbb{Q}(\sqrt{D})$. The modular equation plays a central role in this development, too.

Chapter 7 provides a glimpse of the arithmetic of elliptic curves per se; it is devoted to the theorem of Mordell and Weil stating that the rational points of such curves form a module of finite rank over $\mathbb{Z}$.

Prerequisites. Not much is needed besides a first acquaintance with complex function theory: primarily, Cauchy's theorem, Liouville's theorem, power series, residues, and such. Copson [1935], Hurwitz and Courant [1964], or Ahlfors [1979] would be fine; each of these contains, as well, a nice presentation of elliptic functions from the viewpoint of complex function theory, covering part of Chapters 2 and 3. A first acquaintance with the topology of curves and surfaces and with algebraic number fields and Galois theory would be helpful, but is not really necessary. What is needed is reviewed on the spot and you can either believe it or look it up.

Exercises with occasional hints are placed throughout the text. Please take them as an important part of the discussion and do them faithfully. It will pay off.

Acknowledgments. It is a pleasure to thank the audiences at MIT (1958–66), Rockefeller University (1966–9), NYU (1969–82) [H. McKean], and Tulane [V. Moll] (1986,1994) who have listened patiently while we learned the subject. Now one cannot remember who objected to this or clarified that, but any teacher will know how much we owe to them. It is an additional and very particular pleasure to thank P. Sarnak who has kindly read the whole book, with particular attention to the arithmetic parts, and corrected a number of mistakes and misapprehensions. The text owes its elegant appearance to the expertise of Meredith Mickel and the superb copyediting by G. M. Schreiber. The partial support of the National Science Foundation (summers 1981,1994 under grant nos. NSF-MCS 7900813 [H. McKean] and LEQSF(1991–4)-RD-A-31 [V. Moll] is gratefully acknowledged, too.

So. Landaff and New York — *Henry McKean*
New Orleans — *Victor Moll*

1

First Ideas: Complex Manifolds, Riemann Surfaces, and Projective Curves

This chapter presents some elementary (and not so elementary) ideas in continual use throughout the book. For more details and further information see, for example, Ahlfors [1979], Bliss [1933], Clemens [1980], Farkas and Kra [1992], Hurwitz and Courant [1964], Kirwan [1992], Reyssat [1989], Springer [1981], and/or Weyl [1955]. These are all perfectly accessible to beginners; further references will be given as we go along.

1.1 The Riemann Sphere

Let $\mathbb{R}^3$ be the 3-dimensional (real) space of points $x = (x_1, x_2, x_3)$ and let $|x| = \sqrt{x_1^2 + x_2^2 + x_3^2}$ be the distance from x to the origin $o = (0, 0, 0)$. $\mathcal{M}$ is the unit sphere $|x| = 1$ and $\mathbb{C}$ is its equatorial plane $x_3 = 0$, identified as the complex numbers via the map $(x_1, x_2, 0) \mapsto x_1 + \sqrt{-1}x_2$. $\mathcal{M}$ is temporarily punctured at the north pole $n = (0, 0, 1)$ and the rest ($x_3 < 1$) is mapped 1:1 onto $\mathbb{C}$ by the projection p depicted in profile in Fig. 1.1. The rule is: Sight from n through the point $x \in \mathcal{M}$, the projection $p(x)$ being the intersection of this line of sight with $\mathbb{C}$. Obviously, $p(x)$ and $x_1 + \sqrt{-1}x_2$ lie on the same ray of $\mathbb{C}$; also the triangles $n, o, p(x)$ and $n, q = (0, 0, x_3), x$ are similar, so

$$|p(x)| = \frac{\text{distance}[o, p(x)]}{\text{distance}[o, n]} = \frac{\text{distance}[q, x]}{\text{distance}[q, n]} = \frac{\sqrt{x_1^2 + x_2^2}}{1 - x_3}.$$

In short,

$$p(x) = \frac{x_1 + \sqrt{-1}x_2}{1 - x_3}.$$

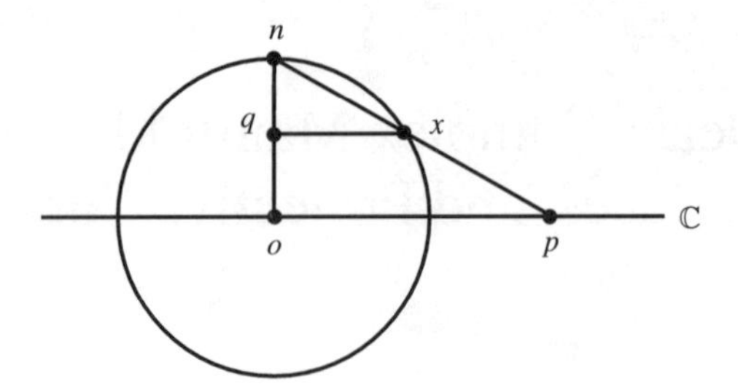

Figure 1.1. The Riemann sphere.

This is the **stereographic projection** of the cartographers. Denote it by p_+ to distinguish it from the analogous projection

$$p_-(x) = \frac{x_1 + \sqrt{-1}x_2}{1 + x_3}$$

of $\mathcal{M} \cap (x_3 > -1)$ produced by sighting from the south pole $(0, 0, -1)$. Now, for $-1 < x_3 < 1$, both maps are available and

$$[p_-(x)]^{-1} = \frac{1 + x_3}{x_1 + \sqrt{-1}x_2} = \frac{1 + x_3}{x_1^2 + x_2^2\,(= 1 - x_3^2)} \times (x_1 - \sqrt{-1}x_2)$$

$$= \frac{x_1 - \sqrt{-1}x_2}{1 - x_3} = [p_+(x)]^*,$$

the star being complex conjugation, so the two images are anticonformally related. Replacing $p_-(x)$ by $[p_-(x)]^*$ produces the following situation: $\mathcal{M}$ is covered by two open patches $U_+ = \mathcal{M} \cap (x_3 < 1)$ and $U_- = \mathcal{M} \cap (x_3 > -1)$, each provided with a **local coordinate**: $z_+ = p_+(x)$ for U_+ and $z_- = [p_-(x)]^*$ for U_-. Most points of $\mathcal{M}$ lie in the overlap $U_- \cap U_+$, and for them the two competing coordinates are conformally related: $z_- = 1/z_+$. This object [$\mathcal{M}+$ patches + projections] is the **Riemann sphere**, alias the extended plane $\mathbb{C} + \infty$, the so-called **point at infinity** being identified with the north pole $n = (0, 0, 1)$.

Exercise 1. Prove that p_+ maps spherical circles into plane circles or lines and vice versa. *Hints*: $x \cdot e = \cos\theta$ marks off a spherical circle for any unit vector e and any angle $0 < \theta \le \pi/2$. Check that $p_+(x) = a + \sqrt{-1}b$ satisfies $(1 - x_3)(ae_1 + be_2) + x_3e_3 = \cos\theta$ and $a^2 + b^2 = (1 - x_3)^{-1}(1 + x_3)$. Then eliminate x_3.

Exercise 2. Prove that the map $p_+: \mathcal{M} \to \mathbb{C}$ is conformal, that is, angle preserving, spherical angles being measured in the natural way.

1.2 Complex Manifolds

A **2-dimensional manifold** or **surface** $\mathcal{M}$ is a geometrical figure that looks in the small like an (open) disk. To be precise, this means three things: (1) $\mathcal{M}$ is a topological space covered by a countable number of open **patches** U. (2) The typical patch U is equipped with a **patch map** $\mathfrak{p} \to x(\mathfrak{p})$ of points $\mathfrak{p} \in U$ to the open unit disk $D = \{(x_1, x_2): x_1^2 + x_2^2 < 1\} \subset \mathbb{R}^2$. This map is 1:1, continuous, and onto; it provides U with **local coordinates** $x = x(\mathfrak{p})$. (3) An ambiguity arises if the point $\mathfrak{p} \in \mathcal{M}$ lies in the overlap $U_- \cap U_+$ of two patches so that two competing coordinates $x_-(\mathfrak{p})$ and $x_+(\mathfrak{p})$ are available; in this case, the composite map $x_-(\mathfrak{p}) \to \mathfrak{p} \to x_+(\mathfrak{p})$, and likewise its inverse, is required to be continuous; see Fig. 1.2.

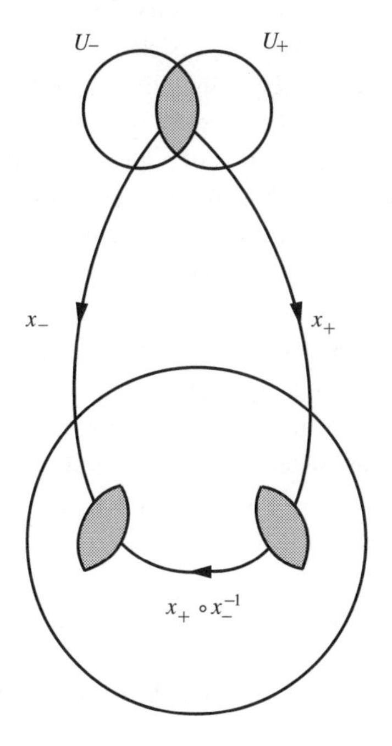

Figure 1.2. Local coordinates on $\mathcal{M}$.

Examples. The plane; the (open) half-plane, disk, or annulus; the sphere or the cylinder; the surface of a doughnut (torus) or a pretzel; the Möbius strip.

- $\mathcal{M}$ is **connected** if it comes in one piece, that is, if any two of its points can be joined by a nice curve; this feature is assumed from now on without further comment.
- $\mathcal{M}$ is **simply connected** if it has no punctures, holes, or handles, that is, if every closed curve (**loop**) can be shrunk to a point in $\mathcal{M}$. This is so for the disk, half-plane, and sphere, but not for the annulus, cylinder, torus, or pretzel.
- $\mathcal{M}$ is **compact** if every cover of it by open sets admits a finite subcover. In this case it is possible to pick a number $0 < r < 1$ so that the images of the closed disk of radius r under a finite number of the patch maps already cover $\mathcal{M}$. This is so for the sphere, torus, and pretzel, but not for the disk, plane, or cylinder.
- $\mathcal{M}$ is **orientable** if the relation between patch maps preserves the sense of (say, counterclockwise) rotation. This is so for the Riemann sphere of Section 1, but is not possible on a Möbius band.
- $\mathcal{M}$ is **smooth** if competing local coordinates x_- and x_+ on overlaps $U_- \cap U_+$ are smoothly related, that is, if $x_-(\mathfrak{p})$ is an infinitely differentiable function of $x_+(\mathfrak{p})$ and vice versa. Then you may speak of smooth functions $f: \mathcal{M} \to \mathbb{R}$, of which you ask that $f(\mathfrak{p})$ be a smooth function of the local coordinate $x(\mathfrak{p})$ on any patch. Plainly, there can be no competition in this regard: On overlaps, $f(\mathfrak{p})$ is a smooth function of both $x_-(\mathfrak{p})$ and $x_+(\mathfrak{p})$ or of neither.
- $\mathcal{M}$ acquires the more subtle structure of a **complex manifold** or **Riemann surface** if the complex local coordinates or **parameters** $z(\mathfrak{p}) = x_1(\mathfrak{p}) + \sqrt{-1}x_2(\mathfrak{p})$ are conformally related on overlaps; orientability is necessary for this. Then it makes sense to speak of the class $\mathbf{K}(\mathcal{M})$ of functions $f: \mathcal{M} \to \mathbb{C} + \infty$ of **rational character** defined by the requirement that in the vicinity of any point $\mathfrak{p}_0$, $f(\mathfrak{p})$ have an expansion $w^d[c_0 + c_1 w + c_2 w^2 + \cdots](d > -\infty, c_0 \neq 0)$ in powers of $w = z(\mathfrak{p}) - z(\mathfrak{p}_0)$. Naturally the expansion changes if the local parameter is changed, but the number d does not, so it is permissible to speak of a **root** of multiplicity d if $d > 0$ and of a **pole** of multiplicity $-d$ if $d < 0$; d is the **degree** of f at $\mathfrak{p}_0$.

Exercise 1. $\mathbf{K}(\mathcal{M})$ is a field.

Exercise 2. Check the statement that the degree d is independent of the local parameter.

Now for some easy examples.

Example 1. It is needless to pause over the complex structure of the plane $\mathbb{C}$ except to note that it has a *global parameter* $z(\mathfrak{p}) = x_1 + \sqrt{-1}x_2$. This example is too simple, as is the disk, half-plane, or annulus, or any other open part of $\mathbb{C}$ which obtains a complex structure by mere inheritance.

Example 2. The cylinder is the quotient of $\mathbb{C}$ by its (arithmetic) subgroup $\mathbb{Z}$, so it, too, obtains a complex structure by inheritance, and likewise the (square) torus, which is the quotient of $\mathbb{C}$ by the lattice $\mathbb{Z} \oplus \sqrt{-1}\mathbb{Z}$; see Fig. 1.3.

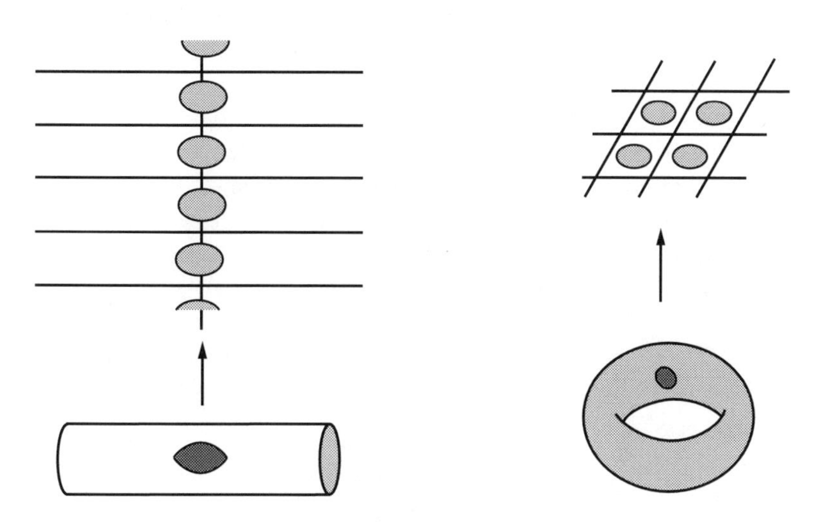

Figure 1.3. Complex structures on the cylinder and the torus.

Example 3. The sphere is more interesting. The stereographic projections of Section 1 provide it with a complex structure after self-evident adjustments; for instance, $z_+(U_+) = \mathbb{C}$ is not the unit disk, but no matter.

Example 4. The **projective line** $\mathbb{P}^1$ is the family of all complex lines in the 2-dimensional complex space $\mathbb{C}^2$: In detail, $\mathbb{C}^2$ is punctured at the origin and two of its points are identified if they lie on the same (complex) line, that is, (a, b) is identified with (a', b') if $a' = ca$ and $b' = cb$ for some nonvanishing complex number c. $\mathbb{P}^1$ is covered by two patches $U_+ = \mathbb{C} \times 1$ and $U_- = 1 \times \mathbb{C}$, provided with self-evident local parameters: $z_+(\mathfrak{p}) = z$ for $\mathfrak{p} = (z, 1) \in U_+$ and $z_-(\mathfrak{p}) = z$ for $\mathfrak{p} = (1, z) \in U_-$; on the overlap $U_- \cap U_+ = \{(a, b) \in \mathbb{C}^2 \colon ab \neq 0\}$, you have the identifications $(a/b, 1) \equiv (a, b) \equiv (1, b/a)$ and so also the relation of local parameters: $z_+(\mathfrak{p}) = [z_-(\mathfrak{p})]^{-1}$. This is the *same rule* as for the Riemann sphere of Section 1. In short, the projective line and the Riemann sphere are identical (as complex manifolds).

Exercise 3. $\mathbb{P}^1$ is compact. That is obvious from its identification with the sphere, but do it from scratch, from the original definition of compactness.

A Little Topology. Let $\mathcal{M}$ be any compact surface, with complex structure or not, and let it be **triangulated** by cutting it up into little (topological) triangles having (in sum) c corners, e edges, and f faces (triangles). Reyssat [1989] has a nice proof that this is always possible. Then (remarkable fact!) the Euler number $\chi = c - e + f$ is always the same: 2 for the sphere, 0 for the torus, -2 for the pretzel, and so forth, that is, it depends only upon the surface and not upon the particular triangulation in hand. This number determines the topology of $\mathcal{M}$ completely. In fact, $\mathcal{M}$ is necessarily a **handlebody**, that is, a (topological) sphere with $g = 1 - (1/2)\chi(\mathcal{M})$ handles attached: 0 for the sphere, 1 for the torus, 2 for the pretzel, and so on; this number is the **genus** of $\mathcal{M}$. Hurwitz and Courant [1964: 497–534] present an elementary proof; see also Coxeter [1980] for more information and Euler [1752] who started it all. The next items are illustrative.

Example. The spherical triangulation seen in Fig. 1.4 has 6 corners, 12 edges, and 8 faces for an Euler number of $6 - 12 + 8 = 2$.

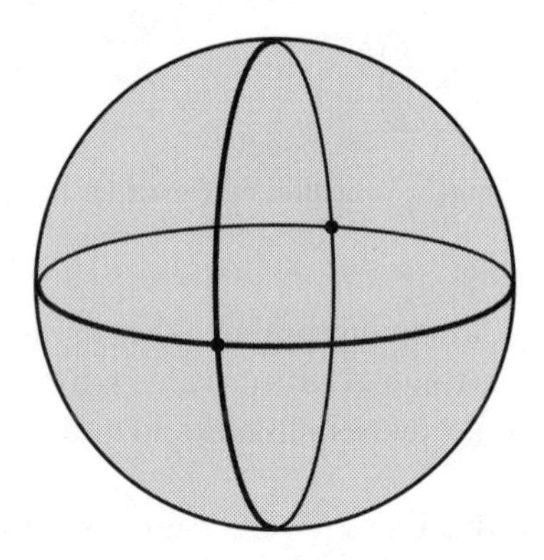

Figure 1.4. Triangulation of the sphere: $c = 6$, $e = 12$, $f = 8$.

Exercise 4. Check by hand that the Euler number and so also the genus of the sphere does not depend upon the triangulation. *Hint*: The sphere can be laid out flat on the plane by cutting all edges that meet at some particular corner. Now count.

Exercise 5. Repeat for higher handlebodies: torus, pretzel, and so on; especially, check that the genus $1 - (1/2)\chi(\mathcal{M})$ really is the handle number.

Preview. It is a fact that any handlebody can be provided with a complex structure, as was already seen for sphere and torus and will appear for the pretzel in Section 12. This can be done in one and only one way for the sphere, but already the torus admits infinitely many (conformally) distinct complex structures; see Section 2.6. It is this unobvious fact that prompted the adjective *subtle* in first speaking of complex manifolds.

1.3 Rational Functions

The function field $\mathbf{K} = \mathbf{K}(\mathbb{P}^1)$ of the projective line is easy to compute: It is just the field $\mathbb{C}(z)$ of rational functions of $z = p_+(x)$.

Proof. $f \in \mathbf{K}$ is a function of rational character of $z_+ = p_+(x)$ on the patch $x_3 < 1$, and likewise of $z_- = [p_-(x)]^*$ on $x_3 > -1$. It follows that f has, in the first patch, a finite number of poles $\mathfrak{p}_i (1 \le i \le m)$, repeated according to their multiplicity, and the possibility of an extra pole of multiplicity n at $\infty =$ the north pole for a total count of $n+m = d$. View $f(\mathfrak{p})$ as a function of $z = p_+(x)$ and let $z_1, \dots, z_m$ be the projections of $\mathfrak{p}_1, \dots, \mathfrak{p}_m$. Then the product $Q(z)$ of $f(\mathfrak{p})$ and $P(z) = (z - z_1) \times \cdots \times (z - z_m)$ is pole-free in $\mathbb{C}$ and of limited growth at ∞: $|Q(z)| \le$ a constant multiple of $|z|^d$ far out. Now use Cauchy's formula for a big circle of radius R:

$$\frac{D^p Q(0)}{p!} = \frac{1}{2\pi\sqrt{-1}} \oint \frac{Q(z)dz}{z^{p+1}}$$

to check that

$$|D^p Q(0)| \le \text{ a constant multiple of } R^d \times R^{-p-1} \times 2\pi R = o(1)$$

for $R \to \infty$ if $p > d$. The upshot is that Q is a polynomial of degree $\le d$ and f is a ratio Q/P, as advertised.

Exercise 1. Check the estimate of Q at ∞.

The **degree** of $f \in \mathbf{K}$ is the total number of its poles, counted according to multiplicity, ∞ included, that is, $\deg f = d = n + m$, and it is plain from its representation as rational function that f has the same number of roots, counted likewise, according to multiplicity. But also $f - c \in \mathbf{K}$ has the same number of poles as f for any complex number c, so f takes on every complex value, ∞ included, d times. In short, d is also the **topological degree** of f as a map of $\mathbb{P}^1$ to itself, taking $f(\mathfrak{p}) = \infty$ at poles. This is a general principle for functions of rational character on compact Riemann surfaces $\mathcal{M}$: As maps of $\mathcal{M}$ to $\mathbb{P}^1$, they take on every value the same number of times; see Section 16.

Exercise 2. Clarify the statement: $f \in \mathbf{K}(\mathbb{P}^1)$ is an analytic map of $\mathbb{P}^1$ to itself.

Exercise 3. Check that the roots and poles of $f \in \mathbf{K}(\mathbb{P}^1)$ can be placed any way you like, provided only that they are the same in number. This is not the case for any other compact Riemann surface: It is already false for the torus; see Section 2.7, item 4.

Exercise 4. Let $\mathfrak{p}_1, \ldots, \mathfrak{p}_n$ be any collection of points on $\mathbb{P}^1$, repetition permitted, and let $\mathfrak{L}$ be the space of functions $f \in \mathbf{K}(\mathbb{P}^1)$ having these poles or softer; for example, f is permitted a pole at $\mathfrak{p}_1$, of degree no more than the number of its repetitions. $\mathfrak{L}$ is a vector space over $\mathbb{C}$. Prove that its (complex) dimension is $n+1$.

Besides its functions of rational character, $\mathbb{P}^1$ also carries **differentials of rational character**. These are the objects ω expressible patchwise as $c(z)dz$ with coefficients c of rational character in the local parameter $z = z(\mathfrak{p})$. If the parameter is changed from $z_+ = z$ to $z_- = w$ on the overlap $U_- \cap U_+$, then the coefficient changes in the natural way, from c to $c \times (dz/dw)$. The differential ω has a root or pole of degree d at the point $\mathfrak{p}_0$ if its coefficient does so. The residue of ω at $\mathfrak{p}_0$ is the integral $(2\pi\sqrt{-1})^{-1} \oint \omega$ taken about a small circle enclosing $\mathfrak{p}_0$. ω is a **differential of the first kind** if it is pole-free, of the **second kind** if it has poles but only vanishing residues, and of the **third kind** otherwise.

Exercise 5. dz is a differential of the second kind on $\mathbb{P}^1$: It has 2 poles at ∞. $df = f'(z)dz$ is likewise of the second kind for any $f \in \mathbf{K}(\mathbb{P}^1)$. $z^{-1}dz$ is different, being of the third kind, in agreement with the fact that the logarithm is not single-valued. Check all that.

Exercise 6. $\mathbb{P}^1$ has no differentials of the first kind besides $\omega = 0$.

Exercise 7. Check that the total degree (roots $-$ poles) of a differential of rational character on $\mathbb{P}^1$ is necessarily -2.

1.4 Luroth's Theorem*

The star means that you may skip this section, but *do* note the following fact which will be useful later: *Any subfield of* $\mathbf{K} = \mathbf{K}(\mathbb{P}^1)$ *containing more than the constant field* $\mathbb{C}$ *is isomorphic to* $\mathbf{K}$ *itself.* This is Luroth's theorem [1876].

Example. The subfield $\mathbf{K}_0$ of functions $f \in \mathbf{K}$ invariant under the involution $z \mapsto 1/z$ is the field $\mathbb{C}(w)$ of rational functions of $w = z + 1/z$. The latter is viewed as a map from one projective line (the cover) to a second projective line (the base); it is of degree 2. The field $\mathbf{K}$ of the cover is likewise of degree 2 over the field $\mathbf{K}_0$ of the base in view of $z = [w \pm \sqrt{w^2 - 4}]/2$. It is this type of counting that is the key to the present proof. It is not the usual proof in that it mixes standard field theory with nonstandard geometric considerations. It is precisely this type of mixture that we want to emphasize in this book. Van der Waerden [1970] presents the standard proof; see also Hartshorne [1977].

A Little Algebra. Not much is needed. The letter $\mathbf{K}$ denotes a field over the rational numbers $\mathbb{Q}$. The degree of a big field $\mathbf{K}$ (the **extension**) over a smaller field $\mathbf{K}_0$ (the **ground field**) is the dimension of $\mathbf{K}$ as a vector space over $\mathbf{K}_0$, denoted by $[\mathbf{K}\colon \mathbf{K}_0] \le \infty$. If the degree is not infinite, then the powers y^n, $n \ge 0$, of an element $y \in \mathbf{K}$ cannot be independent over $\mathbf{K}_0$, so y is a root of some polynomial $P(x) = x^n + c_1 x^{n-1} + \cdots + c_n$ with coefficients from $\mathbf{K}_0$, and y is **algebraic** over $\mathbf{K}_0$. $\mathbf{K}_0[x]$ is the ring of such polynomials. The **field polynomial** of y over $\mathbf{K}_0$ is the irreducible polynomial $P(x) = x^d + c_1^{d-1} + \cdots$ of class $\mathbf{K}_0[x]$ that it satisfies. The extended field $\mathbf{K}_1 = \mathbf{K}_0(y)$ of rational functions of y with coefficients from $\mathbf{K}_0$, obtained by adjunction of y to $\mathbf{K}_0$, is spanned by the d powers $1, y, \ldots, y^{d-1}$; in particular, $[\mathbf{K}_1\colon \mathbf{K}_0] = d$. The roots $x_1 = y, x_2, \ldots, x_d$ of $P(x) = 0$ are necessarily simple. They are adjoined to the ground field $\mathbf{K}_0$ to produce the **splitting field** $\mathbf{K}_2 = \mathbf{K}_0(x_1, \ldots, x_d)$ of $P(x)$. This is the smallest extension of $\mathbf{K}_0$ in which $P(x)$ splits into factors of degree 1: $P(x) = (x - x_1) \cdots (x - x_d)$; it can be realized as the quotient field $\mathbf{K}_0[x]$ modulo $P(x)$. The simplicity of the roots implies that the **discriminant** $\Delta = \prod_{i<j}(x_i - x_j)^2$ does not vanish. This quantity, together with any other symmetric polynomial in the roots, belongs to the ground field $\mathbf{K}_0$. The only other fact that will be needed is that if the extended field $\mathbf{K}$ is obtained from the ground field by the adjunction of n such algebraic elements $y_i (1 \le i \le n)$, then there is a single **primitive element** y_0 that does the job at one stroke: $\mathbf{K} = \mathbf{K}_0(y_0)$. Artin [1953], Lang [1984], Pollard [1950], and/or Stillwell [1994] are recommended as refreshers and for more information.

Exercise 1. Prove directly that the discriminant Δ is, itself, a polynomial in the so-called **elementary symmetric functions**

$$\sigma_1 = \sum x_i, \sigma_2 = \sum_{i<j} x_i x_j, \ldots, \sigma_d = x_1 \cdots x_d.$$

Exercise 2. Deduce $\Delta \in \mathbf{K}_0$.

Exercise 3. What is the discriminant of the general cubic $x^3 + ax^2 + bx + c$?

Aside. Luroth's theorem illustrates, in the simplest circumstances, an important theme of complex geometry: The *complex structure* of a compact Riemann surface $\mathcal{M}$ is determined by the *algebraic structure* of its function field $\mathbf{K}(\mathcal{M})$; see Section 15 under **rational curves** for more information and also Section 2.13 for the case of the torus. The characteristic feature of the rational function field $\mathbf{K} = \mathbb{C}(z)$ is that it is of infinite degree over the ground field $\mathbb{C}$ and isomorphic to any proper intermediate field. As to the geometry of $\mathbb{P}^1$, if $\mathcal{M}$ is a compact complex manifold and if $\mathbf{K} = \mathbf{K}(\mathcal{M})$ is a copy of $\mathbb{C}(z)$ then $\mathbf{K} = \mathbb{C}(f)$ for some distinguished $f \in \mathbf{K}$. Now view f as a map of $\mathcal{M}$ to $\mathbb{P}^1$: It has a degree d just like an ordinary rational function as expounded in Section 3. Besides, it is a fact that $\mathbf{K}$ separates points of $\mathcal{M}$, so $d = 1$ and f maps $\mathcal{M}$ 1:1 onto $\mathbb{P}^1$. In short, as a complex manifold, $\mathcal{M}$ *is* $\mathbb{P}^1$.

Proof of Luroth's theorem. The first item of business is to check that if f_0 is any nonconstant rational function, then the *algebraic degree* of $\mathbf{K} = \mathbb{C}(z)$ over $\mathbf{K}_0 = \mathbb{C}(f_0)$ is the same as the *topological degree* d_0 of f_0. The first degree is finite because $f_0 = a_0/b_0$ with coprime $a_0, b_0 \in \mathbb{C}[z]$ and $P(x) = a_0(x) - f_0(x)b_0(x) \in \mathbf{K}_0[x]$ has $x = z$ as a root; moreover,

$$\begin{aligned} d = [\mathbf{K}\colon \mathbf{K}_0] \le \deg P &= \text{ the larger of the degrees of } a_0 \text{ and } b_0 \\ &= \deg f_0 = d_0. \end{aligned}$$

Now let $P_0 = x^d + s_1x^{d-1} + \cdots + s_d \in \mathbf{K}_0[x]$ be the field polynomial of z over $\mathbf{K}_0$ and observe that *for most values c of f_0* three things happen: (1) c is not a pole of any coefficient $s_n (n \le d)$; (2) $f_0(z) = c$ has d_0 simple roots in $\mathbb{C}$; (3) $z^d + s_1(c)z^{d-1} + \cdots + s_d(c)$ vanishes at each of these. But that makes $d_0 \le d$ and equality prevails: $d_0 = d$.

Now comes the proof of Luroth's theorem itself. Let the intermediate field $\mathbf{K}_1$ lie properly above the constant field $\mathbb{C}$ so that $n = [\mathbf{K}\colon \mathbf{K}_1] \le \min\{\deg f_0\colon f_0 \in \mathbf{K}_1\} < \infty$ and let $P_1(x) = x^n + r_1x^{n-1} + \cdots + r_n \in \mathbf{K}_1[x]$ be the field polynomial of z over $\mathbf{K}_1$. Then r_1 (or some other of its coefficients) is not constant, z being of infinite degree over $\mathbb{C}$. It is to be proved that $\mathbf{K}_1 = \mathbb{C}(r_1)$; see Fig. 1.5.

Now write $r_1 = a_1/b_1$ with $a_1, b_1 \in \mathbb{C}[z]$, and so on, and clear denominators in $P_1(x)$ to produce $P_2(x) = c_0x^n + c_1x^{n-1} + \cdots + c_n \in \mathbb{C}[z][x]$. This divides $P_3(x) = a_1(x)b_1(z) - a_1(z)b_1(x) \in \mathbb{C}[z][x]$ and comparison of degrees with

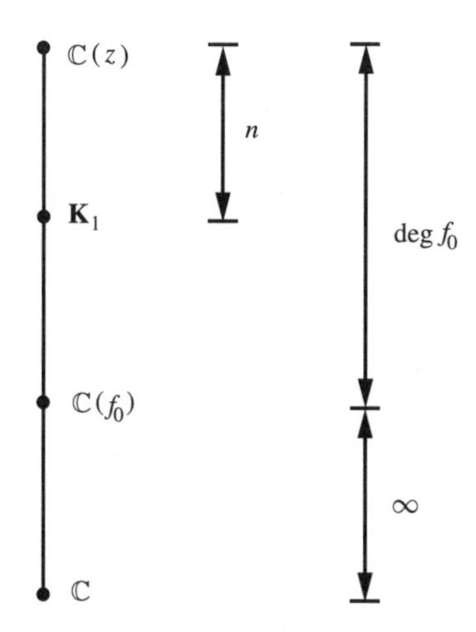

Figure 1.5. The extensions.

regard to z produces

$$\begin{aligned} \deg P_3 &= \text{the larger of the degrees of } a_1 \text{ and } b_1 \\ &= \deg r_1 \\ &\le \text{the larger of the degrees of } c_0 \text{ and } c_1 \\ &\le \deg P_2, \end{aligned}$$

whereupon the divisibility of P_3 by P_2 implies $P_3 = P_2 \times P_4$ with $P_4 \in \mathbb{C}[x]$ independent of z. But $P_4(x)$ divides P_3 only if it divides both $a_1(x)$ and $b_1(x)$, and as these are coprime, so P_4 must be constant, whereupon comparison of degrees with regard to x produces

$$\begin{aligned} [\mathbf{K}{:}\, \mathbb{C}(r_1)] &= \deg r_1 \quad \text{(by the first item of business)} \\ &= \deg P_3 \\ &= \deg P_2 \\ &= [\mathbf{K}{:}\, \mathbf{K}_1]. \end{aligned}$$

The proof is finished: The degrees of $\mathbf{K}$ over $\mathbf{K}_1$ and over $\mathbb{C}(r_1) \subset \mathbf{K}_1$ match only if $\mathbb{C}(r_1) = \mathbf{K}_1$.

Exercise 4. Check that r_1 is of minimal positive degree in $\mathbf{K}_1$. *Hint*: Any other such primitive element of $\mathbf{K}_1$ must be of the same degree n.

1.5 Automorphisms of $\mathbb{P}^1$

An automorphism of a complex manifold $\mathcal{M}$ is a 1:1 analytic map of $\mathcal{M}$ onto itself. The inverse of such a map is likewise analytic, so the automorphisms form a group $\Gamma(\mathcal{M})$. $\Gamma(\mathcal{M})$ is easily identified for $\mathcal{M} = \mathbb{P}^1$: It comprises all the rational functions of degree 1, alias the **fractional linear substitutions** or **Möbius transformations**, of the form

$$z \to \frac{az+b}{cz+d} \quad \text{with} \quad a, b, c, d \in \mathbb{C} \quad \text{and} \quad ad - bc \neq 0.$$

Exercise 1. Explain the proviso $ad - bc \neq 0$.

The numbers a, b, c, d can be scaled to make $ad - bc = 1$ without changing the map, so you may associate to each automorphism the 2×2 complex matrix $[ab/cd]$ of determinant 1. This effects an isomorphism between $\Gamma(\mathbb{P}^1)$ and the **special linear group** $SL(2, \mathbb{C})$ of all such matrices. Actually, that is not quite right: The map is unaffected by the substitution $a, b, c, d \mapsto -a, -b, -c, -d$, so what you really have is an isomorphism with the **projective special linear group** $PSL(2, \mathbb{C})$ which is the quotient of $SL(2, \mathbb{C})$ by its center $(\pm 1)\times$ the identity. The symbol $[ab/cd]$ signifies either the 2×2 matrix or the associated map, as the context requires, $[ab/cd]$ and its negative being identified until further notice.

Exercise 2. Check all that. Chiefly it is required to prove that the association of $\Gamma(\mathbb{P}^1)$ to $PSL(2, \mathbb{C})$ respects the group operations.

Exercise 3. What is $\Gamma(\mathbb{C})$?

Exercise 4. $\Gamma(\mathbb{P}^1)$ is generated by (1) **translations** $z \mapsto z + a$, (2) **magnifications** (including rotations) $z \mapsto bz$, and (3) the **inversion** $z \mapsto -1/z$. Check it.

Exercise 5. Deduce that $\Gamma(\mathbb{P}^1)$ preserves the class of circles and lines.

Exercise 6. $SO(3)$ is the group of proper (i.e., orientation-preserving) rotations of $\mathbb{R}^3$, realized as (real) 3×3 orthogonal matrices of determinant $+1$. Prove that every such rotation is an automorphism of $\mathbb{P}^1$.

$\Gamma(\mathbb{P}^1)$ moves any three distinct points to any other three points, as you will check: For example, $z \mapsto [(b-a)/(c-b)] \times [(c-z)/(z-a)]$ moves a, b, c to $\infty, 1, 0$ with the natural interpretation of the map if one of the points is at infinity.

Table 1.5.1. *Action of* S_3 *on the reduced cross ratio*

(123)	(132)	(321)	(231)	(312)	(213)
x	$1/x$	$1-x$	$1/(1-x)$	$(x-1)/x$	$x/(x-1)$

Exercise 7. The action of an automorphism on three distinct points specifies it completely; especially, it is the identity if it fixes three points. Why?

The (one and only) automorphism $z \mapsto w$ that moves z_1, z_2, z_3 to w_1, w_2, w_3 can be expressed as

$$\frac{z_1 - z_2}{z_2 - z_3} \frac{z_3 - z}{z - z_1} = \frac{w_1 - w_2}{w_2 - w_3} \frac{w_3 - w}{w - w_1}.$$

It moves the additional point z_4 to w_4 if and only if the zs and ws have the same **cross ratio**:

$$\frac{z_1 - z_2}{z_2 - z_3} \frac{z_3 - z_4}{z_4 - z_1} = \frac{w_1 - w_2}{w_2 - w_3} \frac{w_3 - w_4}{w_4 - w_1},$$

with the natural interpretation if any point is at infinity.

Exercise 8. Check that $\Gamma(\mathbb{P}^1)$ preserves cross ratios.

The cross ratio is changed under the action of the **symmetric group** S_4 of permutations of the four "letters" 1, 2, 3, 4. It is invariant under the subgroup

$$(1234) = \text{id}, \ (2143), \ (3412), \ (4321),$$

so to understand the action, it is permissible to place z_4 at infinity and to study the action of S_3(= the symmetric group on three letters) on the reduced ratio $x = (z_1 - z_2)/(z_3 - z_2)$. This is seen in Table 1.5.1.

Exercise 9. $x \neq 0, 1, \infty$. Why?

Exercise 10. Check the table.

The substitutions $x \mapsto x, 1/x, 1-x, 1/(1-x), (x-1)/x, x/(x-1)$ comprise the **group of anharmonic ratios** $\mathfrak{H}$, which is isomorphic to S_3. **Harmonic ratios** arise when two anharmonic ratios of x coincide; the nomenclature goes back to Chasles (1852). The group will play a small but important role later; see Section 7 under **Platonic solids** and also Chapter 4.

Exercise 11. The harmonic ratios are of two kinds: $x = -1, 1/2, 2$ is the *harmonic proportion* of nineteenth-century projective geometry; it corresponds, for example, to the four points $0, 2/3, 1, 2$. Otherwise, $x^2 - x + 1 = 0$, x being a primitive root of -1 such as $\omega = e^{\pi\sqrt{-1}/3}$; the points may now be taken to form an equilateral triangle $1, \omega, 0$ (and ∞). Check that no other harmonic ratios exist.

Exercise 12. Check that the **Schwarzian derivative** $(f')^{-2}[f'f''' - (3/2)(f'')^2]$ commutes with the action of $\Gamma(\mathbb{P}^1) = PSL(2, \mathbb{C})$ on the independent variable; see Ford [1972: 98–101] for more information and for applications of this intriguing object.

1.6 Spherical Geometry

The sphere inherits from the ambient space $\mathbb{R}^3$ its customary (round) geometry, and it is easy to see that the shortest path **(geodesic)** joining two points is an arc of a great circle. Introduce spherical polar coordinates: $x_1 = \sin\varphi\cos\theta$, $x_2 = \sin\varphi\sin\theta$, $x_3 = \cos\varphi$ in which $0 \le \varphi \le \pi$ is the colatitude measured from the north pole and $0 \le \theta < 2\pi$ is longitude. The line element (of arc length) is $ds = \sqrt{(d\varphi)^2 + \sin^2\varphi (d\theta)^2}$. The rotation group $SO(3)$ acts in a distance-preserving way, so you may as well take the first point to be the north pole $n = (1, 0, 0)$. The second point lies on a great circle passing through n, as seen in Fig. 1.6. Plainly, any longitudinal deviation from the great-circle path makes the journey longer.

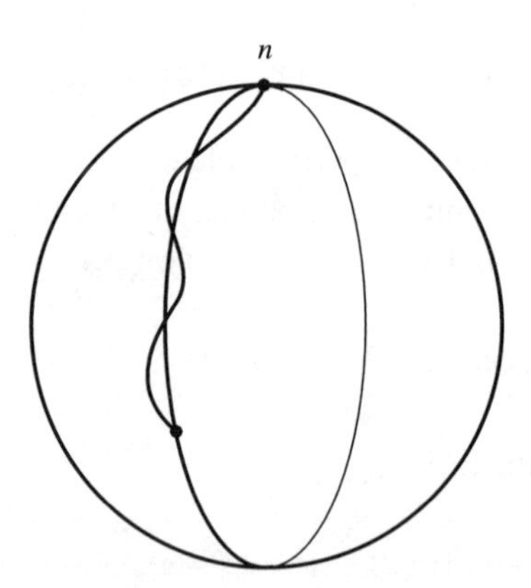

Figure 1.6. Geodesic path on $\mathbb{P}^1$.

Exercise 1. Write this up carefully and find a nice expression for the (shortest) distance.

Exercise 2. Express the spherical distance between two points in the language of the stereographic projection. *Hint*: The line element is $2(1+r^2)^{-1}|dz|$ with $r = |z|$.

Exercise 3. A pair of great circles cuts the spherical surface into two pairs of congruent **lunes**. Check that a lune of angle θ has area 2θ.

Exercise 4. A spherical triangle Δ is formed by joining three points by pieces of great circles. Gauss [1827] found that the area of Δ is the sum of its three interior angles diminished by π. Check this. *Hint*: Each interior angle is marked off by two great circles, determining a lune; see Fig. 1.7.

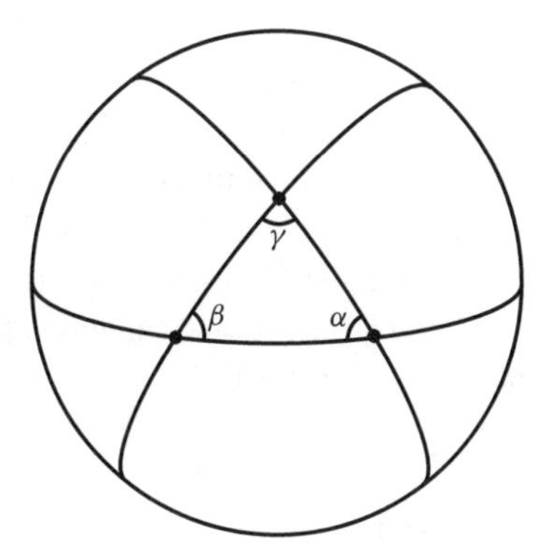

Figure 1.7. Geodesic triangles on $\mathbb{P}^1$.

Exercise 5. Find all the coverings or **tessellations** of the spherical surface by f congruent equilateral triangles, e triangles meeting at each corner. *Hint*: $4/f = 6/e - 1$, the only solutions being $(e, f) = (3, 4), (4, 8), (5, 20)$. To what Platonic solids do these correspond? Compare Section 7.

Exercise 6. Prove that the area of a spherical disk of (geodesic) radius r satisfies

$$A(r) = \pi r^2 \times [1 - r^2/12 + O(r^4)].$$

The number

$$k = \lim_{r\downarrow 0} \frac{12}{\pi} r^{-2}[\pi r^2 - A(r)] = +1$$

measures the deviation of the (round) geometry of the sphere from the (flat) geometry of the plane. It is the **Gaussian curvature** of the sphere; obviously, the plane has curvature $k = 0$; compare Section 9 for the (hyperbolic) geometry of the half-plane with curvature $k = -1$.

1.7 Finite Subgroups and the Platonic Solids

The finite subgroups of $\Gamma = \Gamma(\mathbb{P}^1)$ were already known to Kepler [1596] through their connection with the five **Platonic solids** of antiquity. This will be explained in a moment. The present section serves also as a prototype for the considerations of Chapter 4 about arithmetic subgroups of $PSL(2, \mathbb{R})$; compare Ford [1972: 127–36] and Coxeter [1963].

Let Γ_0 be such a subgroup of order $d < \infty$ and suppose (what is no loss) that no element of Γ_0 fixes ∞, the identity excepted, of course. This can always be achieved by preliminary conjugation of Γ_0 in the ambient group: In fact, the nontrivial elements of Γ_0 have (in sum) at most $2d - 2$ fixed points and almost any conjugation will move them all away from ∞. Let $c \in \mathbb{P}^1 - \infty$ be distinct from any of these fixed points. Then the **orbit** $\Gamma_0 c = \{gc: g \in \Gamma_0\}$ is simple, that is, it is comprised of d distinct points. Pick two such finite points a and b with different orbits. The product $j(z)$ of $(gz - a)(gz - b)^{-1}$, taken over the substitutions $g \in \Gamma_0$, is a rational function of z, of degree d, with three properties. (1) It is invariant under the action of Γ_0. (2) It separates orbits; especially, it has d simple roots and/or poles at the orbit of a/b. (3) Any other (rational) invariant function of Γ_0 is a rational function of j. In short, the field of invariant functions of Γ_0 is $\mathbb{C}(j)$; for this reason, j is called the **absolute invariant** of Γ_0.

Proof of (1). This is self-evident.

Proof of (2). The rational function j is of degree d as it has (simple) poles at the orbit of b and no others. The rest follows from the fact that j takes each value d times and so takes distinct values at distinct orbits of Γ_0.

Proof of (3). Let $\mathbf{K}_0 = \mathbf{K}(\Gamma_0)$ be the field of invariant functions. It is a subfield of $\mathbf{K} = \mathbb{C}(z)$, properly including the constant field $\mathbb{C}$ because j is in it. By Luroth's theorem, $\mathbf{K}_0 = \mathbb{C}(j_0)$ for some $j_0 \in \mathbf{K}_0$ of minimal degree d_0. But this function takes the same value d times on any simple orbit, so $d_0 \geq d = \deg j$. In short, j is of minimal degree and j_0 is a rational function of it, of degree 1.

Exercise 1. The derivative j' of j is of degree $2d$. Why? Note that its roots come from exceptional (nonsimple) orbits.

The next step in the classification of the subgroups of Γ_0 is to describe the family of points fixed by some nontrivial element of Γ_0. It is divided into $h < \infty$ **classes k** according to the action of Γ_0; $n(\mathbf{k}) < \infty$ is the number of points in the class. The function j is invariant, so it takes the same value $j(\mathbf{k})$ at

Table 1.7.1. *Multiplicities*

$m(\mathbf{k}_1)$	$m(\mathbf{k}_2)$	$m(\mathbf{k}_3)$	d
2	2	m	$2m$
2	3	3	12
2	3	4	24
2	3	5	60

each point of $\mathbf{k}$, the latter being a full orbit of Γ_0; moreover, this value is taken each time with the same multiplicity $m(\mathbf{k})$.

Exercise 2. Check that.

The aim of the next few lines is to recompute the degree $2d$ of j' with the aid of these numbers $n(\mathbf{k})$ and $m(\mathbf{k})$ so as to obtain a relation between them. $d = n(\mathbf{k}) \times m(\mathbf{k})$ is plain, j being of degree d. The situation at ∞ has now to be clarified: ∞ is not a pole of j by choice of b, so you have an expansion in powers of the local parameter $1/z$: $j(z) = c_0 + c_1 z^{-1} + \cdots$ in which $c_1 \neq 0$ because the orbit of ∞ is simple. This means that j' has a double root at ∞ for a new count of its degree as per its roots:

$$2d = \deg j' = 2 + \sum_{\mathbf{k}} n(\mathbf{k})[m(\mathbf{k}) - 1],$$

which is to say

$$h - 2 + \frac{2}{d} = \sum_{\mathbf{k}} \frac{1}{m(\mathbf{k})},$$

as you will check using $d = n \times m$. This relation gives rise to a complete list of the possible values of d, h, n, and m. The fact is that *either $d \geq 2$ is arbitrary, $h = 2$, and $m(\mathbf{k}) = d$, or else $h = 3$ and the numbers m fall into one of the four patterns of Table 1.7.1.*

Proof. $m(\mathbf{k}) \geq 2$ (why?) so $h - 2 + 2/d \leq h/2$, and this is contradictory unless $h \leq 3$, that is, $h = 2$ or 3, $h = 1$ being the case $d = 1$, which is trivial. (Why is that?)

Case 1. $h = 2$. There are two classes and $[m(\mathbf{k}_1)]^{-1} + [m(\mathbf{k}_2)]^{-1} = 2/d$. Now $m \leq d$ divides d, and if $m < d$, then it is $\leq d/2$ already and the identity cannot balance. In short, $m = d$ for both classes and each comprises just one point.

Table 1.7.2. *Platonic solids*

Solid	Faces	Edges	Corners	m
tetrahedron	4 triangles	6	4	3
cube	6 squares	12	8	3
dodecahedron	12 pentagons	30	20	5
octahedron	8 triangles	12	6	4
icosahedron	20 triangles	30	12	3

Case 2. $h = 3$. Now $[m(\mathbf{k}_1)]^{-1} + [m(\mathbf{k}_2)]^{-1} + [m(\mathbf{k}_3)]^{-1} = 1 + 2/d$ and if all three multiplicities were three or more, you would have $1 + 2/d \le 1$, so that is out, and *some* class, say the first, has multiplicity two: $m(\mathbf{k}_1) = 2$. $m(\mathbf{k}_2) \le m(\mathbf{k}_3)$ can also be assumed, and you have $[m(\mathbf{k}_2)]^{-1} + [m(\mathbf{k}_3)]^{-1} = 1/2 + 2/d$.

Case 2−. $m(\mathbf{k}_2) = 2$. Then $d = 2m(\mathbf{k}_3)$, which is line 1 of Table 1.7.1.

Case 2+. $m(\mathbf{k}_2) \ge 3$. $m(\mathbf{k}_2) = 4$ or more is contradictory in view of $[m(\mathbf{k}_2)]^{-1} + [m(\mathbf{k}_3)]^{-1} \le 1/2 < 1/2 + 2/d$, so you have $m(\mathbf{k}_2) = 3$ and $[m(\mathbf{k}_3)]^{-1} = 1/6 + 2/d$. It follows that $m(\mathbf{k}_3) = 3, 4$, or 5 producing lines 2, 3, 4 of Table 1.7.1 with $d = 12, 24, 60$. The proof is finished.

Table 1.7.1 displays the possibilities. Now they must be realized concretely. The role of absolute invariant may be played by $(aj+b)(cj+d)^{-1}$ for any substitution $[ab/cd]$ of $\Gamma = PSL(2, \mathbb{C})$, so you may assign the values $j(\mathbf{k}) = 0, 1, \infty$ (or $0, \infty$) to the $h = 3$ (or 2) classes; also, Γ_0 can be conjugated in Γ, permitting you to distribute all (or two) of the points $0, 1, \infty$ among the distinct classes as you will. This freedom permits the group and its absolute invariant to be brought to standard form, with the final result that *each pattern of multiplicities is realized by just one subgroup* $\Gamma_0 \subset \Gamma$, *up to conjugation; in particular, lines 2–4 of Table 1.7.1 are realized by the proper (= orientation-preserving) symmetries of the five Platonic solids of antiquity listed in Table 1.7.2*; compare Fig. 1.8. The idea will now be illustrated in the two simplest cases.

Example 1. $h = 2, m(\mathbf{k}) = d$. The two classes $\mathbf{k}_1$ and $\mathbf{k}_2$ are single points at which j takes its value d-fold; they may be placed at 0 and ∞ and assigned the values $j(\mathbf{k}_1) = 0$ and $j(\mathbf{k}_2) = \infty$. Then $j(z) = z^d$ and the substitutions of Γ_0 fixing, as they do, both 0 and ∞, must be of the form $g\colon z \mapsto \omega z$, ω being a dth root of unity because of the invariance of j, and every one of these roots

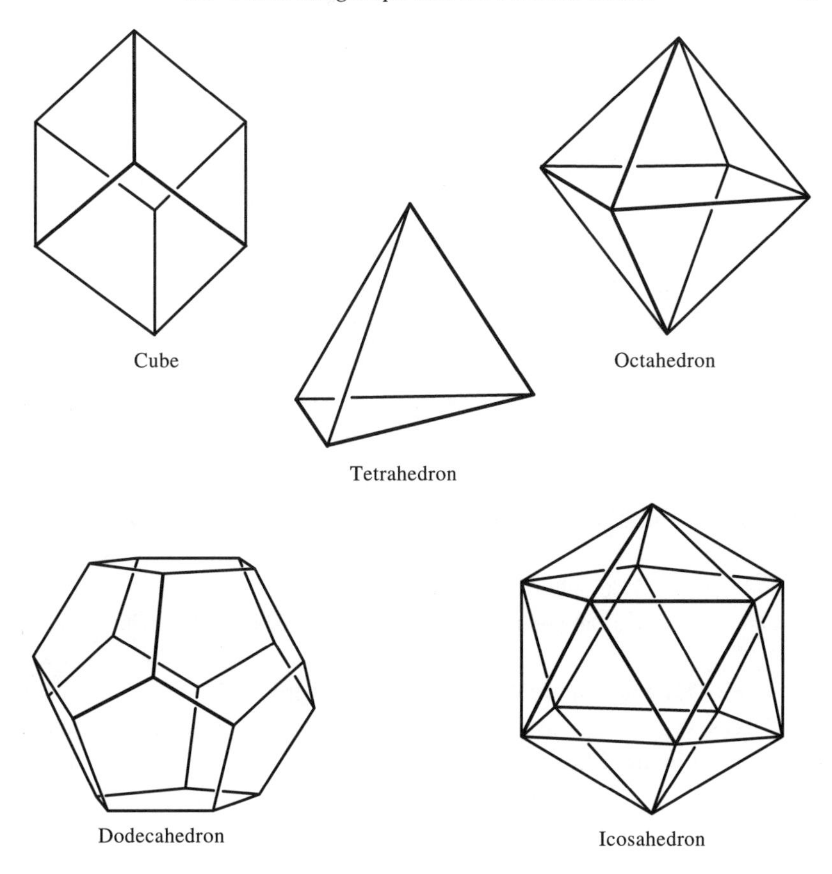

Figure 1.8. The five Platonic solids.

must be employed, Γ_0 having d elements. In short, Γ_0 *is the cyclic group of rotations about the north pole by the* d*th roots of unity.*

Example 2. $h = 3, m(\mathbf{k}_1) = 2, m(\mathbf{k}_2) = 2, m(\mathbf{k}_3) = m, d = 2m$. This is line 1 of Table 1.7.1. Let $j(\mathbf{k}_1) = 0$, $j(\mathbf{k}_2) = 1$, and $j(\mathbf{k}_3) = \infty$ and let $\mathbf{k}_3$, which has just $2 = d/m$ points, be the pair $0, \infty$, so that j has m poles at each of these points. Γ_0 contains a substitution $g \neq$ the identity fixing ∞, and this must also fix 0, $\mathbf{k}_3$ being an orbit of the group. Then g is of the form $z \mapsto \omega z$, and ω can only be an mth root of unity, the invariant function j having m poles at ∞. Any substitution of Γ_0 not of this type still preserves $\mathbf{k}_3$ and so must exchange 0 and ∞, that is, it is of the form $z \mapsto \omega/z$, ω being an mth root of unity, as before. Now $d = 2m$ forces every such substitution to appear in Γ_0. In short, Γ_0 is the *so-called dihedral group comprised of the north-pole rotations by* m*th roots of unity, with the involution* $z \mapsto 1/z$ *adjoined*; it is the symmetry group of a degenerate polyhedron formed by pasting together two copies of the regular m-sided polygon, each divided into m slices of the pie.

Exercise 3. Check that the absolute invariant is $j(z) = z^m + z^{-m}$, up to trivialities.

Exercise 4. The group $\mathfrak{H}$ of anharmonic ratios $x, 1/x, 1-x, 1/(1-x)$, $(x-1)/x, x/(x-1)$ of Section 5 illustrates the case $d = 6$. Use Luroth's theorem to check that

$$j_6(x) = \frac{4}{27} \frac{(x^2 - x + 1)^3}{x^2(1-x)^2}$$

is an absolute invariant and reduce it to $(x^3 + x^{-3} + 2)/4$ by conjugation. The function j_6, harmless as it may look, plays an important role in several subjects discussed later; see especially, Section 2.12 on the classification of complex tori and Chapter 4 on automorphic functions of $PSL(2, \mathbb{Z})$. Quite a big role for such a little function; compare Ford [1972: 127–36] and also Klein [1884].

Example 3: The Platonic solids. A regular polyhedron in 3-dimensional space has f faces, say, and if each face has n sides and if m faces meet at each corner, then it has $e = nf/2$ edges and $c = nf/m$ corners for an Euler number of $2 = f - e + c = f \times (1 - n/2 + n/m)$.

Exercise 5. Check that $n \leq 5$; after all, for $n = 6$, you have the hexagonal tessellation of the plane and do not get a 3-dimensional object at all.

Case 1. $n = 3$. $4m = f(6-m)$ so $3 \leq m \leq 5$, and these choices are all permitted: $m = 3$ with $f = 4$ (for the tetrahedron), $m = 4$ with $f = 8$ (octahedron), and $m = 5$ with $f = 20$ (icosahedron).

Case 2. $n = 4$. $2m = f(4-m)$, the only possibility being $m = 3$ with $f = 6$, that is, a cube.

Case 3. $n = 5$. $4m = f(10 - 3m)$ and only $m = 3$ with $f = 12$ will do (dodecahedron).

Coxeter [1963] presents a beautiful discussion of these issues. Now come the symmetry groups of the solids. Let S_n be the symmetric group of permutations of n letters $\{1, 2, 3, \dots, n\}$ and A_n the alternating group of permutations of parity $+1$, preserving the sign of the square root of the discriminant $\sqrt{\Delta} = \prod_{i<j}(x_i - x_j)$. Line 2 of Table 1.7.2 is realized by the tetrahedral group A_4, line 3 by the group S_4 of the cube or of its dual the octahedron, and line 4 by the icosahedral group A_5 which is also the group of the (dual) dodecahedron. Here *duality* is effected by placing a corner at the center of each face of the original solid.

Exercise 6. Make yourself a cardboard icosahedron and check its group A_5 by hand.

The orders of the groups are easy to compute. If the solid has f faces and each face has e edges, then you can map face 1 to any other and rotate the displaced face in e different ways for a total count of $e \cdot f = 3 \cdot 4 = 12$ for the tetrahedron, $6 \cdot 4 = 24$ for the cube, and $5 \cdot 12 = 60$ for the icosahedron. The tetrahedron has four corners which any symmetry permutes; odd permutations such as (2134) are improper, so the group can only be A_4. The cube has 24 symmetries. These can be enumerated as follows: Corner 1 can be sent to its own (front) face or to the back, and to the bottom or to the top, for a count of four; there are three further possibilities for corner 2 and two for corner 3, for a total count of $4 \cdot 3 \cdot 2 = 24$. This mode of counting identifies the group as S_4. The dodecahedron is a bit more complicated. There are 20 corners which may be divided into four families of five, each family belonging to a single face and labeled 12345. A symmetry maps corner 1 of family 1 to one of the other four families and ascribes to it a new label; this can be done in $4 \cdot 5 = 20$ ways. The two corners of family 1 adjacent to corner 1 can then be placed in three ways for a total count of 60. It follows that the group is part of S_5; it can only be A_5.

Exercise 7. Why?

Now imagine such a Platonic solid inscribed in a sphere and project its edges outward, from the center to the surface, to obtain a tessellation of the sphere. Its symmetries appear as elements of $\Gamma = PSL(2, \mathbb{C})$, so these Platonic groups appear among the subgroups Γ_0; indeed, you now see that these exhaust the possibilities.

Proof for line 2 of Table 1.7.1. $n(\mathbf{k}_3) = 4$ and Γ_0 permutes these four points, its action being determined by what it does to any three of them. $\Gamma_0 \subset S_4$ follows and $d = 12$ identifies it as A_4, S_4 having no other subgroup of index 2.

Proof for line 3. This is the same, only now $d = 24$ is the order of S_4.

Proof for line 4. This is the same as for line 2, only now with $A_5 \subset S_5$.

Klein [1884: 110–43] computed the absolute invariants for the most symmetrical disposition of the solids:

$$j_2(z) = \frac{(z^4 - 2\sqrt{-3}z^2 + 1)^3}{(z^4 + 2\sqrt{-3}z^2 + 1)^3} \qquad \text{for the tetrahedron,}$$

$$j_3(z) = \frac{(z^8 + 14\,z^4 + 1)^3}{108\,z^4(z^4 - 1)^4} \qquad \text{for the cube,}$$

$$j_4(z) = \frac{(-z^{20} + 228\,z^{15} - 494\,z^{10} - 228\,z^5 - 1)^3}{1728\,z^5(z^{10} + 11\,z^5 - 1)^5} \qquad \text{for the dodecahedron.}$$

The phrase *symmetrical disposition* is explained for the dodecahedron: j_4 takes the value 1 with multiplicity 2 at the 30 midpoints of the edges, 0 with multiplicity 3 at the 20 corners, and ∞ with multiplicity 5 at the midpoints of the 12 faces, in accord with line 4 of Table 1.7.1. The claim is that line 4 can be realized only in this way, up to conjugation.

Proof for the tetrahedron. The class $\mathbf{k}_1$ contains six points, of multiplicity 2 each, with assigned value $j(\mathbf{k}_1) = 1$. Let $0, 1, \infty$ be placed in $\mathbf{k}_1$. Then Γ_0 contains a nontrivial substitution $z \mapsto a+bz$ fixing ∞, and you must have $b^2 = 1$ to fix the form of $j(z) = 1+c/z^2+\cdots$. The possibility $b = +1$ is excluded by the fact that the substitution is of order at most 12, so $b = -1$, and conjugation of Γ_0 by the substitution $z \mapsto z - a/z$ reduces a to 0, with the result that j is invariant under the substitution $z \mapsto -z$ and so a function of z^2. Now Γ_0 also contains a substitution $z \mapsto a + b/z$ that sends $0 \in \mathbf{k}_1$ to ∞. This means that, besides 0 and ∞, $\mathbf{k}_1$ also contains the six points $\pm a$, $\pm(a+b/a)$, and $\pm(a-b/a)$ for a total count of eight, which would be too many if they were distinct. Take $a \neq 0$. Then $0 = a \pm b/a$ is one possible conjunction, in which case $b = \pm a^2$, and the class-preserving substitution $z \mapsto a + b/(\pm z) = a + a^2/z$ sends a to $a \times 2, 3/2, 5/3, 8/5, 3/8, 21/13, \ldots$, which is too many points. Otherwise, $a = -(a \pm b/a)$, in which case $b = \pm 2a^2$, and $z \mapsto a + b/(\pm z) = a + 2a^2/z$ sends a to $a \times 3, 5/3, 11/5, 21/4, 43/21, 85/43, \ldots$, which is also wrong, so $a = 0$ and a further conjugation of Γ_0 reduces b to $+1$, with the result that $j(z) = j(1/z)$. Now assign the values $j(\mathbf{k}_2) = 0$ and $j(\mathbf{k}_3) = \infty$ and reflect that each of these classes contains four points of multiplicity 3, whence $j(z)$ is of the form $(z^4 + 2\alpha z^2 + 1)^3/(z^4 + 2\beta z^2 + 1)^3$. It remains to pin down the numbers α and β. Let $z \mapsto (z+a)(z+b)^{-1}$ be a substitution of Γ_0 mapping

∞ to $1 \in \mathbf{k}_1$. Then $a/b \in \mathbf{k}_1$ and of the eight points $0, \infty, \pm 1, \pm a/b, \pm b/a$ not more than six can be distinct.

Case 0. $ab = 0$. Then Γ_0 contains a substitution of the form $z \mapsto 1 + cz$, the number c being a primitive nth root of unity, and this leads to a contradiction in every case $n = 2, 3, 4, 6$ (n divides 12).

$n = 2$: $c = -1$ and that is impossible since, together with $z \mapsto 1 - z$, $z \mapsto -z \mapsto 1 + z$ also belongs to Γ_0.

$n = 3$ produces, from ± 1, four additional points $\pm(1 + c)$, $\pm(1 - c)$, and this is too many.

$n = 4$: $c = \pm\sqrt{-1}$ and $z \mapsto \pm z \mapsto 1 + \sqrt{-1}z$ produces, from ± 1, the distinct points $\pm 1 \pm \sqrt{-1}$.

$n = 6$ produces, from 1, five additional distinct points $1 + c + \cdots + c^m$ $(m \leq 5)$.

Case 1. $1 = a/b$ is not possible, but $1 = -a/b$ *is*, in which case the substitutions known to date produce new points $\pm(1+a)(1-a)^{-1}$ and $\pm(1-a)(1+a)^{-1}$ of $\mathbf{k}_1$ which cannot be distinct. This forces $a^2 = -1$ so, besides $0, \infty$, and ± 1, $\mathbf{k}_1$ also contains $\pm(1 + \sqrt{-1})(1 - \sqrt{-1})^{-1} = \pm\sqrt{-1}$.

Case 2. $1 \neq -a/b$. Then $a/b = \pm b/a$ implies $a = \pm\sqrt{-1}b$, and the same result is obtained: $\mathbf{k}_1$ contains $a/b = \pm\sqrt{-1}$.

It follows that

$$1 = j(\pm 1) = \left(\frac{1+\alpha}{1+\beta}\right)^3 \quad \text{and} \quad 1 = j(\pm\sqrt{-1}) = \left(\frac{1-\alpha}{1-\beta}\right)^3$$

from which follow $\alpha^2 = \beta^2$ and $3\alpha + \alpha^3 = 3\beta + \beta^3$. But $\alpha \neq \beta$ since $j \not\equiv 1$, so $\alpha = -\beta$ and $\alpha^2 = -3$, that is, $\alpha = \pm\sqrt{-3}$ and $\beta = \mp\sqrt{-3}$, as per the formula for $j = j_2$ previously displayed.

Proof for the dodecahedron. The values $j(\mathbf{k}_1) = 1$, $j(\mathbf{k}_2) = 0$, and $j(\mathbf{k}_3) = \infty$ are assigned and the point ∞ is placed in the class $\mathbf{k}_3$. Γ_0 contains a nontrivial substitution A fixing ∞. This must be of the form $z \mapsto a + bz$ with $b^5 = 1$ to fix the pole of $j(z) = cz^5 + \cdots$. Now A is of finite order n and $A^n z = a(1 + b + \cdots + b^{n-1}) + b^n z$ so $b \neq 1$ can only be a primitive fifth root of unity, $b = \omega = e^{2\pi\sqrt{-1}/5}$, say, and you can conjugate the group by $z \mapsto z - a(1-\omega)^{-1}$ to bring A to its simplest form: $z \mapsto \omega z$. Then j is a function of z^5. Now place the point 0 in the class $\mathbf{k}_3$. The group Γ_0 contains a substitution B: $z \mapsto a + b/z$ mapping 0 to ∞. If $a \neq 0$, B will map ∞ to a new point a of $\mathbf{k}_3$. Then $a, \omega a, \ldots, \omega^4 a$ belong to $\mathbf{k}_3$, and new applications of the substitutions A and B produce further points of $\mathbf{k}_3$: $\omega^i a + \omega^j b/a (0 \le i, j < 5)$. But $\mathbf{k}_3$ contains just 12 points, including ∞, so at most 11 of these new points are distinct, and a picture will convince you that this is not possible. The upshot is that $a = 0$ and a further conjugation reduces b to 1 and j to the form

$$\frac{(z^{10} - az^5 + b + a/z^5 + 1/z^{10})^3}{f(z^5 + c - 1/z^5)^5} = \frac{(z^{20} - az^{15} + bz^{10} + az^5 + 1)^3}{fz^5(z^{10} + cz^5 - 1)^5}$$

with undetermined constants a, b, c, and f, in which the top accounts for the 20 points of the class $\mathbf{k}_2$ together with the pole at ∞, and the bottom for $\mathbf{k}_3 - \infty$. The final step is to require that j take the value 1 with multiplicity 2 at each of the 30 points of the class $\mathbf{k}_1$. Now the derived function j' is of degree 70, having four poles at ∞ and six more at each of the other points of $\mathbf{k}_3$; it vanishes 40-fold on class $\mathbf{k}_2$, so the condition to be imposed is that the square of j', with its poles and its roots of the class $\mathbf{k}_2$ removed, should be a constant multiple of

$$(z^{20} - az^{15} + bz^{10} + az^5 + 1)^3 - fz^5(z^{10} + cz^5 - 1)^5.$$

The rest of the computation elicits the values $a = 228$, $b = 494$, $c = 11$, $f = -1728$. This is omitted, as there seems to be no slick way to do it.

1.8 Automorphisms of the Half-Plane

The group $\Gamma(\mathbb{H})$ of automorphisms of the open upper half-plane $\mathbb{H} = \{z = x_1 + \sqrt{-1}x_2 : x_2 > 0\}$ may be identified with the real special linear group $PSL(2, \mathbb{R})$.

Proof. Let g be an automorphism of $\mathbb{H}$ mapping $\sqrt{-1}$ to $a + \sqrt{-1}b\,(b > 0)$ and observe that $g_1 = b^{-1}(g - a)$ fixes $\sqrt{-1}$. Let h be the standard map $z \mapsto (z - \sqrt{-1})(z + \sqrt{-1})^{-1}$ of $\mathbb{H}$ to the disk $|h| < 1$. Then $g_2 = h \circ g_1 \circ h^{-1}$ is an automorphism of the disk fixing the origin and $|g_2(z)/z| \le 1$ by application of the maximum modulus principle on the perimeter. The same idea applies to the inverse map g_2^{-1}, so the reciprocal modulus $|z/g_2(z)| = |g_2^{-1}(z')/z'|$ with $z' = g_2(z)$ is likewise ≤ 1. The upshot is that $g_2(z)/z$ is of modulus $\equiv 1$ and so must be constant. This pretty trick is due to H. A. Schwartz. The rest is computation if you like. A better way is to note that $g = [ba/01] \circ h^{-1} \circ g_2 \circ h$ is a fractional linear substitution preserving the completed line $\mathbb{R} + \infty$ bordering $\mathbb{H}$ and to deduce that it belongs to $PSL(2, \mathbb{R})$.

Exercise 1. Do it.

Exercise 2. Check that a nontrivial element $[ab/cd]$ of $\Gamma(\mathbb{H}) = PSL(2, \mathbb{R})$ is conjugate to (1) a magnification ($d = 1/a, b = c = 0, a > 0$), (2) a translation ($a = d = 1, c = 0, b \in \mathbb{R}$), or (3) a rotation ($a = d = \cos\theta, -b = c = \sin\theta, \theta \neq 0, \pi$) according as the absolute value of its trace is $> 2, = 2$, or < 2.

Exercise 3. Use ex. 2 to prove that any subgroup of $PSL(2, \mathbb{R})$ isomorphic to $\mathbb{Z}^2$ comes as close to the identity as you like. *Hint*: What commutes with a translation *is* a translation.

Exercise 4. Show that $\Gamma = PSL(2, \mathbb{R})$ will move one point of $\mathbb{H}$ to $\sqrt{-1}$ and any *second* point to some position on the imaginary half-line.

1.9 Hyperbolic Geometry

This is a model of the non-Euclidean geometry discovered by Bolyai, Lobatchevsky, and Gauss about 1820; it is connected in a beautiful way to $PSL(2, \mathbb{R})$ and $\mathbb{H}$. The latter can be equipped with the **hyperbolic line element** $ds = x_2^{-1}\sqrt{dx_1^2 + dx_2^2}$ of Liouville and Beltrami (1868) and Klein (1870), which was rediscovered by Poincaré [1882] and usually called by his name, relative to which the vicinity of each point looks like a mountain pass or saddle point: two ridges rising, one on either hand, two valleys falling away. The geodesics of this geometry are semicircles with centers on the bordering line $x_2 = 0$ and/or vertical lines, and $\Gamma(\mathbb{H}) = PSL(2, \mathbb{R})$ is its group of proper (= orientation-preserving) rigid motions, playing the same role as the rotation group $SO(3)$ for the (round) spherical geometry of Section 6 or the Euclidean motion group (translations and rotations) for the (flat) geometry of $\mathbb{R}^2$. See Milnor [1982] for historical remarks, Pogorelov [1967] and Beardon [1983] for further information, and Section 4.9 for a deep connection to complex function theory.

Exercise 1. Check that the hyperbolic line element is invariant under the action of $\Gamma(\mathbb{H})$.

Now let $\sqrt{-1}a$ and $\sqrt{-1}b$ ($0 < a < b$) be two points of the imaginary half-line. Inspection of the line element shows that any deviation from the vertical path joining them makes the journey longer, so *that* is the geodesic, and a self-evident application of ex. 8.4 confirms the statement made before that the general geodesic is either a semicircular arc meeting $\mathbb{R}$ at 90° or else a vertical line. The identification of $\Gamma(\mathbb{H})$ as the rigid motions is a by-product.

Exercise 2. The hyperbolic distance from $\sqrt{-1}a$ to $\sqrt{-1}b$ is $\int_a^b x_2^{-1}dx_2 = \log(b/a)$. Confirm the general formula for the distance between any two points of $\mathbb{H}$: $\cosh(d(p,q)) = 1 + (1/2)\times$ the flat distance between p and q, squared, divided by the product of their heights. *Hint*: The right-hand side had better be invariant under the action of $\Gamma(\mathbb{H})$.

Exercise 3. Check that the circle of hyperbolic radius r, centered at $\sqrt{-1}$, is the same as the flat circle of radius $\sinh r$, centered at $\sqrt{-1}\cosh r$.

The hyperbolic area inside this circle is

$$A(r) = \int\int \frac{d(\text{flat area})}{(\text{height})^2} = \int_0^{\sinh r}\int_0^{2\pi} \frac{\rho d\rho d\theta}{(\cosh r + \rho \sin\theta)^2}$$

and

$$k = \lim_{r\downarrow 0} \frac{12}{\pi} r^{-2}[\pi r^2 - A(r)] = -1$$

is the **curvature**; compare ex. 6.6. This completes the zoo of geometries: the flat plane ($k = 0$), the round sphere ($k = +1$), and the half-plane like a mountain pass ($k = -1$).

Exercise 4. Do the computation of $k = -1$.

A down-to-earth picture of curvature $k = -1$ is obtained from the simple surface of revolution $y = h(x)$ seen in Fig. 1.9. The so-called principal curvatures

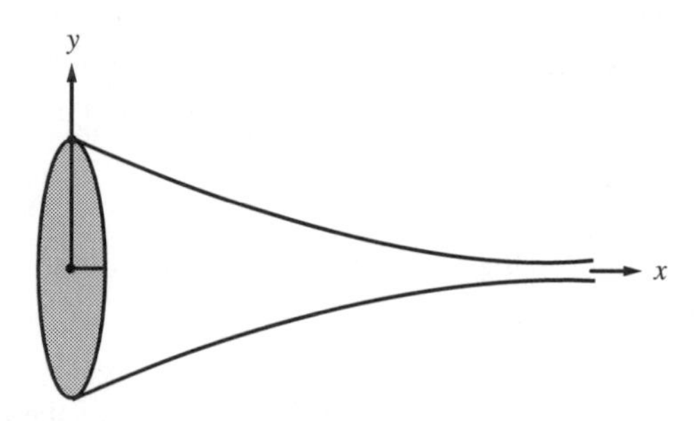

Figure 1.9. A surface with curvature -1.

are $k_1 = -h''[1 + (h')^2]^{-3/2}$ in the plane of the paper and $k_2 = 1$ in the perpendicular plane. Their product is the Gaussian curvature k, so the condition $k = -1$ becomes $h'' = [1 + (h')^2]^{3/2}$. The value $k = -4$ is more convenient:

It produces a convex function starting at $h(0) = 1$ and vanishing at ∞, expressible as

$$x = -\frac{1}{2}\log\left(\frac{1-\sqrt{1-h^2}}{h}\right) - \sqrt{1-h^2}.$$

Exercise 5. Check this.

Exercise 6. Compute the hyperbolic area of a geodesic triangle with interior angles α, β, γ.

Pogorelov [1967: 127–67] is recommended as an introduction to this circle of ideas; see also Stillwell [1992] which is nice and elementary.

1.10 Projective Curves

Clemens [1980] presents a splendid account of this subject; Kirwan [1992] is fine, too. The projective line $\mathbb{P}^1$ you know. The projective plane $\mathbb{P}^2$ is similarly defined: It is the space of complex lines in $\mathbb{C}^3$ obtained by identifying two points of $\mathbb{C}^3 - 0$ that differ by a nonvanishing complex multiplier. It is covered by three patches $U_1 = \mathbb{C} \times \mathbb{C} \times 1$, $U_2 = \mathbb{C} \times 1 \times \mathbb{C}$, and $U_3 = 1 \times \mathbb{C} \times \mathbb{C}$, and, with a self-evident extension of the terminology of Section 2, it is a compact manifold of (complex) dimension 2. The higher projective spaces $\mathbb{P}^d$ $(d \geq 3)$ follow the same pattern.

Exercise 1. A *line* in $\mathbb{P}^2$ is determined by the vanishing of a linear form $ax + by + cz = 0$ with fixed $(a, b, c) \in \mathbb{C}^3 - 0$. Check that it is nothing but a copy of the projective line $\mathbb{P}^1$.

To explain projective curves, think first of the *real* quadratic curves (conic sections) of the schoolroom. They come in three varieties: circle, hyperbola, and parabola, typified by $x^2 + y^2 = 1$, $xy = 1$, and $x^2 = y$; see Fig. 1.10. A different and, in many respects, a superior picture is obtained by complexifying x and y; it is even better to compactify everything, by taking a projective standpoint in $\mathbb{P}^2$, so as to take into account what is happening at *infinity*. The idea is due to Bezout [1779]. The case of the circle will convey the idea. The old curve $x^2 + y^2 = 1$ in $\mathbb{R}^2$ is replaced by the locus $x^2 + y^2 + z^2 = 0$ in $\mathbb{C}^3 - 0$. This makes projective sense, each term being of the same degree, and so defines a **projective curve** in $\mathbb{P}^2$ of 1 complex (= 2 real) dimension. The old curve (and more) is seen in the patch $\mathbb{C}^2 \times \sqrt{-1} = \mathbb{C}^2 \times 1$, projectively, but

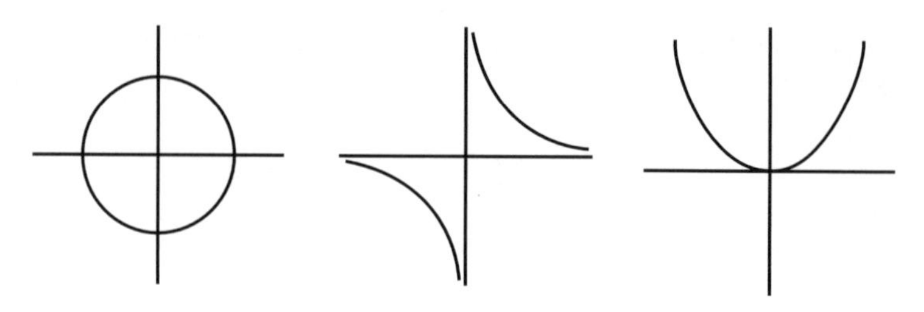

Figure 1.10. Quadratic curves.

now you pick up two new (compactifying) points "at infinity": $(1, \pm\sqrt{-1}, 0)$ in $(1 \times \mathbb{C}^2) \cap (\mathbb{C} \times 1 \times \mathbb{C})$. The same recipe applies to the projective hyperbola $xy - z^2 = 0$ and to the parabola $x^2 - yz = 0$, which are evidently the same in $\mathbb{P}^2$. In fact, *all three types of curves fall together into a single projective class.* The (1:1 projective) substitution

$$x \mapsto x + \sqrt{-1}y, \ \ y \mapsto x - \sqrt{-1}y, \ \ z \mapsto \sqrt{-1}z$$

converts $xy = z^2$ into $x^2 + y^2 + z^2 = 0$. In short, up to such substitutions, the conic sections are indistinguishable in $\mathbb{P}^2$. But what does the circle $\mathbf{C}$: $x^2 + y^2 + z^2 = 0$ really look like in $\mathbb{P}^2$? The pretty answer is: *the projective line*. The proof is easy. $\mathbb{P}^1$ is viewed as $\mathbb{C} + \infty$ and provided with the parameter w. Then $x = (1/2)(w + 1/w)$ and $y = (1/2\sqrt{-1})(w - 1/w)$ solve $x^2 + y^2 = 1$ provided $w \neq 0, \infty$. This presents, in a 1:1 manner, the finite part of $\mathbf{C}$ that lies in the patch $\mathbb{C}^2 \times \sqrt{-1}$: in fact, $w = x + \sqrt{-1}y$. To cope with the points at ∞, the correspondence $[x, y, z] = [(w + 1/w)/2, \sqrt{-1}(w - 1/w)/2, \sqrt{-1}]$ is expressed in the projectively equivalent forms $[(1 + 1/w^2)/2, (1 - 1/w^2)/2\sqrt{-1}, \sqrt{-1}/w]$ and $[(w^2 + 1)/2, (w^2 - 1)/2\sqrt{-1}, \sqrt{-1}w]$: The first places the north pole $w = \infty$ in correspondence with $[1/2, 1/2\sqrt{-1}, 0] = [1, -\sqrt{-1}, 0]$ projectively; the second places the south pole $w = 0$ in correspondence with the second point at infinity $[1, \sqrt{-1}, 0]$.

Moral. The recipe places the whole projective circle in faithful (rational) correspondence with the projective line; in particular, *the totality of real solutions of* $x^2 + y^2 = 1$ *is obtained by numerical specialization of the rational functions* $x = \frac{1}{2}(w + 1/w)$ *and* $y = \frac{1}{2\sqrt{-1}}(w - 1/w)$. Do you recognize them?

More generally, any irreducible polynomial $P \in \mathbb{C}[x, y]$ defines, by its vanishing, a projective curve $\mathbf{X}$ in $\mathbb{P}^2$: The individual terms $x^m y^n$ of P are brought up to a common (minimal) degree d by powers of a new variable z. Then the vanishing of P makes projective sense, and the rest is as before; in particular, the original curve $\mathbf{X}_0 = \mathbb{C}^2 \cap \{P(x, y) = 0\}$ is seen in the patch $\mathbb{C}^2 \times 1$. $\mathbb{P}^2$ is compact and $\mathbf{X}$ inherits that; it is also connected, automatically, but that lies deeper.

Amplification 1. $\mathbf{X}$ need not be a projective line; that is the exception, not the rule. For example, if e_1, e_2, e_3 are distinct complex numbers, then the cubic $y^2 = (x - e_1)(x - e_2)(x - e_3)$ is a torus in $\mathbb{P}^2$ and cannot be identified with $\mathbb{P}^1$ for topological reasons; see Section 1.12 for pictures and Section 2.11 for a full explanation of this striking geometric fact.

Amplification 2. $\mathbf{X}$ inherits from $\mathbb{P}^2$ the structure of a complex manifold, with exceptions at a few places. Let $P \in \mathbb{C}[x, y, z]$ be the homogeneous polynomial that defines $\mathbf{X}$ by its vanishing. Then, with the notation $P_1 = \partial P/\partial x$, and so forth, the form $P_1 dx + P_2 dy + P_3 dz$ vanishes on $\mathbf{X}$, so if, for example, you place yourself in the patch $\mathbb{C}^2 \times 1$ and if (P_1, P_2) does not vanish, then you can get rid of $dz\,(z = 1)$ and use the implicit function theorem in the small to solve for y in terms of x (or vice versa), placing a whole patch of neighboring points $\mathfrak{p} = (x, y) \in \mathbf{X}$ in faithful correspondence with a little disk by means of the *local parameter* $x = x(\mathfrak{p})$; see Section 13 for more details from another point of view. The exceptions alluded to are the **singular points** of $\mathbf{X}$ at which the gradient (P_1, P_2, P_3) vanishes. If the gradient never vanishes the curve is termed **nonsingular**.

Exercise 2. Check that the circle $x^2 + y^2 + z^2 = 0$ is nonsingular.

Exercise 3. The cubic $\mathbf{X}$: $y^2 = x^2 + x^3$ has two singular points: the origin and also $(0, 0, 1)$, as you will check.

This example is more typical. At $x = 0$, the two analytic branches $y = \pm x\sqrt{1 + x}$ of $\mathbf{X}$ cross, accidentally so to speak, so, in the small, the curve looks like two complex lines touching at one point and not like a disk; see Fig. 1.11. This can be cured. The branches are distinguished at $x = 0$ by their slopes $y' = \pm 1$, which suggests a new attitude toward $\mathbf{X}$, described afresh as the common roots (in $\mathbb{P}^3$) of the *two* relations $xz = y$ and $z^2 = 1 + x$; in fact, $z = y/x$ takes distinct values ± 1 at $x = 0$, depending on the branch, so the old point $(0, 0)$ splits into two separate points $(0, 0 \pm 1)$. The present discussion is just to whet the appetite; it is continued in Section 11, and also in Section 13. For further information, Bliss [1933], Walker [1978], Kirwan [1992], Shafarevich [1977], and Mumford [1976] are recommended, in order of sophistication.

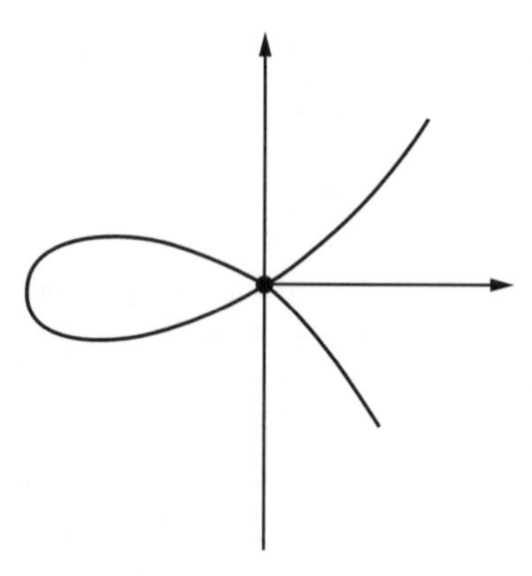

Figure 1.11. Crossing of branches.

1.11 Covering Surfaces

We review a few topological facts needed for the further study of projective curves and/or Riemann surfaces. The ideas go back to Poincaré. Ahlfors [1973], Forster [1981], and Massey [1991] are recommended for more information.

Covering Spaces. Let $\mathcal{M}$ and $\mathcal{K}$ be surfaces, that is, topological manifolds of real dimension 2. $\mathcal{K}$ is an (unramified) **cover** of $\mathcal{M}$ if it admits a **projection** p onto $\mathcal{M}$ with the characteristic feature that every point $\mathfrak{p} \in \mathcal{M}$ has an open neighborhood U such that $p^{-1}(U)$ breaks up into disjoint open pieces V of $\mathcal{K}$, finite or countable in number, the restrictions $p: V \to U$ being homeomorphisms; see Fig. 1.12. The cardinality of the **fiber** $p^{-1}(\mathfrak{p})$ is independent of the point $\mathfrak{p} \in \mathcal{M}$; it is the **degree** or **sheet number** of the cover.

Exercise 1. Check that $p(z) = z^k (k \in \mathbb{N})$ is a projection from $\mathbb{C} - 0$ to itself. What happens if 0 is included?

Exercise 2. Check that the exponential $\exp: \mathbb{C} \to \mathbb{C} - 0$ is a projection.

Exercise 3. Let ω be a complex number of positive imaginary part and let $\mathbb{L}$ be the lattice $\mathbb{Z} \oplus \omega\mathbb{Z}$. The map $p: \mathbb{C} \to \mathbf{X} = \mathbb{C}/\mathbb{L}$, reducing the plane modulo the lattice $\mathbb{L}$, induces a topology on the torus $\mathbf{X}$. Check that p is a projection.

Universal Cover. Among all the surfaces that cover $\mathcal{M}$, there is a largest one. This is its **universal cover** $\mathcal{K}$, distinguished by the property that if $\mathcal{K}_0$ is any other cover of $\mathcal{M}$, then $\mathcal{K}$ covers $\mathcal{K}_0$. $\mathcal{K}$ is unique up to homeomorphisms and

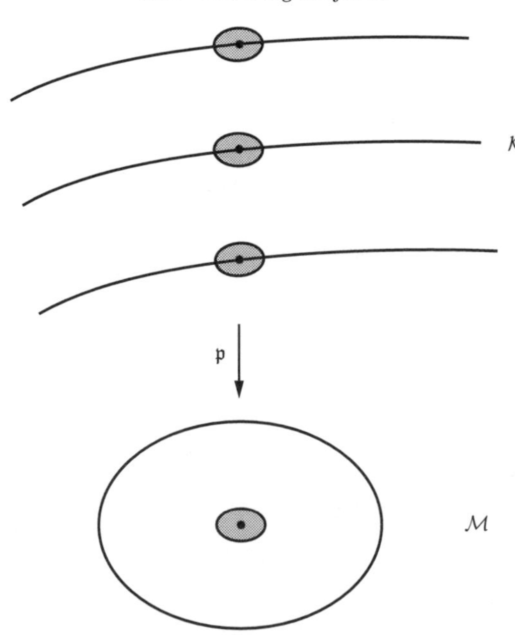

Figure 1.12. Covering spaces.

simply connected, that is, any closed loop in $\mathcal{K}$ can be shrunk to a point. This feature, too, distinguishes the universal cover among its competitors.

Exercise 4. The universal cover of the torus $\mathbf{X} = \mathbb{C}/\mathbb{L}$ is the plane $\mathbb{C}$, by ex. 3. What is the universal cover of the disk, annulus, sphere, cylinder, and the once-punctured plane?

Lifting and Covering Maps. The universal cover $\mathcal{K}$ is provided with a group $\Gamma(\mathcal{K})$ of **covering maps**. These are homeomorphisms of $\mathcal{K}$ that commute with the projection $p: \mathcal{K} \to \mathcal{M}$; in particular, covering maps preserve fibers. Let $\mathfrak{o}$ be a point of $\mathcal{M}$, fix a point $\mathfrak{o}_1$ of $\mathcal{K}$ covering it, and select a loop in the base starting and ending at $\mathfrak{o}$, as in Fig. 1.13. The beginning of the loop lifts unambiguously to a patch about $\mathfrak{o}_1$ via the inverse projection, and this lifting can be continued without obstruction until the moving base point returns to $\mathfrak{o}$. The lifted loop ends at a point $\mathfrak{o}_2$ covering $\mathfrak{o}$ and the map $\mathfrak{o}_1 \mapsto \mathfrak{o}_2$ may be extended to a covering map of the whole of $\mathcal{K}$ by a self-evident continuation.

Exercise 5. Check all that by means of pictures. Think of an example in which the lifted curve is not closed.

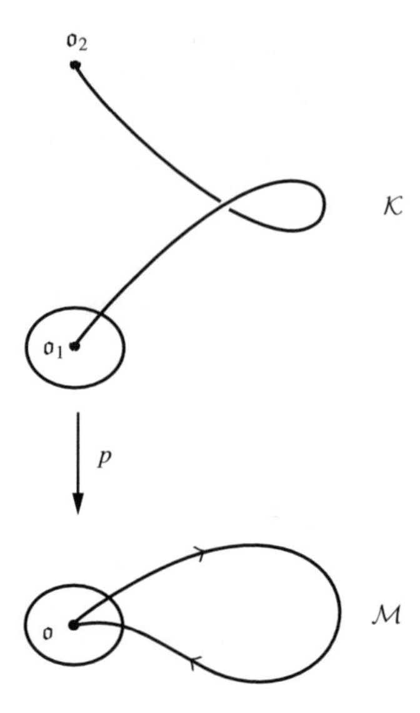

Figure 1.13. Lifting of loops.

Exercise 6. Prove that every covering map arises in this way.

Exercise 7. Prove that covering maps have no fixed points, the identity excepted.

Fundamental Group. The loops of $\mathcal{M}$, starting and ending at $\mathfrak{o}$, fall into deformation classes; and it is easy to see that these classes form a group: The formation of classes respects the composition of loops effected by passing first about loop 1 and then about loop 2; the identity is the class of the trivial loop = the point $\mathfrak{o}$ itself; the inverse is the class of the loop run backward; and so on. This is the **fundamental group** $\pi_1(\mathcal{M})$ of the surface.

Exercise 8. What is the fundamental group of the annulus, sphere, torus, once-punctured plane, and twice-punctured plane? *Answer*: $\mathbb{Z}$, id, $\mathbb{Z}^2$, $\mathbb{Z}$, the free group on two generators.

Exercise 9. The covering map attached to a loop depends only upon the class of the latter. Why? Deduce that the group $\Gamma(\mathcal{K})$ of the cover is isomorphic to the fundamental group of the base.

More Structure. $\mathcal{K}$ may be equipped with any extra structure, smooth or complex, that $\mathcal{M}$ enjoys: Just lift it up via the inverse projection. Then the covering maps appear as diffeomorphisms of $\mathcal{K}$ if $\mathcal{M}$ is smooth, and as conformal automorphisms if $\mathcal{M}$ is a complex manifold.

Exercise 10. Give all the necessary details in the complex case.

Monodromy. The universal cover leads to a simple proof of the **monodromy theorem** of classical function theory. Let $\mathfrak{f}_0$ be a function element, that is, a convergent power series in the local parameter of a small patch of a complex manifold $\mathbf{X}$, and suppose its continuation along paths of $\mathbf{X}$ is unobstructed. This leads to a cover $\mathcal{K}_0$ of $\mathbf{X}$ whose points are pairs comprising a point $\mathfrak{p}$ of $\mathbf{X}$ and a function element $\mathfrak{f}$ at $\mathfrak{p}$ obtained by continuation of $\mathfrak{f}_0$. Now let $\mathbf{X}$ be simply connected, so that it is its own universal cover. Then $\mathbf{X} = \mathcal{K}_0$, which is to say that the process of continuation of $\mathfrak{f}_0$ leads to a single-valued function on $\mathbf{X}$. This is the monodromy theorem. The more conventional proof is to note that a function element produced by continuation is insensitive to small (and so also to large) deformations of the path as long as the endpoints of the latter are fixed.

Exercise 11. Why?

1.12 Scissors and Paste

The most general compact orientable surface is a sphere with handles, as noted in Section 2. The number of handles is the genus g: 0 for the sphere, 1 for the torus, 2 for the pretzel, and so on. It is a fact that each of these can be equipped with a complex structure, and that in many distinct ways if $g \geq 1$. This will now be made plausible by the familiar informal method of *scissors and paste*.

Take the simple two-valued function $\sqrt{z}$. It is desired to make a 2-sheeted Riemann surface S on which it can live comfortably as a single-valued function. Let $\mathbb{P}^1$ be the projective line with parameter z. The radical branches at $z = 0$ and at $z = \infty$, but may be made single-valued by cutting $\mathbb{P}^1$ from 0 to ∞ in view of the fact that a circuit enclosing both 0 and ∞ produces two changes of sign and so no change at all. It has, however, different signs at the two banks of the cut, the endpoints 0 and ∞ excepted, and now it is clear what to do. Take two copies of the cut $\mathbb{P}^1$ with the cuts opened up to make holes and paste them together as in Fig. 1.14, matching the handles according to the sign of the radical to produce a copy S of $\mathbb{P}^1$. The two original copies of $\mathbb{P}^1$ are the **sheets** of S; naturally, upon erasing the cuts, it will not be clear where one sheet ends and the other begins. The discussion is sloppy: For example, careless pasting could leave a crease in S; presumably, it can be ironed out, but the more

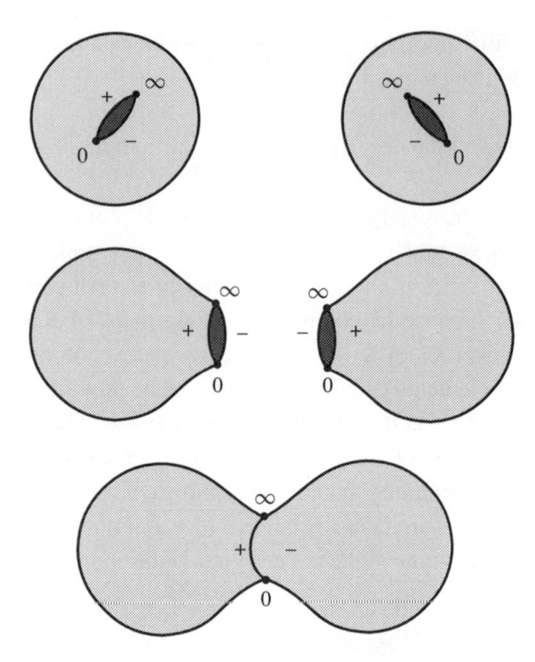

Figure 1.14. The Riemann surface for $\sqrt{z}$.

subtle (because less visible) complex structure of S is best treated from another standpoint. Now look at the function z^2 as a 2:1 map of one copy of $\mathbb{P}^1$ to another, with exceptions at 0 and ∞ where it is 1:1 (but still of degree 2 in that 0 is a double root and ∞ a double pole). The map is seen in Fig. 1.15. The great circle seen in the upper sphere represents the inverse stereographic projection of $\mathbb{R} \subset \mathbb{C}$, cutting $\mathbb{P}^1$ into two hemispheres, labeled $+$ and $-$, and z^2 opens up each of these into a full copy of $\mathbb{P}^1$. The inverse map is $\sqrt{z}$ whose Riemann surface S, seen in Fig. 1.14, may now be identified with the covering projective line of Fig. 1.15. The complex structure of S is clarified thereby: Plainly, it is compatible with that of the base except over 0 and ∞ where the branching of the radical takes place. There the cover is **ramified** over the base, its two sheets touching as in Fig. 1.16, or, more realistically, as in Fig. 1.17, in which you see that one revolution about 0 carries you from sheet 1 to sheet 2, and a second revolution (not shown) brings you back to sheet 1. Figures 1.16 and 1.17 hint that the complex structure of the cover goes bad at the ramifications, but *this is not so*: Both cover and base are complex manifolds in themselves; it is just that their complex structures are not the same: At 0, z is local parameter downstairs and $\sqrt{z}$ is local parameter upstairs; at ∞, you must use $1/z$ downstairs and $1/\sqrt{z}$ upstairs. Forster [1981] and Springer [1981] provide more details.

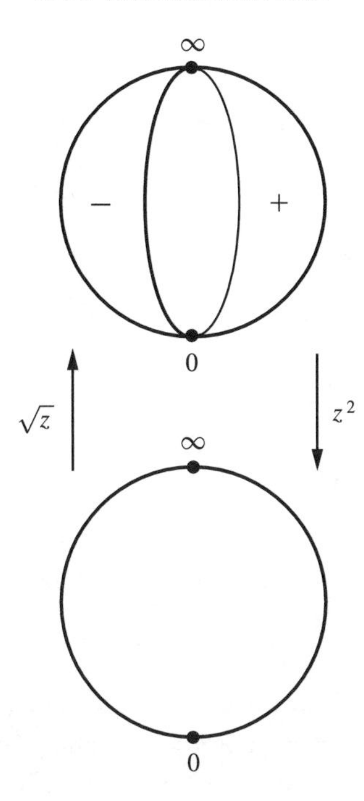

Figure 1.15. The maps $z \mapsto z^2$ and $z \mapsto \sqrt{z}$.

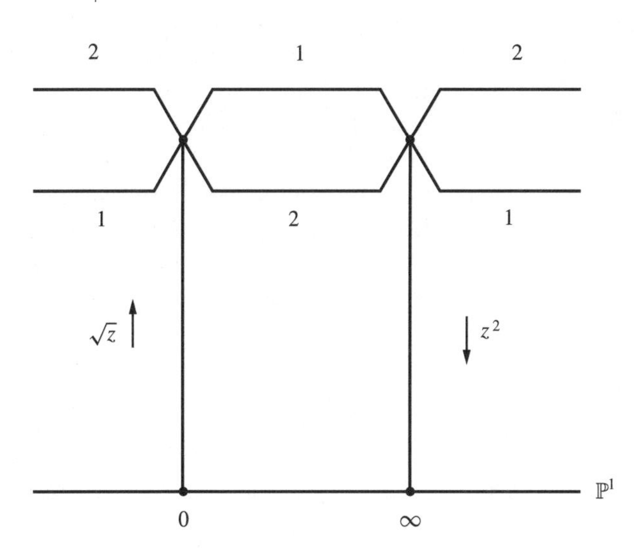

Figure 1.16. The two sheets touch at certain points.

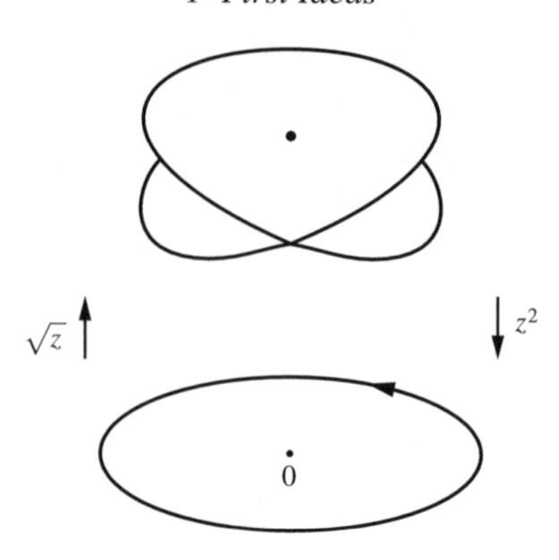

Figure 1.17. The two sheets.

Exercise 1. Give a parallel discussion of the Riemann surface for $\sqrt[d]{z}$ as a d-fold ramified cover of $\mathbb{P}^1$ for $d \geq 3$.

Handlebodies. The same idea applies to the radical $\sqrt{(z-e_1)\cdots(z-e_n)}$ with distinct branch points $e_i (i \leq n)$. For $n = 2$, nothing changes; it is only that e_1 and e_2 play the roles of 0 and ∞. For $n = 3$, the radical branches at e_1, e_2, e_3, and ∞ since a loop enclosing e_1, e_2, and e_3 produces $3(=1)$ changes of sign. To make the radical single-valued on $\mathbb{P}^1$ requires two cuts, one from e_1 to e_2, say, and one from e_3 to ∞, as in Fig. 1.18. Two copies of this cut $\mathbb{P}^1$ are now pasted together in the manner of Fig. 1.14, but with the very different outcome seen in Fig. 1.18. The complex structure of the torus so produced may be clarified as in Fig. 1.16; for example, $\sqrt{z-e_1}$ is local parameter over e_1 and so on. For $n = 4$, you get the same figure with e_4 in place of ∞, but for $n = 5$ or 6 a pretzel appears, and for general $n = 2g+1$ or $2g+2$, a handlebody of genus (= handle number) g. The moral is that every handlebody appears as a Riemann surface; in particular, they all admit a complex structure. More complicated examples abound. The projective curve $y^3 = x - 1/x$ is pretty typical: y branches triply over $x = 0, \infty$, and over the roots ± 1 of $x^2 = 1$, as in Fig. 1.19; in detail

$$\begin{aligned} y &= x^{1/3}\left[1 + \text{ powers of } 1/x\right] && \text{at} \quad x = \infty \\ &= x^{-1/3}\left[-1 + \text{ powers of } x\right] && \text{at} \quad x = 0 \\ &= \sqrt[3]{x-e}\left[\sqrt[3]{2} + \text{ powers of } x-e\right] && \text{at} \quad x = e = \pm 1, \end{aligned}$$

so a little counterclockwise circuit about $x = \infty, 0, -1$, or $+1$ multiplies y by $e^{-2\pi\sqrt{-1}/3}$ for $x = \infty$ or 0 and by $e^{2\pi\sqrt{-1}/3}$ for $x = \pm 1$. Otherwise, y has

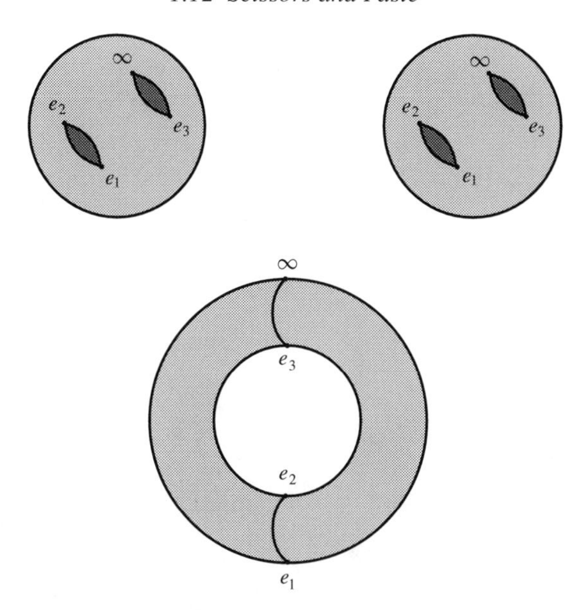

Figure 1.18. The Riemann surface for $\sqrt{(z-e_1)(z-e_2)(z-e_3)}$.

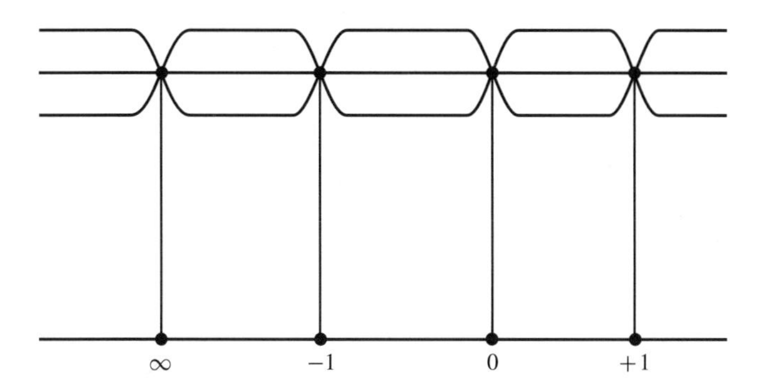

Figure 1.19. The projective curve $y^3 = x - 1/x$.

three distinct values over every value of $x \in \mathbb{P}^1$. Now take three copies of the projective line cut from ∞ to -1 and from 0 to 1 so as to account for the three different determinations of $y = \sqrt[3]{x - 1/x}$. These are seen in Fig. 1.20I. They must be pasted by the numbers. A preliminary pasting of 3 to 3 and 4 to 4 produces Fig. 1.20II; the final handlebody (III) is of genus 2.

Exercise 2. Check the numbering of Fig. 1.20. *Hint*: As you pass just below the real line from $-\infty$ to 1 and just above it to $-\infty$, y changes by the factor

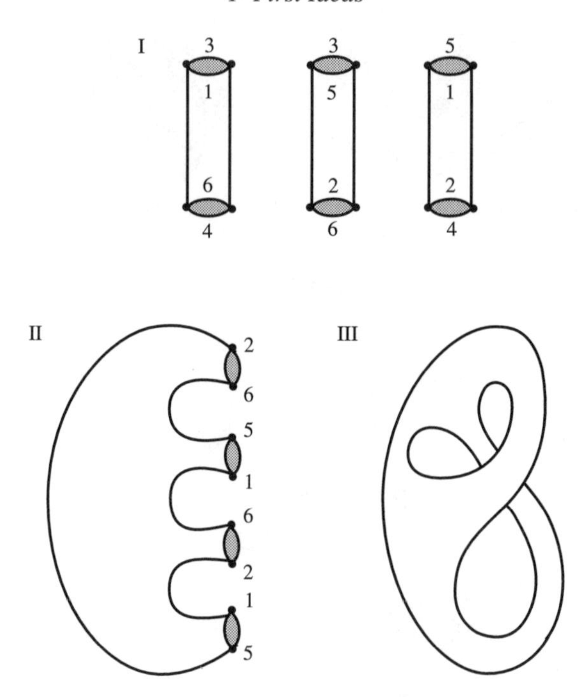

Figure 1.20. Cutting and pasting for $y^3 = x - 1/x$.

$\omega = e^{\pi\sqrt{-1}/3}$ at -1, ω^2 at 0, and ω^2 at 1, and again, on the return trip, by ω^2 at 0 and ω at -1, for a net change of $\omega^8 = \omega^2$.

Exercise 3. Check that the Riemann surface of $y^3 = x^2 - x^{-2}$ has genus 4.

The Helix. The Riemann surface of the logarithm should also be mentioned. The function $w = e^z$ maps each horizontal strip $\mathbb{R} + 2\pi\sqrt{-1} \times [n, n+1)$ of $\mathbb{C}$ faithfully onto the punctured plane $\mathbb{C} - 0$, as in Fig. 1.21, so the inverse function $z = \log w$, defined on the base $\mathbb{C} - 0$, has the covering plane $\mathbb{C}$ as its Riemann surface, each strip constituting a sheet of the latter. A counterclockwise circuit about the puncture downstairs raises the covering point to the next sheet up (by addition of $2\pi\sqrt{-1}$), suggesting that the cover is better viewed as the infinite helix, with $\log z$ itself filling the office of global parameter, opening up the punctured disk $0 < r < 1$ downstairs into the half-plane $(-\infty, 0) \times \sqrt{-1}\mathbb{R}$ upstairs.

Ramified Covers. The Riemann surface of $z^{1/3}$ provides an example of a **ramified cover**: It is a copy of $\mathbb{P}^1$ covering the projective line three-fold, except over 0 and ∞ which are covered once; compare Fig. 1.22. The anomalous covering

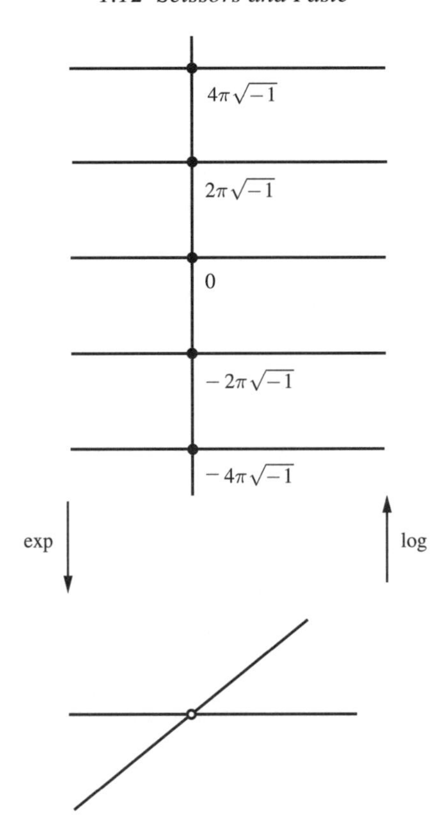

Figure 1.21. The exponential and logarithm.

points are the **ramifications**. More generally, a cover is ramified if $d \geq 2$ sheets meet at a point in the manner of the Riemann surface $z^{1/d}$, $|z| < 1$. The idea of a cover is now widened to admit such ramifications; it is even convenient to admit infinite ramifications as in the Riemann surface of the logarithm. The **index** of a ramification is 1 less than the degree, that is, 1 less than the number of adjacent sheets.

Riemann–Hurwitz Formula. Let $\mathcal{K}$ be a compact orientable manifold (= a handlebody) covering the projective line $\mathbb{P}^1$ with total ramification index r, d sheets, and g handles. Then $r = 2(d + g - 1)$. This is the **Riemann–Hurwitz formula**.

Example 1. The formula precludes unramified covers unless $d = 1$ and $g = 0$ ($\mathcal{K} = \mathbb{P}^1$).

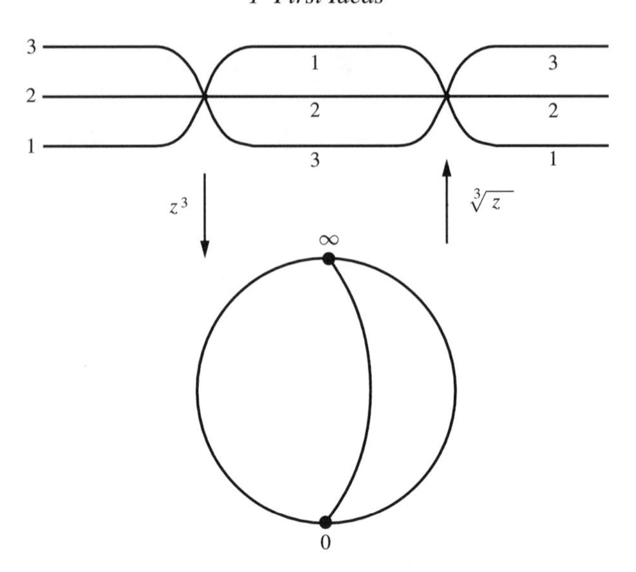

Figure 1.22. Ramified covers.

Example 2. The Riemann surface of $y^3 = x - x^{-1}$ seen in Fig. 1.19, viewed as a ramified cover of the projective line with parameter x, has 4 ramifications of degree 3 apiece for a total index of $4(3-1) = 8$, 3 sheets, and 2 handles: $8 = 2(3+2-1)$.

Proof of the Riemann–Hurwitz formula. The compactness of $\mathcal{K}$ implies that both the ramification index and the sheet number d are finite. Let the base be provided with a fine triangulation in which the projections of the ramified points of the cover appear as corners, and lift it up by the inverse projection to obtain a triangulation of $\mathcal{K}$. The alternating sum: corners − edges + faces is the Euler number: Upstairs its value is $2-2g$; downstairs its value is $2(g=0)$. Now the Euler number of the cover should be $2-2g = 2d$ because everything downstairs appears d-fold upstairs, *but not quite*: A base point that lifts to one or more ramified points of $\mathcal{K}$, of total index m, is covered only $d-m$ times, so there is an imbalance of $-\sum m = -r$ to the right, the true relation being $2-2g = 2d - r$. This is the formula.

Exercise 4. Extend the Riemann–Hurwitz formula to the case of a ramified cover of a general compact complex manifold.

1.13 Algebraic Functions

The idea of a Riemann surface applies to the projective curves of Section 10. Let $P(x, y)$ be an irreducible polynomial in y with coefficients from $\mathbb{C}[x]$. The roots y of $P(x, y) = 0$ are viewed temporarily as elements of the splitting field of P over the ground field $\mathbb{C}(x)$. They are simple, so their discriminant is a nonvanishing element of $\mathbb{C}(x)$; see Section 4. The totality of these roots is an **algebraic function y** of the **indeterminate x**. Riemann's idea provides a geometrical picture of this, as will now be explained in stages.

The Covering. Fix a point x_0 of the projective line $\mathbb{P}^1$ punctured at ∞ and at the points where either the discriminant or the top coefficient of $P(x, y) = c_0(x)y^d + c_1(x)y^{d-1} + \cdots$ vanishes. Near x_0, you have d distinct numerical roots $y_1, \ldots, y_d$, each of which is capable of being expanded in powers of $z = x - x_0$: $\mathbf{y} = k_0 + k_1 z + \cdots$. The resulting pairs $\mathfrak{p}_0 = (\mathbf{x}_0, \mathbf{y}_0)$ with $\mathbf{x}_0 = x_0$ and $\mathbf{y}_0 = y_1, \ldots, y_d$ are displayed in Fig. 1.23 stacked up over x_0. The latter is the **base**

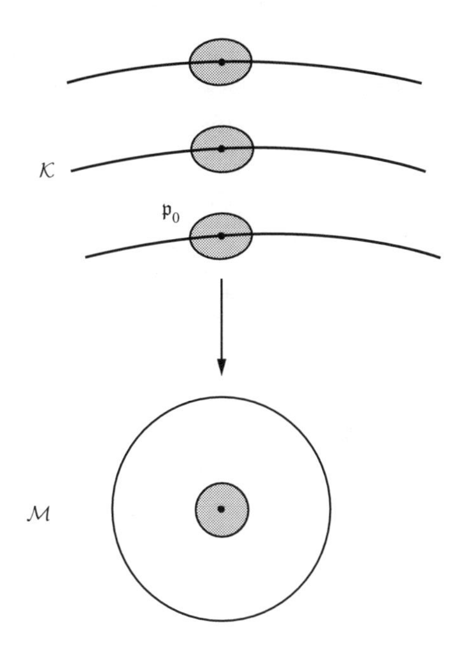

Figure 1.23. Function elements.

point; **y** is a **function element**; the map $\mathfrak{p}_0 \mapsto x_0$ is the **projection**. Now any such function element $\mathbf{y}_0$ can be reexpanded about any center $\mathbf{x}$ close to x_0. This produces a patch of points $\mathfrak{p} = (\mathbf{x}, \mathbf{y})$ about $\mathfrak{p}_0 = (\mathbf{x}_0, \mathbf{y}_0)$, and as this patch is faithfully mirrored by the projection $x = \mathbf{x}(\mathfrak{p})$, so it may be equipped with a complex structure by use of the **local parameter** $z(\mathfrak{p}) = \mathbf{x}(\mathfrak{p}) - x_0$. This makes the totality of points $\mathfrak{p}$ into a complex manifold $\mathcal{K}$ covering the punctured base $\mathbb{P}^1$.

Connectivity. The possible disconnectivity of the cover is easily disproved. Any function element $\mathbf{y}_0$ can be continued without obstruction over the punctured base, producing a subfamily $\mathbf{y}_1, \ldots, \mathbf{y}_n$ of the function elements over each base point $\mathbf{x}$, of fixed cardinality $n \leq d$. The coefficients of the polynomial $P_1(\mathbf{y}) = (\mathbf{y} - \mathbf{y}_1) \cdots (\mathbf{y} - \mathbf{y}_n)$ are single-valued functions of $\mathbf{x}$. These must be *rational* since any function element obeys an estimate $|\mathbf{y}| \leq C_1 \times |\mathbf{x} - \mathbf{x}_0|^{-m}$ at a finite puncture or $|\mathbf{y}| \leq C_2 \times |\mathbf{x}|^m$ at ∞. It follows that $P_1(\mathbf{y})$ divides $P(\mathbf{x}, \mathbf{y})$ over $\mathbb{C}(\mathbf{x})$, violating the irreducibility of the latter unless $n = d$; in short, continuation of $\mathbf{y}_0$ produces the full family of function elements over every base point, which is to say that $\mathcal{K}$ is connected.

The Punctures Filled In. This is a pretty application of the monodromy theorem of Section 11. Fix a puncture $\mathbf{x} = 0$, say, and let $0 < |\mathbf{x}| < r_0$ be puncture-free. The left half-plane $\mathcal{H}$ of the Riemann surface of $\log \mathbf{x}$ covers this punctured disk, and any function element $\mathbf{y}$ over a point of the latter can be lifted to the former and continued there without obstruction. This produces a single-valued function of the logarithm because $\mathcal{H}$ is *simply connected.* Now each of the points $\log \mathbf{x} + 2\pi\sqrt{-1}\mathbb{Z}$ covers x and each of the associated function elements is a root of $P(x, y) = 0$. It follows that only a finite number of different function elements appear at $\log \mathbf{x} + 2\pi\sqrt{-1}\mathbb{Z}$ and that the original branch repeats itself after a continuation upward by $2\pi\sqrt{-1}n$ units with minimal $n \leq d$ independent of $\mathbf{x}$, the intervening branches being distinct. This means that the continued function element $\mathbf{y}$ may be viewed as a single-valued function of $\mathbf{x}^{1/n}$; it is even of rational character in this parameter at $\mathbf{x} = 0$ in view of the estimate $|\mathbf{y}| \leq C_1 \times |\mathbf{x}|^{-m}$ used before. The upshot is that the totality of function elements so produced can be obtained from a single fractional expansion $\mathbf{y}_0 = c_{-k}\mathbf{x}^{-k/n} + \cdots + c_0 + c_1\mathbf{x}^{1/n} + \cdots$ by reexpansion about centers $0 < |\mathbf{x}| < r_0$. $\mathcal{K}$ is now completed at punctures by the insertion of points of this new type $\mathfrak{p}_0 = (0, \mathbf{y}_0)$, and the complex structure is extended by attributing the local parameter $z(\mathfrak{p}) = [\mathbf{x}(\mathfrak{p})]^{1/n}$ to an ambient patch. The new points have the (ramified) aspect of Fig. 1.24, and the completed surface $\mathcal{K}$ is a compact *ramified* cover of the unpunctured base $\mathbb{P}^1$ with a full complex structure: It is *the Riemann surface of the algebraic function* $\mathbf{y}$. Bliss [1933] and Springer [1981] present more details and additional information. Weyl [1955] and Narasimhan [1992] are recommended for a more sophisticated view. $\mathcal{K}$ is a **nonsingular** model of the projective curve defined by $P(x, y) = 0$; compare

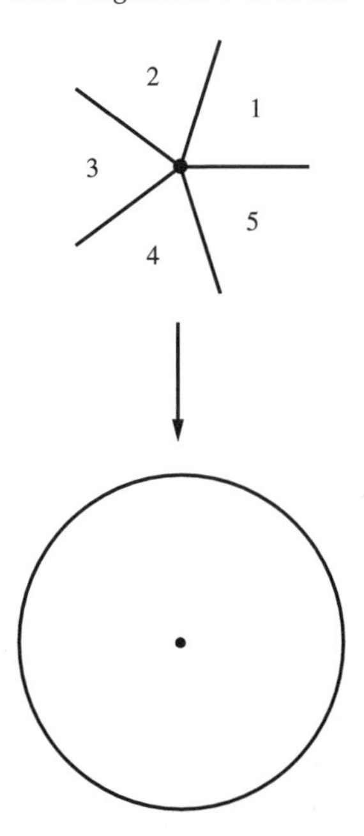

Figure 1.24. The ramified cover.

amplification 2 in Section 10. The present desingularization is an overkill as it is realized only in the infinite-dimensional space of points $\mathfrak{p} = (\mathbf{x}, \mathbf{y})$. Actually, a nonsingular model of any projective curve can always be accommodated in $\mathbb{P}^3$; see Shafarevich [1977].

Uniformization. The universal cover $\mathcal{K}$ of a complex manifold can be equipped with the natural complex structure lifted up from the base; it is also simply connected. What are the possibilities? The answer is contained in the celebrated theorem of Klein [1882], Poincaré [1907], and Koebe [1909–14]: Up to conformal equivalence, *$\mathcal{K}$ is either the sphere $\mathbb{P}^1$, the plane $\mathbb{C}$, or else the disk $\mathbb{D}$: $|x| < 1$.* Note that $\mathbb{P}^1$ is inequivalent to $\mathbb{C}$ or $\mathbb{D}$ on topological grounds already and that $\mathbb{C}$ is inequivalent to $\mathbb{D}$ because an analytic map of $\mathbb{C}$ to $\mathbb{D}$ has a one-point image, by Liouville's theorem: In short, sphere, plane, and disk are genuinely different. The statement includes the earlier Riemann mapping theorem [1851]: *A simply connected region of the sphere omitting two or more points is conformally equivalent to a disk.* It is important to understand what is involved in the general statement, though the proof is not so simple and we refer you to Ahlfors [1973], Springer [1981], or Weyl [1955] for full details. Poincaré himself did not find a fully successful proof. The following informal discussion may be helpful.

Some Hydrodynamics. A pretty hydrodynamical picture was introduced by Klein in his Cambridge lectures [1893]. Let U be a patch of $\mathcal{K}$ equipped with a local parameter $z(\mathfrak{p}) = x_1(\mathfrak{p}) + \sqrt{-1}x_2(\mathfrak{p})$. A steady irrotational flow of an incompressible fluid is specified by a velocity field $\mathbf{v}(\mathfrak{p}) = (v_1, v_2)$ subject to (1) $\mathrm{div}(\mathbf{v}) = \partial v_1/\partial x_1 + \partial v_2/\partial x_2 = 0$ for the incompressibility and (2) $\mathrm{curl}(\mathbf{v}) = \partial v_2/\partial x_1 - \partial v_1/\partial x_2 = 0$ for the irrotational character. (2) implies the existence of a **potential function** p, producing the velocity field $\mathbf{v} = \mathrm{grad} p$, subject to (3) $\Delta p = 0$ in place of (1), and conversely, if $\Delta p = 0$ in U, then $\mathbf{v} = \mathrm{grad}\ p$ is the velocity field of a steady irrotational flow of an incompressible fluid: $\dot{\mathbf{x}}(\mathfrak{p}) = \mathbf{v}(\mathfrak{p})$ with $\mathbf{x}(\mathfrak{p}) = (x_1(\mathfrak{p}), x_2(\mathfrak{p}))$. Let $z'(\mathfrak{p}) = x_1'(\mathfrak{p}) + \sqrt{-1}x_2'(\mathfrak{p})$ be a new local parameter. The new velocity field is also incompressible and irrotational, and the new flow appears in the old coordinates as $\dot{\mathbf{x}} = JJ^\dagger \partial p/\partial \mathbf{x} = c\mathbf{v}$ with Jacobian $J = \partial \mathbf{x}/\partial \mathbf{x}'$, its transpose $J^\dagger$, and the positive factor $c = |\det J|^2 = |dz/dz'|^2$, as you will check by means of the Cauchy–Riemann equations $\partial x_1/\partial x_1' = \partial x_2/\partial x_2'$, $\partial x_1/\partial x_2' = -\partial x_2/\partial x_1'$. The new streamlines are the same as the old; it is only the speed that is changed. This ambiguity of speeds is unavoidable: $\mathcal{K}$ has only a conformal structure, so there is no *preferred* local parameter on any patch.

Exercise 1. Check the new flow.

A Global Flow. $\mathcal{K}$ is equipped with the streamlines of a steady incompressible irrotational flow produced by a single *source* at a point $\mathfrak{o}$, with the understanding that if $\mathcal{K}$ is compact, it will be necessary to destroy fluid at a complementary *sink* $\mathfrak{o}'$. The source is modeled by the potential function $p = \log|z(\mathfrak{p})|$ with local parameter $z(\mathfrak{p})$ vanishing at $\mathfrak{p} = \mathfrak{o}$, but care is needed: If the local parameter is poorly chosen, it may not be possible to extend the flow patchwise over the whole of $\mathcal{K}$ without unpleasant singularities besides the necessary sink in the compact case. What is needed, and this is the hard part of the proof, is the existence of a global function p with $\Delta p = 0$ at ordinary points of $\mathcal{K}$, the singularity $\log|z(\mathfrak{p})|$ at the source, and in the compact case, a second singularity $-\log|z'(\mathfrak{p})|$ at the sink, $z'(p)$ being a local parameter there. This defines a flow over the whole of $\mathcal{K}$ with velocities $\mathbf{v} = (v_1, v_2) = \mathrm{grad}\ p$. The conjugate flow with velocities $(v_2, -v_1)$ is at right angles; it has a (local) potential function q determined up to an additive constant. The associated **circulation** $= \oint dq$ taken about any of the (closed) level lines of p is independent of the particular level line, by Stokes's theorem, and may be evaluated as 2π by shrinking the level line to $\mathfrak{o}$. The new parameter $z(\mathfrak{p}) = \exp\left[p(\mathfrak{p}) + \sqrt{-1}q(\mathfrak{p})\right]$ is now seen to be a single-valued function of rational character on $\mathcal{K}$ with a simple root at the source and, in the compact case, a simple pole at the sink. This function

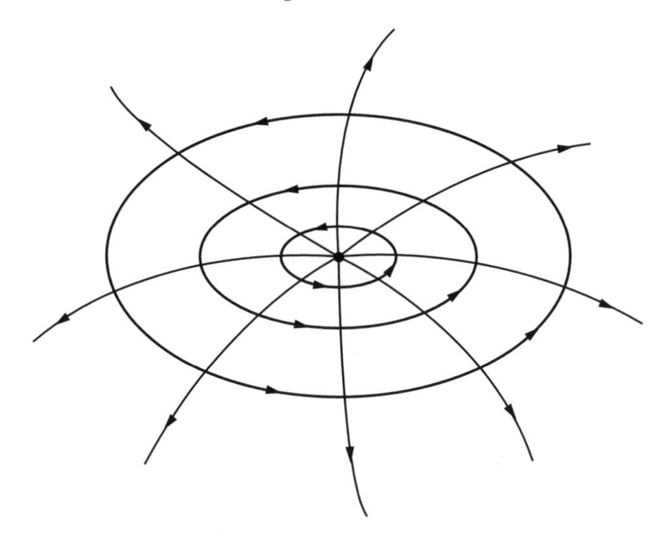

Figure 1.25. Global parameters.

is 1:1 as is plain from the picture (Fig. 1.25) but not so easy to prove: The streamlines issuing from $\mathfrak{o}$ cover the punctured surface simply, except that they come together again at the sink in the compact case; p serves as coordinate *along* the streamline and q tells *which* streamline it is. Three cases are now distinguished according as $\mathcal{K}$ is compact or not and, in the noncompact case, according as p is bounded or not.

Case 1. $\mathcal{K}$ is compact. Then it is a topological sphere and $z: \mathcal{K} \to \mathbb{P}^1$ is a conformal equivalence between $\mathcal{K}$ and the projective line.

Case 2. $\mathcal{K}$ is noncompact and p is unbounded. Then $z: \mathcal{K} \to \mathbb{C}$ is a conformal equivalence between $\mathcal{K}$ and the whole complex plane.

Case 3. $\mathcal{K}$ is noncompact and p tends to the finite number $p(\infty)$. Then z maps $\mathcal{K}$ 1:1 onto a disk of radius $r = \exp[p(\infty)]$. This is Riemann's case.

Idea of the proof. Springer [1981] explains the actual construction of p: Let $\bigwedge^1$ be the class of smooth 1-forms $\omega = \omega_1 dx_1 + \omega_2 dx_2$ on $\mathcal{K}$ and let $*\omega = -\omega_2 dx_1 + \omega_1 dx_2$. Weyl [1955] proved that $\bigwedge^1$ splits into three pieces, mutually perpendicular relative to the natural quadratic form $\int \omega \wedge^* \bar{\omega}$,[1] of which the first is differentials of smooth functions ($\omega = df = (\partial f/\partial x_1)dx_1 + (\partial f/\partial x_2)dx_2$),

[1] $\bar{\omega}$ is the complex conjugate of ω.

the second is codifferentials ($\omega = *df$), and the third is the **harmonic differentials** which are simultaneously **closed** ($d\omega = (\partial\omega_2/\partial x_1 - \partial\omega_1/\partial x_2)dx_1 \wedge dx_2 = 0$), like differentials of functions, and also **coclosed** ($d * \omega = 0$), like codifferentials. Now place yourself in the noncompact case and cook up a smooth exact differential ω that imitates $d\log z(\mathfrak{p})$ near $\mathfrak{p} = \mathfrak{o}$ and vanishes at a little distance from it. Then $\omega - \sqrt{-1} * \bar{\omega}$ vanishes near 0 and ∞ and, being smooth, may be split into its three parts: $df_1 + *df_2 + \omega_3$. The new differential $\omega - df_1 = \sqrt{-1} * \omega + *df_2 + \omega_3$ is plainly smooth, exact, and coclosed away from $\mathfrak{o}$, and its real part is the differential of a single-valued harmonic function p on $\mathcal{K} - \mathfrak{o}$ having the required singularity $\log|z(\mathfrak{p})|$ at $\mathfrak{p} = \mathfrak{o}$. The simple connectivity of $\mathcal{K}$ is now used to confirm that $dq = *dp$ is also exact modulo 2π so $z(\mathfrak{p}) = \exp\left[p(\mathfrak{p}) + \sqrt{-1}q(\mathfrak{p})\right]$ is single-valued too. The final point is that $z\colon \mathcal{K} \to \mathbb{C}$ is 1:1. This is more subtle, so we stop here.

Exercise 2. The map of $\mathcal{K}$ to a sphere, plane, or disk (= half-plane) may be standardized by fixing its values at any three points as you will. Why?

1.14 Examples

The deep content of the Koebe–Poincaré theorem is plain from elementary examples.

Spheres. This is already instructive. The statement is that a complex structure on a topological sphere can be described by a *global parameter* $z\colon \mathcal{K} \to \mathbb{P}^1$. This is what is meant by saying that *the sphere has just one complex structure.*

Annuli. The sphere is peculiar in this respect; for instance, two annuli equipped with the natural complex structure they inherit from $\mathbb{C}$ are conformally equivalent if and only if they have the same ratio r of inner to outer radii. In short, the inequivalent complex structures of a topological annulus are in faithful correspondence with the numbers $0 < r < \infty$.

Exercise 1. Why? *Hint*: A map of annuli extends by circular reflection to a map of punctured planes.

Punctured Spheres. The once-punctured sphere is the plane, so the twice-punctured sphere is a punctured plane, and its universal cover $\mathcal{K}$ may be viewed as the Riemann surface of the logarithm; in short, $\mathcal{K}$ is a plane. The thrice-punctured sphere (= the doubly punctured plane) is different: *Its universal cover is the disk.*

Proof. The cover is not compact (why?), so only $\mathcal{K} = \mathbb{C}$ needs to be ruled out. But if $\mathcal{K} = \mathbb{C}$, then the covering group is populated by substitutions of $PSL(2, \mathbb{C})$ fixing ∞; see Section 5. These are of the form $z \mapsto az + b$ and have fixed points if $a \neq 1$, whereas covering maps do not; see ex. 11.7. It follows that the covering group is commutative, contradicting the fact that the fundamental group of the twice-punctured plane is not. A very different proof will be found in Section 4.9.

Picard's Little Theorem [1879]. A very pretty bonus is a proof of the fact that a *nonconstant integral function takes on every complex value with at most one exception*. The exception is real: *the exponential does not vanish.*

Proof. Let the integral function f omit the values a and b. Then $(b-a)^{-1}(f-a)$ omits 0 and 1, so it is permissible to take $a = 0$ and $b = 1$. Now use f to map a small disk into the punctured plane $\mathbb{C} - 0 - 1$, lift the map to the (universal) covering half-plane via any branch of the inverse projection, and map the lift into the unit disk. The composite map $\mathbb{C} \to \mathbb{C} - 0 - 1 \to \mathbb{H} \to \mathbb{D}$ may be continued without obstruction along paths of the plane. This produces a single-valued function in that plane by the monodromy theorem of Section 11, and as its values are confined to the disk, so it must be constant. But the projection is not constant nor is the map from the half-plane to the disk. The only way out is for f to be constant.

Ahlfors [1973: 19–21] presents a beautiful elementary proof, not employing such transcendental aids; see also Nevanlinna [1970: 248–9] for a thorough geometric discussion and Section 4.9 for a reprise.

Hyperbolic Geometry. Another amusing consequence is that the thrice-punctured sphere admits a geometry of constant curvature -1. The cusps of this **horned sphere** are modeled in ex. 9.5; see Fig. 1.26. The point is that the covering group of the punctured sphere is realized by substitutions of $PSL(2, \mathbb{R})$ acting upon the covering half-plane, and as these are rigid motions of the cover, so this geometry drops down to the punctured sphere; see Section 8 and Pogorelov [1967: 166–7] for such matters. It is not possible to do this for the unpunctured sphere. The obstacle is expressed by the Gauss–Bonnet formula which states that if a handlebody $\mathbf{X}$ of genus g is equipped with a geometry with (possibly variable) curvature κ and surface element $d\sigma$, then the value of the **curvatura integra** $\int_{\mathbf{X}} \kappa\, d\sigma$ is 2π times the Euler number $2 - 2g$ of $\mathbf{X}$; for example, it is $+4\pi$ for the sphere. Pogorelov [1967: 164–6] explains this well.

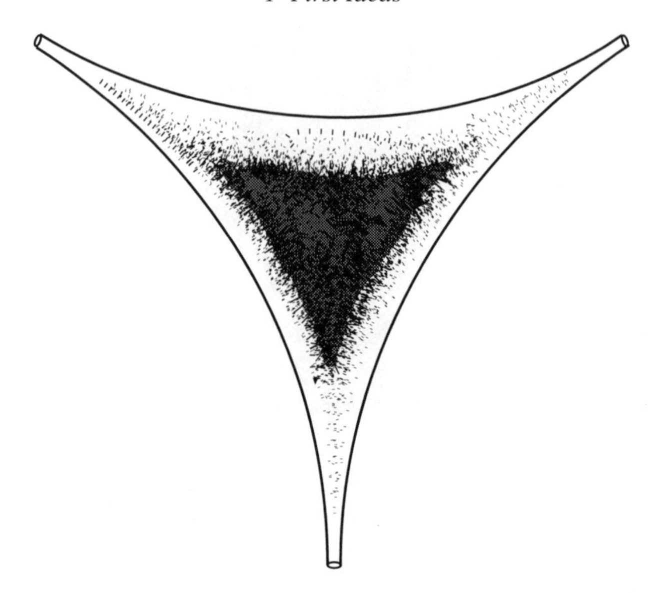

Figure 1.26. The horned sphere.

Higher Handlebodies. The idea behind the horned sphere applies, as well, to higher handlebodies of genus 2 or more: *Their universal covers are always half-planes* so the covering maps appear as substitutions of $PSL(2,\mathbb{R})$ and the geometry of curvature -1 drops down. Note, also, that the covering group must be noncommutative. This is evident from Fig. 1.28 by consideration of the commutator $aba^{-1}b^{-1}$.

Exercise 2. Check the cover. *Hint*: Two handles produce a nontrivial commutator $aba^{-1}b^{-1}$ in the fundamental group; see Fig. 1.27.

Poincaré [1898] had the attractive idea of producing the universal cover of a higher handlebody by endowing the body with a geometry of curvature -1, lifting this geometry to the cover, and identifying the latter with the hyperbolic half-plane; see Kazdan [1985] for such matters.

Tori. These are different; for example, the curvatura integra vanishes, so a geometry of constant curvature ± 1 is not possible. Let ω be a complex number of positive imaginary part and let $\mathbb{L}$ be the lattice $\mathbb{Z} \oplus \omega\mathbb{Z}$. The torus $\mathbf{X} = \mathbb{C}/\mathbb{L}$ inherits the complex structure of $\mathbb{C}$ and has this plane as its universal cover. The latter divides naturally into cells as in Fig. 1.28. The **fundamental cell** is shaded. The projection $\mathbb{C} \to \mathbf{X}$ is *identification modulo* $\mathbb{L}$ and the covering

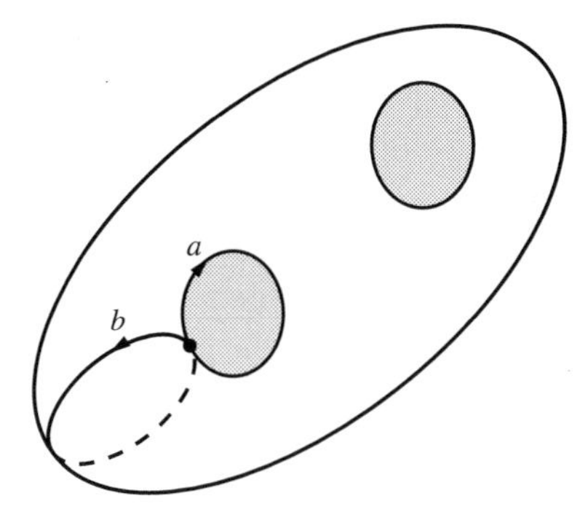

Figure 1.27. Handles produce noncommutativity.

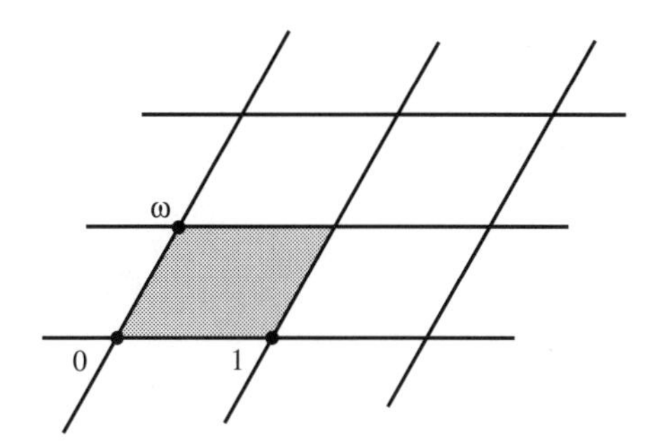

Figure 1.28. The lattice.

group is a copy of the fundamental group $\mathbb{Z}^2 = \mathbb{Z} \oplus \mathbb{Z}$, as it had to be. The fact is that, up to conformal equivalence, every complex torus arises in this way.

Proof. The universal cover of the torus is the plane; intuitively, it is just the torus rolled out vertically and horizontally. Now the covering maps are fixed-point-free conformal self-maps of $\mathbb{C}$, that is, translations $z \mapsto z + c$, as noted under the heading "punctured spheres." The numbers c form a sublattice $\mathbb{L}$ of $\mathbb{C}$, isomorphic to $\mathbb{Z}^2$ and so of the form $\omega_1\mathbb{Z} \oplus \omega_2\mathbb{Z}$ with noncollinear ω_1 and ω_2, as will be verified in Section 2.6. $\mathbf{X}$ is now identified with the quotient $\mathbb{C}/\mathbb{L}$. A trivial map $z \mapsto z/\omega_1$ and, if need be, a change of sign of the ratio $\omega = \omega_2/\omega_1$ brings $\mathbb{L}$ to the standard form $\mathbb{Z} \oplus \omega\mathbb{Z}$ with ω of positive imaginary part.

The identification $\mathbf{X} = \mathbb{C}/\mathbb{L}$ leads to a simple proof of the fact that, unlike the unpunctured sphere, but like the annulus of ex. 1, *the topological torus has many inequivalent conformal structures*. The same is true of the higher handlebodies.

Proof. Let $\mathbf{X}_1 = \mathbb{C}/\mathbb{L}_1$ and $\mathbf{X}_2 = \mathbb{C}/\mathbb{L}_2$ be complex tori with a conformal map between them and lift this map up to a self-map of their common universal cover $\mathbb{C}$; see Fig. 1.29. The lifted map is single-valued by the monodromy theorem, the continuation being unobstructed; moreover, it is of linear growth at ∞ since a closed path that goes around or through the hole of $\mathbf{X}_1$ leads from a point in one cell of the cover to a new point in an *adjacent* cell. But then the lifted map is a linear function $az + b$ of the parameter z of $\mathcal{K}_1 = \mathbb{C}$, and as it commutes with projections, so you must have $b \in \mathbb{L}_2$ and $a\mathbb{L}_1 \subset \mathbb{L}_2$; indeed,

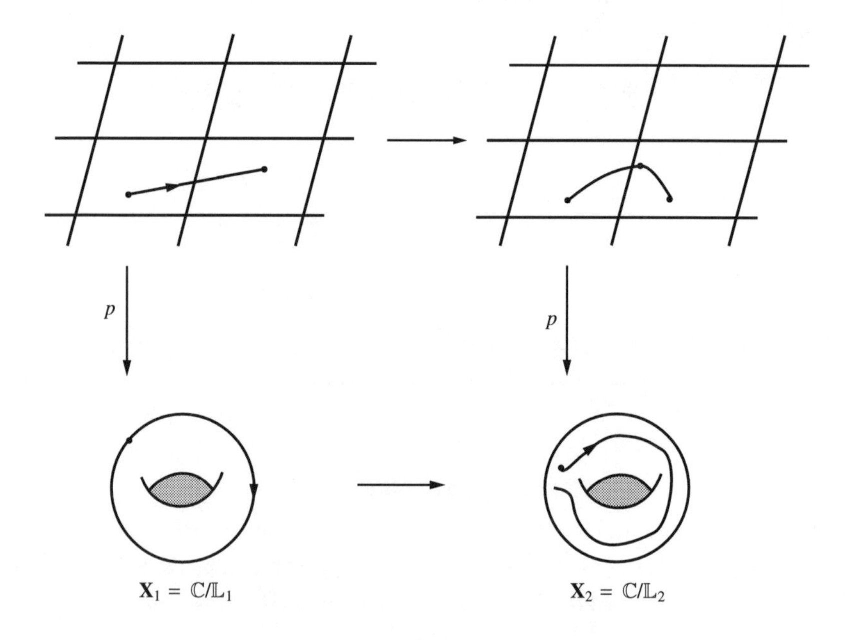

Figure 1.29. Conformal maps between tori.

$a\mathbb{L}_1 = \mathbb{L}_2$ by consideration of the inverse map. The moral is that the tori cannot be conformally equivalent otherwise. For example, the period ratios $\omega = \sqrt{-1}$ and $\omega = (1 + \sqrt{-3})/2$ produce inequivalent tori.

Geometric Explanation. The fact that tori admit many different complex structures is the subject of Section 2.6 and of the first part of Chapter 4. A geometric explanation can be given right now. Let the fundamental cell be

$$\mathfrak{F} = \{x = x_1 + sx_2 + \sqrt{-1}x_2 : 0 \le x_1 < 1, 0 \le x_2 < h\}$$

with fixed base $0 \le x_1 < 1$. This standardization does not affect the number of complex structures. There are two degrees of freedom left: the **slant** s and the **height** $h > 0$. The slant can be made to satisfy $0 \le s < 1$, 1 being a lattice point. Now the corresponding torus is made in two steps from the rectangular cell of height $h(s = 0)$: First paste together the vertical sides to make a right cylinder; then twist the upper circle by sh and paste it to the bottom. This produces a shear in the amount sx_2 at level $0 \le x_2 < h$ and changes the complex structure; a change of height usually changes the structure, too.

Example. $h = 1$. If the slants s_1 and s_2 produce equivalent tori, then the corresponding lattices

$$\mathbb{L}_1 = \mathbb{Z} \oplus (\sqrt{-1} + s_1)\mathbb{Z} \quad \text{and} \quad \mathbb{L}_2 = \mathbb{Z} \oplus (\sqrt{-1} + s_2)\mathbb{Z}$$

are related by a multiplication $a\mathbb{L}_1 = \mathbb{L}_2$, as you know, and $|a| = 1$ by comparison of areas of fundamental cells. But then $a \times 1 = i + j\sqrt{-1}s_2$ with $i, j \in \mathbb{Z}$ and $1 = (i + js_2)^2 + j^2$, so either $j = 0$, $a = i \pm 1$, and $\mathbb{L}_1 = \mathbb{L}_2$, or else $j = \pm 1$ and $s_2 = 1$, which is not permitted.

Exercise 3. Discuss the case when two heights produce equivalent tori. Prove that, for slant = 1, the height determines the structure.

1.15 More on Uniformization

The statement of Koebe–Poincaré is called the **uniformization theorem** in the older literature. The name refers to the nineteenth-century usage uni-form = one-valued. To clarify matters let **X** be a projective curve defined by the vanishing of an irreducible polynomial: $P(x, y) = 0$. Its points are triples $(x, y, z) \in \mathbb{C}^3 - 0$ with projective identifications, so $\mathbf{x} = x/z$ and $\mathbf{y} = y/z$ make projective sense; in fact, they are functions of rational character on **X**, and the vanishing of P expresses a relation between them, specifying **y** as a many-valued function of **x** and vice versa. These functions may be promoted to the universal cover $\mathcal{K}$ of **X** where they appear as functions **x**, **y** of rational character, invariant under the action of the covering group. In this way, **X** is uniformized by $\mathcal{K}$ in that the totality of points **X** is displayed by means of single-valued functions on $\mathcal{K}$(= sphere, plane, or half-plane).

Genus 0. If **X** has no handles, then it is a projective line $\mathbb{P}^1$ and is its own universal cover, so **x** and **y** appear as rational functions of the parameter w of $\mathbb{P}^1$ and the map $w \mapsto (\mathbf{x}, \mathbf{y})$ is a conformal equivalence between $\mathbb{P}^1$ and **X**. The simplest nontrivial example is provided by the projective circle **X**: $\mathbf{x}^2 + \mathbf{y}^2 = 1$

of Section 10 and its uniformization by

$$\mathbf{x} = \frac{1}{2}(w + w^{-1}), \quad \mathbf{y} = \frac{1}{2\sqrt{-1}}(w - w^{-1}).$$

Rational Curves. Here is a deeper fact: If $\mathbf{X}$ is a **rational curve** in the sense that there exists a nonconstant map of rational character of $\mathbb{P}^1$ into $\mathbf{X}$, then $\mathbf{X}$ *is of genus* 0 *already and so itself a projective line*, as before; compare Section 14. The idea will be plain from a picture: $\mathbb{P}^1$ is simply connected, so a map of rational character of $\mathbb{P}^1$ into $\mathbf{X}$ lifts to a map of $\mathbb{P}^1$ into $\mathcal{K}$, by a self-evident application of the monodromy theorem. This is impossible if $\mathcal{K}$ is not a projective line since nonconstant rational functions take all complex values, ∞ included.

Tori. The complex torus $\mathbf{X}$ cannot be uniformized by **rational functions** since its universal cover $\mathcal{K}$ is not a projective line; see ex. 2.11.4. Indeed, $\mathcal{K} = \mathbb{C}$, $\mathbf{X}$ is its quotient by the lattice $\mathbb{L}$ as in Section 14, and $\mathbf{x} = x/z$ and $\mathbf{y} = y/z$ may be promoted to functions of rational character on $\mathbb{C}$ having every complex number $\omega \in \mathbb{L}$ as a period. These functions uniformize $\mathbf{X}$ in the former style, but are not rational. The whole of Chapter 2 is devoted to such **elliptic functions**.

Higher Handlebodies. $\mathcal{K}$ is a half-plane by Section 14. Now the uniformizing functions $\mathbf{x}$ and $\mathbf{y}$, promoted to $\mathcal{K}$, are invariant under a subgroup of $PSL(2, \mathbb{R})$: the so-called **automorphic functions**. The subject lies mostly outside the scope of this book, but see Ford [1972] and Terras [1985] for more information, and Chapters 4 and 5 for a number of special instances and some general information.

1.16 Compact Manifolds as Curves: Finale

The methods sketched in Section 14 can be used to prove the existence of nonconstant functions of rational character on any compact complex manifold $\mathcal{M}$; see, for example, Hurwitz and Courant [1964], Springer [1981], or Weyl [1955]. Let $\mathbf{x}$ be such a function: It has a finite number of poles and an equal number of roots, as may be seen by integrating $(2\pi\sqrt{-1})^{-1} d \log[\mathbf{x}(\mathfrak{p})]$ about the edges of a triangulation of $\mathcal{M}$, every edge being traversed twice, in opposite directions. This number is its degree d. It follows that $\mathbf{x}$ takes on every complex value d times, so $\mathcal{M}$ appears as a d-fold ramified covering of $\mathbb{P}^1$ with projection $\mathfrak{p} \mapsto \mathbf{x}(\mathfrak{p})$. Now two cases arise according as $d = 1$ or $d \geq 2$.

Case 1. If $d = 1$, then the projection is a conformal equivalence of $\mathcal{M}$ to $\mathbb{P}^1$; indeed, $\mathbf{x}$ fills the office of global parameter on $\mathcal{M}$.

Case 2. If $d \geq 2$, a further construction is necessary to produce a function $\mathbf{y}$ of rational character taking d distinct numerical values $\mathbf{y} = \mathbf{y}_1, \ldots, \mathbf{y}_d$ at the d points of the fiber $x = \mathbf{x}(\mathfrak{p})$, for most values of the base point $x \in \mathbb{P}^1$. Then it is easy to see that $(\mathbf{y} - \mathbf{y}_0) \cdots (\mathbf{y} - \mathbf{y}_d)$ is of rational character in $\mathbf{x}$ and so represents a polynomial $P \in \mathbb{C}(x)[y]$ satisfied by $\mathbf{x}$ and $\mathbf{y}$; moreover, P is necessarily irreducible over the ground field $\mathbb{C}(x)$, as you will see by continuation of the identity $P(\mathbf{x}, \mathbf{y}) = 0$ over the necessarily connected manifold $\mathcal{M}$. Now comes the punch line: The vanishing of $P(x, y)$ defines a nonsingular projective curve $\mathbf{X}$, as in Section 10, and the map $\mathfrak{p} \mapsto (\mathbf{x}, \mathbf{y})$ of $\mathcal{M}$ to $\mathbf{X}$ is a conformal equivalence; in short, *every compact complex manifold is a projective curve.* The discussion has come full circle.

2

Elliptic Integrals and Functions

The purpose of this chapter is to study the field **K** of functions of rational character on the complex torus $\mathbf{X} = \mathbb{C}/\mathbb{L}$. These are the **elliptic functions** and one could begin with them, but we prefer a more historical approach starting from the **elliptic integrals**. The incomplete elliptic integral of the first kind of Jacobi [1829] is typical. It is

$$\int_0^x [(1 - x^2)(1 - k^2x^2)]^{-1/2} dx.$$

Gauss [1797] and Abel [1827] made the remarkable discovery that *the inverse function of such an integral belongs to an elliptic function field.* The idea of investigating the integral with modulus $k^2 = -1$ occurred to Gauss on Jan. 8, 1797, in connection with the rectification of the lemniscate.[1] This circle of ideas is one strand of the story. A second strand, intimately entwined with the first, is the fact that *the class of complex tori* $\mathbb{C}/\mathbb{L}$ *is the same as the class of projective quartics* $y^2 = (1 - x^2)(1 - k^2x^2)$ *with complex moduli* k^2 *different from* 0 *or* 1. The third strand reflects the second: *Every (nonsingular) projective quartic (or cubic) may be equipped with a commutative law of addition.* Thus, three strands are interwoven – function theory, geometry, and arithmetic – in the most beautiful creation of nineteenth-century mathematics. The viewpoint adopted in the rest of the present chapter is mainly that of Weierstrass (1870), as recorded in the Berlin lecture notes of H. A. Schwarz [1893]. Jacobi's viewpoint [1829] was different; it is explained episodically here, and at length in Chapter 3. Ahlfors [1979], Copson [1935], and/or Whittaker and Watson [1963] may be consulted for the elementary part. Hurwitz and Courant [1964] and Siegel [1969–73] take the matter further. Lang [1987] may be preferred by the more algebraically inclined. The splendid classical accounts of the subject

[1] Gauss [1870–1929 (10): 483–574].

such as Fricke [1916, 1922] and Weber [1891] are recommended too; see also Houzel [1978] and Weil [1975] for historical information, and Bateman [1953: 295–322] and Gradshteyn and Ryzhik [1994] for a very complete collection of formulas.

2.1 Elliptic Integrals: Where They Come From

The integral

$$\int_0^1 (1-x^2)^{-1/2}dx = \int_0^{\pi/2} (1-\sin^2\theta)^{-1/2}\cos\theta d\theta = \pi/2$$

is elementary, but the equally innocent-looking lemniscatic integral of Gauss

$$\int_0^1 (1-x^4)^{-1/2}dx$$

is not. The latter is a (complete) elliptic integral, meaning that it is the integral of a rational function of x and y in which y^2 is a polynomial in x, of degree 3 or 4, having simple roots. Most integrals of this type are not elementary. Legendre [1811, 1825], Gauss [1797], Abel [1827], and Jacobi [1829] studied them exhaustively. The next examples illustrate how they come up in practice. Greenhill [1892] and Lawden [1989] have good collections of other interesting examples.

Example 1: Mapping the half-plane to a rectangle. This example is central to our version of the story, so don't miss it even if you skip the rest. Let $0 < k < 1$ be fixed and consider the map f of the upper half-plane defined by Jacobi's (incomplete) elliptic integral of the first kind:

$$x \longmapsto \int_0^x [(1-x^2)(1-k^2x^2)]^{-1/2}dx.$$

The radical is taken to be real and positive for $-1 < x < 1$ and extended by continuation to the closed upper half-plane, avoiding the branch points $\pm 1, \pm 1/k$ by infinitesimal semicircles as in Fig. 2.1. The integrand turns by a factor of $\sqrt{-1}$ as you pass 1 and $1/k$ from left to right, so as x runs from 0 to ∞, its image moves, first from 0 to

$$K = \int_0^1 [(1-x^2)(1-k^2x^2)]^{-1/2}dx;$$

then it turns 90° and moves up to $K + \sqrt{-1}K'$ with

$$K' = \int_1^{1/k} [(x^2-1)(1-k^2x^2)]^{-1/2}dx;$$

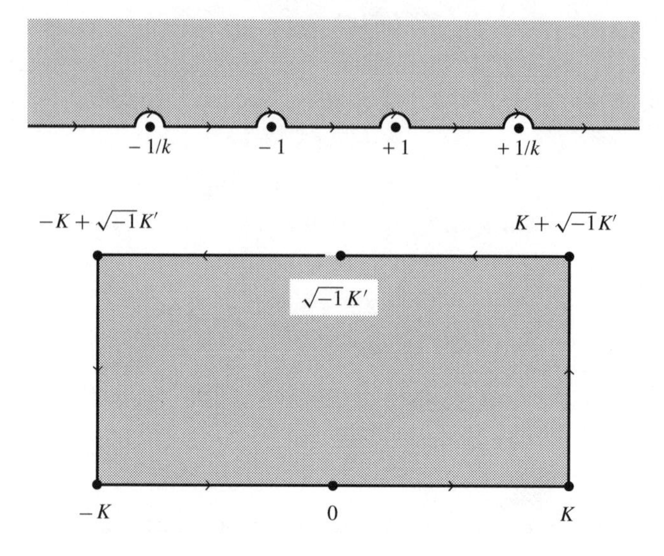

Figure 2.1. Map of the upper half plane by the incomplete integral.

finally, it makes another 90° turn and moves back to the left, as in Fig. 2.1. The image of the left half-line is just the reflection in the vertical of the image of $[0, \infty)$. The nice thing is that the top closes up precisely at $x = \pm\infty$, so the image of the whole line is the complete perimeter of a rectangle. To spell it out, the leftward-moving image of $x > 1/k$ fills out a line segment of length

$$\int_{1/k}^{\infty} [(x^2 - 1)(k^2x^2 - 1)]^{-1/2} dx,$$

and this is just K = half the length of the bottom, as you can see by means of the substitution $x \mapsto 1/kx$:

$$\int_{1/k}^{\infty} [(x^2 - 1)(k^2x^2 - 1)]^{-1/2} dx$$

$$= \int_0^1 \left[\left(\frac{1}{k^2x^2} - 1\right)\left(\frac{1}{x^2} - 1\right)\right]^{-1/2} \frac{dx}{kx^2} = K,$$

or, better still, by observing that the integral of $[(1-x^2)(1-k^2x^2)]^{-1/2}dx$ about a big semicircle of radius R is small like $R^{-2} \times \pi R$, so

$$\int_{-\infty}^{\infty} [(1 - x^2)(1 - k^2x^2)]^{-1/2} dx$$

vanishes. The fact that the line is mapped 1:1 onto the perimeter of the rectangle implies that the whole of the (open) upper half-plane is mapped 1:1 onto the inside. K is Jacobi's so-called **complete elliptic integral of the first kind**, of **modulus** k. K' is the **complementary integral of the first kind** formed with the **complementary modulus** $k' = \sqrt{1-k^2}$:

$$K' = \int_1^{1/k} [(x^2-1)(1-k^2x^2)]^{-1/2} dx = \int_0^1 [(1-x^2)(1-(k')^2x^2)]^{-1/2} dx.$$

Exercise 1. Prove this by use of the substitution $x \mapsto [1-(k')^2x^2]^{-1/2}$.

Exercise 2. Show that rectangles of any proportion, length to width, are obtained, that is, $K\colon K'$ takes all values between 0 and ∞.

Exercise 3. For $k = 0^+$, $K = \pi/2$, $K' = \infty$, and the map is just the elementary $\sin^{-1}(x) = \int_0^x (1-x^2)^{-1/2} dx$ taking the closed half-plane to the strip $|x_1| \le \pi/2$, $0 \le x_2 < \infty$. Think it over.

Exercise 4. Discuss the mode of divergence of $K(k)$ at $k = 1^-$. *Hint*: For $k \downarrow 0$,

$$K(k') = \int_1^{1/k} \left(\frac{1}{\sqrt{x^2-1}} - \frac{1}{x} \right) \frac{dx}{\sqrt{1-k^2x^2}} + \int_k^1 \left(\frac{1}{\sqrt{1-x^2}} - 1 \right) \frac{dx}{x} + \int_k^1 \frac{dx}{x}.$$

Check that the first two integrals contribute $\log 2$ apiece plus $o(1)$; the third integral is just $-\log k$, so

$$K(k) = 2\log 2 - \log\sqrt{1-k^2} + o(1) \quad \text{for} \quad k \uparrow 1.$$

Example 2: Rectification of the lemniscate. The **lemniscate** is realized when the product of the distances of a moving point $Q = (x_1, x_2)$ from two fixed points P_- and P_+ is constant: $d_- \times d_+ = d$. Take $P_\pm = \pm(1/\sqrt{2}, 0)$ and $d = 1/2$. Then $d_- d_+ = \sqrt{(x_1 + 1/\sqrt{2})^2 + x_2^2} \times \sqrt{(x_1 - 1/\sqrt{2})^2 + x_2^2} = 1/2$, so $x_1^2 = (1/2)(r^2 + r^4)$ and $x_2^2 = (1/2)(r^2 - r^4)$ with $r^2 = x_1^2 + x_2^2$, as you will check; see Fig. 2.2. The length of the lemniscate is $4 \times 1/\sqrt{2} \times$ Gauss's lemniscatic integral:

$$\int_0^1 \sqrt{(dx_1)^2 + (dx_2)^2} = \frac{1}{\sqrt{2}} \int_0^1 \frac{dr}{\sqrt{1-r^4}}.$$

This is Jacobi's complete elliptic integral of the first kind with the special modulus $k = \sqrt{-1}$.

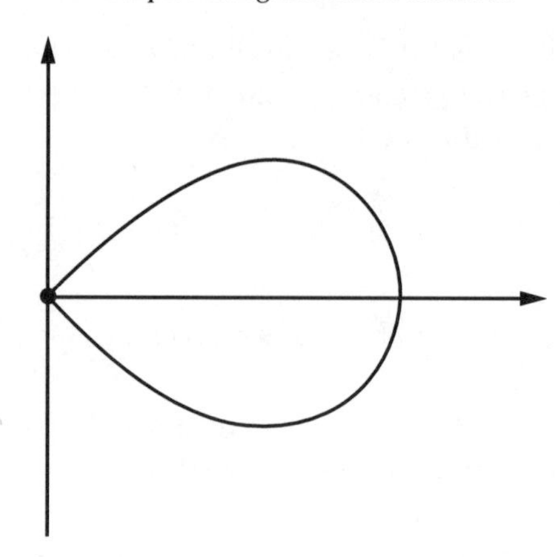

Figure 2.2. The lemniscate.

Exercise 5. Check the computation.

Example 3: Rectification of the ellipse. An ellipse is realized when the sum $d_- + d_+$ is constant $(= d)$, as for $\mathbf{E}$: $x_1^2/a^2 + x_2^2/b^2 = 1 (a > b)$; the fixed points are now the foci $(\pm f, 0)$ with $f = \sqrt{a^2 - b^2}$ and $d = 2a$. The length of $\mathbf{E}$ is 4 times

$$\int_0^a \sqrt{1 + [x_2'(x_1)]^2} dx_1 = a \times \int_0^1 \sqrt{\frac{1 - k^2 x^2}{1 - x^2}} dx$$

with $k^2 = 1 - b^2/a^2$ between 0 and 1, in which you see Jacobi's **complete elliptic integral of the second kind**:[2]

$$E(k) = \int_0^1 \sqrt{\frac{1 - k^2 x^2}{1 - x^2}} dx.$$

This explains the name **elliptic** integral.

Exercise 6. What is the surface area of the 3-dimensional ellipsoid $\mathbf{E}$: $x_1^2/a^2 + x_2^2/b^2 + x_3^2/c^2 = 1$?

Example 4: The simple pendulum. Classical mechanics provides many elegant instances of elliptic integrals. A simple pendulum is depicted in Fig. 2.3. L is its length, m is the mass of the bob, G is the gravitational constant. The vertical force MG has only a tangential effect, of magnitude $MG \sin \theta$; this comes with

[2] Don't panic: There are only three kinds.

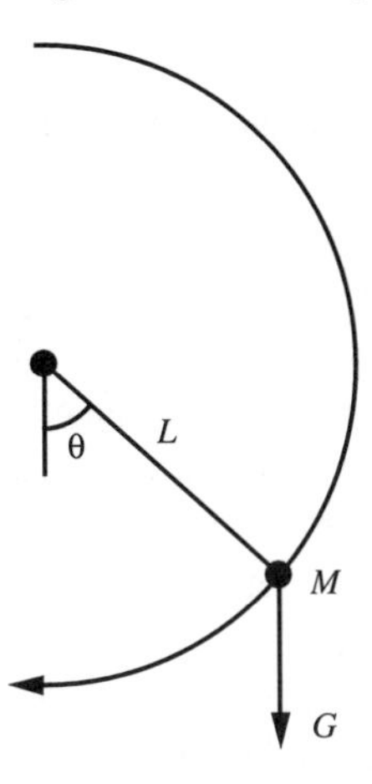

Figure 2.3. The simple pendulum.

a negative sign. The corresponding acceleration is $L\ddot{\theta}$,[3] so Newton's law (force = mass × acceleration) provides the equation of motion $\ddot{\theta} + (G/L)\sin\theta = 0$. Let $L = G$ for simplicity. Then a first integral is $I = \cos\theta - (1/2)(\dot{\theta})^2$, $\cos^{-1} I < 180°$ being the angle of highest elevation. The substitution $x = \sqrt{2/(1-I)} \times \sin(\theta/2)$ converts this identity into $\dot{x} = \pm\sqrt{(1-x^2)(1-k^2x^2)}$ with $k^2 = (1-I)/2$ between 0 and 1, expressing the fact that the time t, in its dependence upon x, is an incomplete elliptic integral of the first kind:

$$t = \int_0^x \left[(1-y^2)(1-k^2y^2)\right]^{-1/2} dy;$$

it is understood that the sign of the radical flips when the highest elevation ($x = \pm 1$) is reached. In particular, the period of motion is 4 times the complete integral: $T = 4K(k)$.

Exercise 7. Carry out the substitution.

Example 5: Capacitance of an ellipsoid. The electrostatic capacitance of the 3-dimensional ellipsoid

$$\mathbf{E}: x_1^2/a^2 + x_2^2/b^2 + x_3^2/c^2 = 1 \quad (a > b > c)$$

is determined by solving Laplace's equation

$$\Delta p = \frac{\partial^2 p}{\partial x_1^2} + \frac{\partial^2 p}{\partial x_2^2} + \frac{\partial^2 p}{\partial x_3^2} = 0$$

[3] The *dot* signifies differentiation by time.

outside **E**, subject to $p = 1$ on the surface and $p = 0$ at ∞. The function $p(x)$ is the (equilibrium) electrostatic potential of **E**. The name will now be explained. **E** is an electrical conductor. A charge placed on **E** distributes itself in a natural way in accord with Coulomb's law of mutual repulsion of like charges: force = product of charges divided by the distance squared.[4] For example, on a sphere the charge distribution is uniform, whereas in Montana, it likes the plains but not the mountains. The quantity $p(x)$ is the work required to bring a unit charge from ∞ to x in the face of the electrostatic field produced by the charge on **E**; it is independent of the path and constant on **E**. Keep loading charge up on **E** until the potential of the surface reaches the value $p = 1$. This is the equilibrium distribution; its total charge is the **capacitance** $C(\mathbf{E})$ of the ellipsoid. Coulomb's law states that the potential produced by the point charge Q at distance D is Q/D. Now reflect that **E** looks like a point from far away, so $p(x) \simeq C(\mathbf{E}) \times |x|^{-1}$ near ∞. The capacitance is read off from that. But what is p? The fact is Δp vanishes outside **E**, this quantity being proportional to charge, by Gauss's divergence theorem suitably applied, so you have to solve $\Delta p = 0$ with $p = 1$ on **E** and $p = 0$ at ∞. The computation is facilitated by **ellipsoidal coordinates**, a favorite trick of Jacobi [1866]. A point x outside **E** determines the parameter $0 \le r = r(x) < \infty$ by means of

$$\frac{x_1^2}{a^2+r} + \frac{x_2^2}{b^2+r} + \frac{x_3^2}{c^2+r} = 1.$$

Now look at

$$p(x) = \frac{1}{2}\int_{r(x)}^{\infty} \left[(a^2+r)(b^2+r)(c^2+r)\right]^{-1/2} dr$$

outside **E**. It vanishes at ∞, by inspection; it is constant on **E** ($r = 0$); and $\Delta p = 0$, by elementary but tiresome computation. Also $r(x) \sim |x|^2$ far out and $p(x) \simeq \frac{1}{2}\int_r^\infty r^{-3/2} dr = r^{-1/2} = |x|^{-1}$, so the capacitance of the ellipsoid is the reciprocal of the complete elliptical integral

$$C^{-1}(\mathbf{E}) = \frac{1}{2}\int_0^\infty \left[(a^2+r)(b^2+r)(c^2+r)\right] dr.$$

Exercise 8. Check that $\Delta p = 0$. *Hint*: $\Delta p(r)$ is proportional to

$$p''(r) + (1/2)\left[(a^2+r)^{-1} + (b^2+r)^{-1} + (c^2+r)^{-1}\right] p'(r).$$

Example 6: Random walk in 2 and 3 dimensions. Let $x_n (n = 0, 1, 2, \ldots)$ be the nth position of a random walker starting at the origin of the 2-dimensional

[4] Irrelevant physical constants are left out.

lattice $\mathbb{Z}^2$. The rule is that at the next step, he is equally likely to move to any of the four neighbors of his present position. The probability that the walker returns to the origin in n steps vanishes if n is odd, while if n is even ($n = 2m$), a return requires an even number $2l$ ($2m - 2l$) of vertical (horizontal) steps of which half must be up (left) and half down (right): In other words,[5, 6]

$$\begin{aligned} P[x_n = 0] &= \sum_{l=0}^{m} \binom{2m}{2l} \binom{2l}{l} \binom{2m-2l}{m-l} 4^{-2m} \\ &= 4^{-2m} \binom{2m}{m} \sum_{l=0}^{m} \binom{m}{l}^2 \\ &= \left[\frac{(2m-1)\cdots 3\bullet 1}{2m(2m-2)\cdots 4\bullet 2} \right]^2 . \end{aligned}$$

It follows that

$$\sum_{n=0}^{\infty} P[x_n = 0]k^n = \frac{2}{\pi} \int_0^{\pi/2} [1 - k^2 \sin^2 \theta]^{-1/2} d\theta,$$

this being the trigonometric form of the integral of the first kind $K(k)$. The formula can then be checked by expanding K in a power series in the disk $|k| < 1$:[7]

$$K(k) = \sum_{0}^{\infty} \binom{-1/2}{n} (-k^2)^n \int_0^{\pi/2} \sin^{2n} \theta \, d\theta.$$

and doing the integrals by hand.

Exercise 9. Do that.

Now for $k \uparrow 1$, the integral, and so also the left-hand sum, is $+\infty$. The latter is recognized as the expected time spent at the origin. Let p be the probability to come back to the origin *some time*. Then p^n is the probability of n returns, so $\infty = 1 + p + p^2 + \cdots$ and $p = 1$, a fact due to Pólya [1921]. The 3-dimensional walk on $\mathbb{Z}^3$ is similarly defined, only now a point has six neighbors and the behavior of the walk is quite different; in particular, $p < 1$

[5] $P(E)$ is the probability of the event E. $\binom{n}{m}$ is "n choose m," that is, $n!/m!(n-m)!$.

[6] $\sum_{l=0}^{m} \binom{m}{l}^2 = \binom{2m}{m}$. Why?

[7] $\binom{-1/2}{n} = \frac{(-1)^n \bullet (2n-1)\cdots 3\bullet 1}{n!}$.

and the sum may be expressed as

$$(1-p)^{-1} =$$
$$\int_0^1\int_0^1\int_0^1 [1-\{3[\cos(2\pi x_1)+\cos(2\pi x_2)+\cos(2\pi x_3)]\}^{-1}]dx_1dx_2dx_3.$$

This is also due to Pólya [1921]. Watson [1939] evaluated the integral in terms of the complete elliptic integral $K(k)$ of modulus $k^2 = (2-\sqrt{3})(\sqrt{3}-\sqrt{2})$:

$$(1-p)^{-1} = \frac{12}{\pi^2}\left(18+12\sqrt{2}-10\sqrt{3}-\sqrt{6}\right)K^2$$

(just the numbers that would spring to your mind!), with the result that $p = .344$, that is, a little more than $1/3$, so the mean number of returns to the origin is about $3/2$.

2.2 The Incomplete Integrals Reduced to Normal Form

This was done first by Legendre [1811]. Here, Jacobi's format is preferred. Bateman [1953: 295–322] is recommended for more information; see also Byrd and Friedman [1954] for a big table of integrals, both complete and incomplete.

Let $Q \in \mathbb{C}[x]$ be of degree $d \leq 4$ and let $F(x, y)$ be rational in x and $y = \sqrt{Q(x)}$. Then the incomplete integral $\int F(x, y)dx$ is either elementary, or else it is an **elliptic integral**, by definition. The integral is visibly elementary if F does not depend upon y; also, if $d = 1$, so $y^2 = ax + b$, it yields to the substitution $x \mapsto a^{-1}(x^2 - b)$, whereas if $d = 2$, so $y^2 = c(x-a)(x-b)$, it yields to the substitution $x \mapsto a - (1/4)(b-a)(x - 1/x)^2$; the integral is also elementary if $d = 3$ or 4 but $Q(x) = 0$ has repeated roots. In all these cases it can be expressed by elementary functions: rational, trigonometric, exponential, or logarithmic. Now let Q be of degree 3 or 4, having only simple roots. Powers of y can be reduced *modulo* $y^2 = Q$, leaving for serious consideration only the integrals $\int F(x)y^{-1}dx$ with $F \in \mathbb{C}(x)$. The rest of the reduction is carried out in easy steps.

Step 1. First dispose of the case $d = 3$. Let $Q(x) = (x-e_0)(x-e_1)(x-e_2)$ with distinct roots e_0, e_1, e_2. The substitution $x \mapsto x^2 + e_0$ converts the differential dx/y into $2dx/\sqrt{(x^2+e_0-e_1)(x^2+e_0-e_2)}$. Now the radicand has *four* simple roots $\pm\sqrt{e_1-e_0}$ and $\pm\sqrt{e_2-e_0}$. In short, it is no loss to take $d = 4$.

Step 2. Reduce Q to Jacobi's standard form $(1-x^2)(1-k^2x^2)$ with a suitable modulus $k^2 \neq 0, 1$. Map the (distinct) roots e_1, e_2, e_3, e_4 of Q into 1,

$-1, 1/k, -1/k$ by means of a fractional linear substitutions $g = [ab/cd]$. As you know from Section 1.5, this can be done if and only if the cross ratios match:

$$\frac{e_1 - e_2}{e_2 - e_3} \bullet \frac{e_3 - e_4}{e_4 - e_1} = \frac{2}{-1 - 1/k} \bullet \frac{2/k}{-1/k - 1} = \frac{4k}{(1+k)^2}.$$

This determines one or two possible values of k, and this number cannot be 0 or ± 1 since the cross ratio is neither 0, 1, nor ∞.

Exercise 1. There are, in fact, two possible choices for k^2 except for the harmonic ratio $r = 2$, this being the lemniscatic case $k = \pm\sqrt{-1}$.

Now the inverse substitution $x \mapsto g^{-1}x$ takes the factor $x - e$ into

$$g^{-1}x - e = \frac{dx - b}{-cx + a} - e = \frac{(ce + d)x - (ae + b)}{-cx + a} = \frac{x - ge}{(-cx + a)(ce + d)^{-1}},$$

and this does not disturb the rational character of F. The rest will be plain.

Step 3. Split $F(x)$ according to parity into $F_1(x^2) + xF_2(x^2)$. The contribution of the second piece is elementary:

$$\int xF_2(x^2)y^{-1}dx = \frac{1}{2}\int F_2(x^2)\left[(1 - x^2)(1 - k^2x^2)\right]^{-1/2} dx^2,$$

which is back to the case $d = 2$. This leaves only $\int F_1(x^2)y^{-1}dx$, which may be reduced, by means of the identities

$$\frac{x^2 - a^2}{x^2 - b^2} = 1 + \frac{b^2 - a^2}{x^2 - b^2}$$

and

$$\frac{1}{(x^2 - a^2)(x^2 - b^2)} = \frac{(x^2 - a^2)^{-1} - (x^2 - b^2)^{-1}}{a^2 - b^2},$$

to two families of integrals:

$$I_n = \int x^{2n}y^{-1}dx \ (n \geq 0), \quad I'_n = \int (x^2 - c^2)^n y^{-1}dx \ (n < 0).$$

Step 4. Reduce the first family to I_0 and I_1. The derivative, with respect to x^2, of $x^{2n}y$ is

$$y^{-1} \times \left[nx^{2n-2}(1 - x^2)(1 - k^2x^2) + x^{2n}\left(k^2x^2 - \frac{1 + k^2}{2}\right)\right]$$

and, integrating back, you find

$$nI_{n-1} - n(1+k^2)I_n + (n+1)k^2 I_{n+1} - \tfrac{1}{2}(1+k^2)I_n = x^{2n}y + \text{a constant},$$

so I_n ($n \geq 2$) may be determined from the values of I_0 and I_1.

Step 5. Finally apply the same trick to the second family. The derivative, with respect to x^2, of $(x^2 - c^2)^n y$ is

$$y^{-1} \times \left[n(x^2-c^2)^{n-1}(1-x^2)(1-k^2x^2) + \frac{1}{2}(x^2-c^2)^n \left[2k^2x^2 - 1 - k^2\right] \right].$$

Now express the bracket solely in powers of $x^2 - c^2$ and integrate back to conclude that I'_{n-1} can be determined from I'_n and I'_{n+1} as long as $n(1-c^2)(1-k^2c^2)$ does not vanish, as is the case for most values of c^2, provided $n < 0$; in detail, if I'_{-1} is known, then I'_{-2} can be found from I'_{-1} and $I'_0 = I_0$, and so on down the line.

Moral. The irreducible (incomplete) integrals are those of the first, second, and third kinds:

$$I_0 = \int \frac{dx}{\sqrt{(1-x^2)(1-k^2x^2)}}, \qquad I_0 - k^2 I_1 = \int \sqrt{\frac{1-k^2x^2}{1-x^2}}\,dx,$$

and

$$I'_{-1} = \int \frac{dx}{(x^2-c^2)\sqrt{(1-x^2)(1-k^2x^2)}}.$$

Exercise 2. Check that the substitution $x \mapsto 1 + \sqrt{3}(1-x)(1+x)^{-1}$ brings the integral $\int_1^\infty (x^3-1)^{-1/2}dx$ into standard form and evaluate it as

$$2(3+2\sqrt{3})^{-1/2} K\left[\sqrt{-1}(2-\sqrt{3})\right].$$

Exercise 3. (Math. Tripos 1911) Use the reduction of step 2 to evaluate $\int_0^2 \left[(2x-x^2)(4x^2+9)\right]^{-1/2} dx$ as $2K(1/\sqrt{5})/\sqrt{15}$.

Exercise 4. The incomplete integral $\int [P_5(x)]^{-1/2}\,dx$ with $P_5 \in \mathbb{C}[x]$ of degree 5 is *not* an elliptic integral, but may be reduced to such in special cases; for example, Jacobi [1832] found that $\int [x(1-x)(1+ax)(1+bx)(1-abx)]^{-1/2}\,dx$ is reducible to a sum of two integrals of the form $\int [P_3(x)]^{-1/2}\,dx$. Check that the substitution $x \mapsto (ab)^{-1}[x + \sqrt{x^2-ab}]$ does the trick. *Hint*: $-\sqrt{2}x^{-3/2}dx \mapsto [(x+\sqrt{ab})^{-1/2} - (x-\sqrt{ab})^{-1/2}]\,dx$ is the key.

The next two exercises explain the geometric background to ex. 4.

Exercise 5. The substitutions $x \mapsto k(x-1)(x+1)^{-1}$ and $y \mapsto y(x+1)^{-3}$ convert the quintic $y^2 = x(x-e_0)(x-e_1)(x-e_2)(x-e_3)$ into $y^2 = (x^2 - 1)(x-f_0)(x-f_1)(x-f_2)(x-f_3)$ with $f/e_0 = (k+e)(k-e)^{-1}$, up to an irrelevant constant multiplier. Check this. Now observe that $f_0 = -f_3$ forces $k = \sqrt{e_0 e_3}$, so, if $e_0 = 1$ and if also $f_1 = -f_2$, then $e_1 e_2 = e_3$. This is Jacobi's case: $e_0 = 1, e_1 = -a, e_2 = -b, e_3 = ab$.

Exercise 6. Use the Riemann–Hurwitz formula to show that the final curve of ex. 5 $y^2 = (x^2-1)(x^2-f_0^2)(x^2-f_1^2)$ is of genus 2. It is a two-sheeted cover of the obvious elliptic curve, with projection $(x, y) \mapsto (x^2, y)$.

This is the geometric reason behind the reduction in ex. 4; see Picard [1882], Belokolos et al. [1994], and Belokolos and Enol'skii [1994] for more details.

2.3 The Complete Integrals: Landen, Gauss, and the Arithmetic–Geometric Mean

The simple transformation

$$K(\sqrt{-1}k) = \frac{1}{\sqrt{1+k^2}} K\left(\frac{k}{\sqrt{1+k^2}}\right)$$

is obtained by making the substitution $x \mapsto x/\sqrt{1+k^2(1-x^2)}$ in the left-hand integral.

Exercise 1. Check it.

This is the first of many more or less deep transformations of the complete elliptic integrals. Landen's transformation [1775]:

$$K(k) = \frac{1}{1+k} K\left(\frac{2\sqrt{k}}{1+k}\right)$$

is the most famous. A variety of proofs of it are presented in what follows; see Sections 17 and 4.12 for proofs in a completely different style. Almkvist and Berndt [1988] and Watson [1933] have brief accounts of Landen's life. The former also cites the interesting proof of Ivory [1796].

Exercise 2. Express Landen's transformation in the form

$$K(k) = (1+k_1)K(k_1)$$

with $k_1 = (1-k')/(1+k')$, k' being the complementary modulus $\sqrt{1-k^2}$. ($k \mapsto k_1$ does it.) Iterate and discuss convergence; in particular, show that $k_1 > k_2(=k_1') > k_3(=k_2')$, and so on. Deduce that $K(k) = \frac{\pi}{2} \times \prod(1+k_n)$.

Legendre's proof [1811]. This employs the substitution

$$x \mapsto y(x) = (1+k')x\sqrt{1-x^2}/\sqrt{1-k^2x^2}$$

with the complementary modulus $k' = \sqrt{1-k^2}$. The function y vanishes at $x = 0$ and at $x = 1$: It has peak value 1 at $x_0 = (1+k')^{-1/2}$, and

$$\left[(1-y^2)(1-k_1^2y^2)\right]^{-1/2} dy = \pm(1+k')\left[(1-x^2)(1-k^2x^2)\right]^{-1/2} dx$$

with $k_1 = (1-k')/(1+k')$ as in ex. 2, as you will verify (with tears). The upshot is

$$K(k_1) = \frac{1+k'}{2} K(k)$$

which you should recognize as Landen's transformation in the form of ex. 2. *Hint*: $1+k' = 2(1+k_1)^{-1}$. Cayley [1895: 180–1] discusses this and related substitutions; they go back to Euler [1760].

Exercise 3. Do the computation.

Gauss and the Arithmetic–Geometric Mean. Gauss [1799] found a highly efficient recipe for computing the complete integrals K for moduli $0 < k < 1$. The substitution $x = \cos\theta$ converts the latter into its trigonometric form, expressed here as

$$G = G(a,b) = K(k)/a = \int_0^{\pi/2} \left[a^2\cos^2\theta + b^2\sin^2\theta\right]^{-1/2} d\theta$$

with $k^2 = 1 - b^2/a^2$ and $1 > a > b > 0$. Gauss hit upon the surprising fact that the integral is unchanged if a is replaced by the arithmetic mean $a_1 = (a+b)/2$ and b by the (smaller) geometric mean $b_1 = \sqrt{ab}$. This is equivalent to Landen's transformation. Now the substitution $a, b \mapsto a_1, b_1$ can be repeated, the successive terms having a common limit $M(a,b) =$ the **arithmetic–geometric mean** of a and b; plainly, $G = \pi/2M$. By May 30, 1799, Gauss had observed that $1/M(1,\sqrt{2})$ and $(2/\pi) \times \int_0^1 (1-x^4)^{-1/2}dx$ agree to 11 places. His diary for that day states that this result "will surely open up a whole new field of analysis." By that date, Gauss had also recognized that the incomplete integral $\int(1-x^4)^{-1/2}dx$ is the inverse of an (elliptic) function with two independent complex periods, but more on that later.

Exercise 4. Check that Gauss's rule is really Landen's transformation in disguise.

Exercise 5. Discuss the speed of convergence of the successive means to their final value M: $a_n - b_n < 2^{-n}$ is plain from $a_1 - b_1 = \frac{1}{2}(\sqrt{a} - \sqrt{b})^2$. Sharpen this to $a_n - b_n < \frac{8}{3} \exp\left[-(3\log 2) \times 2^n\right]$ for $1/3 < b < a < 2/3$, say.

Exercise 6. Compute Gauss's integral G by his recipe for $a = 2/3$ and $b = 1/3$ to three places. *Answer*: 3.235.

Gauss's proof will now be described in brief:

$$M(a, b) = M\left(\frac{a+b}{2}, \sqrt{ab}\right) = \frac{a+b}{2} M\left(1, \frac{2\sqrt{ab}}{a+b}\right),$$

which is to say that $M(1+k, 1-k) = (1+k)M(1+k^*, 1-k^*)$ with $k^* = 2\sqrt{k}/(1+k)$. Gauss deduced the power series

$$\frac{1}{M(1+k, 1-k)} = \frac{1}{M(1, \sqrt{1-k^2})} = \sum_{n=0}^{\infty} \left[\frac{(2n+1)\cdots 5 \cdot 3}{(2n)\cdots 4 \cdot 2}\right]^2 k^{2n}$$

and matched it to $\pi/2 \times$ the series for the complete elliptic integral of the first kind noted in example 1.6. This is a real tour de force.

Exercise 7. Newman [1985] found a very clever proof of the invariance of $G(a, b)$. The substitution $x = b \tan\theta$ converts $2G(a, b)$ into

$$\int_{-\infty}^{\infty} \left[(a^2 + x^2)(b^2 + x^2)\right]^{-1/2} dx.$$

Now make the substitution $x \mapsto x + \sqrt{x^2 + ab}$.

Exercise 8. The substitution $x \mapsto (1+k)x(1+kx^2)^{-1}$ changes $K(k^*)$ into $(1+k)K(k)$. This is Gauss's second proof of Landen's transformation; see Hancock [1958] for this and related substitutions.

Almkvist and Berndt [1988: esp. 593], Borwein and Borwein [1987], and Cox [1984] have further information about the arithmetic–geometric mean and its uses.

2.4 The Complete Elliptic Integrals: Legendre's Relation

Legendre [1825] found an elegant relation between the complete integrals of the first and second kinds. The complementary integral of the first kind is

$$K'(k) = \int_1^{1/k} \left[(x^2-1)(1-k^2x^2)\right]^{-1/2} dx$$
$$= \int_0^1 \left[(1-x^2)(1-(k')^2x^2)\right]^{-1/2} dx = K(k')$$

with $k' = \sqrt{1-k^2}$; compare example 1.1. The complementary integral of the second kind is a bit more complicated:

$$\int_1^{1/k} \sqrt{\frac{1-k^2x^2}{x^2-1}}dx = K(k') - \int_0^1 \sqrt{\frac{1-(k')^2x^2}{1-x^2}}dx;$$

it is the final integral $E(k')$ that is complementary to the integral of second kind

$$E(k) = \int_0^1 \sqrt{\frac{1-k^2x^2}{1-x^2}}dx.$$

Exercise 1. Check this rule by the substitution $x \mapsto k^{-1}\sqrt{1-(k')^2x^2}$.

Differential Equations. The complete integrals solve a variety of differential equations; for example, with $D = d/dk, k^{-1}Dk(1-k^2)DK = K$. Then to begin with, $kDE = E - K$ and $k^{-1}DK = \int_0^1 (1-x^2)^{-1/2}(1-k^2x^2)^{-3/2}x^2dx$. The latter is simplified by integrating

$$\frac{d}{dx}\frac{x\sqrt{1-x^2}}{\sqrt{1-(k')^2}} = \frac{1-x^2}{\sqrt{(1-x^2)(1-k^2x^2)}} - \frac{(1-k^2)x^2}{\sqrt{(1-x^2)(1-k^2x^2)^3}}$$

from $x = 0$ to $x = 1$ to obtain $k(k')^2DK = E - (k')^2K$. Now differentiate with respect to k once more.

Exercise 2. Prove that the differential operator $H = k^{-1}Dk(k')^2D$ commutes with the substitution $k \mapsto k' = \sqrt{1-k^2}$. Conclude that $f = aK + bK'$ is the general solution of $Hf = f$.

Exercise 3. Legendre proved

$$\int_0^1 \frac{dx}{\sqrt{(1-x^2)(1-k^2x^2)^3}} = \frac{1}{1-k^2}\int_0^1 \sqrt{\frac{1-k^2x^2}{1-x^2}}dx.$$

Check this. *Hint*: $y = x\sqrt{1-x^2}/\sqrt{1-k^2x^2}$ vanishes at $x = 0$ and at $x = 1$ so $\int_0^1 dy = 0$.

Exercise 4. Prove that $KE' + EK' - KK'$ is the product of $k(k')^2$ and the Wronskian $KDK' - K'DK$. Now use ex. 2 to conclude that this object is independent of k.

Legendre [1825] proved that $KE' + EK' - KK' = \pi/2$; in extenso,

$$\frac{\pi}{2} = \int_0^1 \frac{dx}{\sqrt{(1-x^2)(1-k^2x^2)}} \times \int_0^1 \sqrt{\frac{1-(k')^2x^2}{1-x^2}}dx$$

$$+ \int_0^1 \frac{dx}{\sqrt{(1-x^2)(1-(k')^2x^2)}} \times \int_0^1 \sqrt{\frac{1-k^2x^2}{1-x^2}}dx$$

$$- \int_0^1 \frac{dx}{\sqrt{(1-x^2)(1-k^2x^2)}} \times \int_0^1 \frac{dx}{\sqrt{(1-x^2)(1-(k')^2x^2)}}.$$

Euler [1782] verified the lemniscatic identity

$$\int_0^1 \frac{dx}{\sqrt{1-x^4}} \times \int_0^1 \frac{x^2dx}{\sqrt{1-x^4}} = \frac{\pi}{4}$$

by the clever manipulations of series of which he was so fond. This is Legendre's relation in the self-complementary case $k = k' = 1/\sqrt{2}$.

Special Values. The value $\pi/2$ in Legendre's relation is elicited from special values of the complete integral. The modulus $k = 1/\sqrt{2}$ is best since it is its own complement. $K(1/\sqrt{2}) = \sqrt{2}K(\sqrt{-1})$ by the transformation at the start of Section 3, and

$$K(\sqrt{-1}) = \int_0^1 \frac{dx}{\sqrt{1-x^4}} = \frac{1}{4}\int_0^1 x^{-3/4}(1-x)^{-1/2}dx$$

$$= \frac{1}{4}\frac{\Gamma(1/4)\Gamma(1/2)}{\Gamma(3/4)} = 2^{-5/2}\pi^{-1/2}\,[\Gamma(1/4)]^2\,.$$

The integral of the second kind is a little different: $\sqrt{2}E(1/\sqrt{2}) = E(\sqrt{-1})$, and

$$E(\sqrt{-1}) = \int_0^1 \frac{1+x^2}{\sqrt{1-x^4}}dx = K(\sqrt{-1}) + \frac{1}{4}\int_0^1 x^{-1/4}(1-x)^{-1/2}dx$$

$$= K(\sqrt{-1}) + 2^{1/2}\pi^{3/2}\,[\Gamma(1/4)]^{-2}\,.$$

The upshot is

$$(2E - K)K \text{ at modulus } 1/\sqrt{2} = 2(E - K) \text{ at modulus } \sqrt{-1}$$
$$= 2 \times 2^{1/2}\pi^{5/2} \times 2^{-5/2}\pi^{-1/2} = \pi/2,$$

as advertised. Cayley [1895] provides lots of special values of this kind; see also Section 17 for more recondite ones.

Exercise 5. Give a second proof using ex. 1.4.

A third proof produces Legendre's relation in an unconventional form as

$$\int_0^1 \sqrt{\frac{1-k^2x^2}{1-x^2}}dx \times \int_1^{1/k} \frac{dx}{\sqrt{(1-x^2)(1-k^2x^2)}}$$
$$- \int_1^{1/k} \sqrt{\frac{1-k^2x^2}{1-x^2}}dx \times \int_0^1 \frac{dx}{\sqrt{(1-x^2)(1-k^2x^2)}} = \frac{\pi}{2}.$$

Let I be the incomplete integral of the second kind

$$\int_0^x \sqrt{\frac{1-k^2x^2}{1-x^2}}dx$$

and form the differential $Iy^{-1}dx$ with $y^2 = (1-x^2)(1-k^2x^2)$. I is single-valued in the plane cut from $-\infty$ to -1 and from $+1$ to $+\infty$. The integral of $Iy^{-1}dx$ about a big circle is now computed in two different ways: At ∞, $Iy^{-1}dx$ looks like $x^{-1}dx$ so the integral is $2\pi\sqrt{-1}$, more or less. Legendre's relation is produced by deforming the two semicircles into the upper and lower banks of the cut, as in Fig. 2.4. The two determinations of y agree on the upper and lower banks, except in $1 < |x| < 1/k$, where they differ in sign; likewise the two determinations of I agree for $|x| < 1$, sum up to

$$2\int_0^1 \sqrt{\frac{1-k^2x^2}{1-x^2}}dx$$

for $1 < |x| < 1/k$, and differ by

$$2\int_1^{1/k} \sqrt{\frac{1-k^2x^2}{x^2-1}}dx$$

for $|x| > 1/k$. Now reevaluate the integral about the big circle as

$$2\int_0^1 \sqrt{\frac{1-k^2x^2}{1-x^2}}dx \times 2\int_1^{1/k} \frac{dx}{\sqrt{(1-x^2)(1-k^2x^2)}}$$

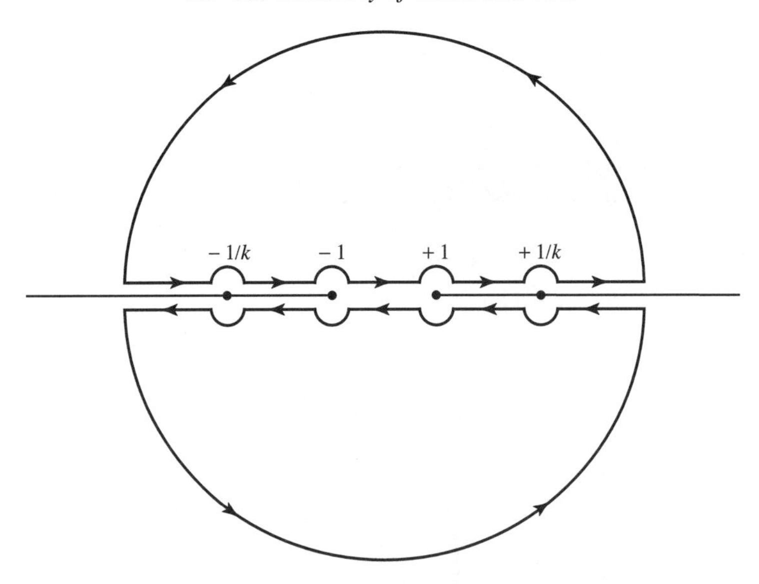

Figure 2.4. The contour.

$$-2\int_1^{1/k}\sqrt{\frac{1-k^2x^2}{1-x^2}}dx \times \int_0^1 \frac{dx}{\sqrt{(1-x^2)(1-k^2x^2)}}.$$

The original form of Legendre's relation comes from the identification of the third integral as $K' - E'$, as noted at the start of this section.

Exercise 6. Think it through, keeping careful track of the phase of $Iy^{-1}dx$.

2.5 The Discovery of Gauss and Abel

The map of the half-plane to a rectangle was presented in example 1.1. A whole new world opens up if you look instead at the inverse map. But first, here is a simpler example to get warmed up.

A Trigonometric Integral. The elementary integral

$$\sin^{-1} x = \int_0^x (1-y^2)^{-1/2} dy$$

maps the upper half-plane 1:1 onto the shaded strip of Fig. 2.5; indeed, by reflection in the punctured plane $\mathbb{C} - (\pm 1)$, it produces a full tiling of the target plane by congruent, nonoverlapping images of the upper (+) and lower (−) half-planes. The viewpoint emphasized here is that the integral is inverted by

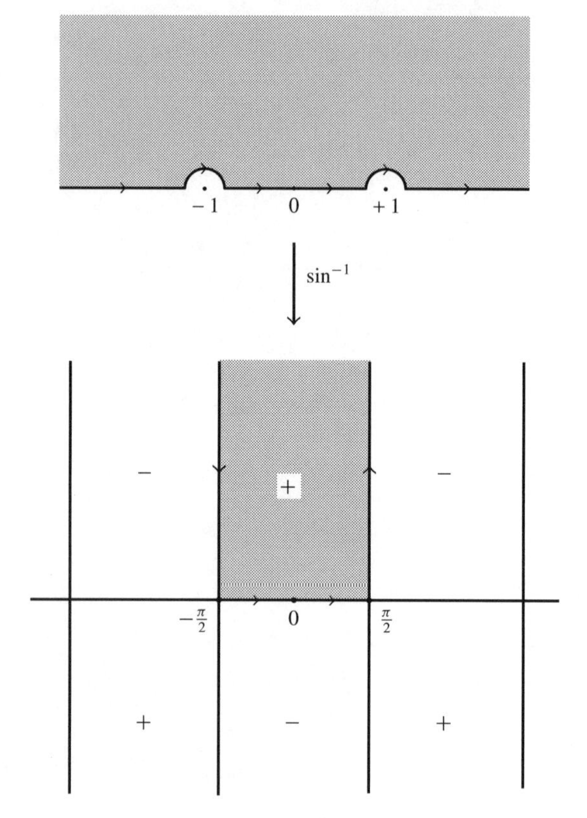

Figure 2.5. The map $y = \sin^{-1} x$.

a function **sinus** $= \sin x$, which is (1) of rational character, and (2) of period

$$2\pi = 4 \times \text{the complete integral} \int_0^1 \frac{dy}{\sqrt{1-y^2}},$$

permitting you to view it as a function on the complex cylinder $\mathbf{X} = \mathbb{C}/\mathbb{L}$ with $\mathbb{L} = 2\pi\mathbb{Z}$:

$$x = \int_0^{\sin x} (1-y^2)^{-1/2} dy \quad \text{up to periods.}$$

Exercise 1. Think this through from the figure.

Exercise 2. The **sinus** may be viewed as the Riemann map of the strip $|x| < \pi/2$, $y > 0$ to the upper half-plane, standardized by the values $0, 1, \infty$ at

$0, \pi/2, \sqrt{-1}\infty$, and subsequently extended to the whole plane by reflection. Think this through too.

Jacobi's Integral of the First Kind. It is a remarkable fact that the same thing happens for the incomplete integral of the first kind:

$$x \mapsto \int_0^x \left[(1-y^2)(1-k^2y^2)\right]^{-1/2} dy.$$

$k = 0$ is the trigonometric case just discussed. The novel point for $k^2 \neq 0, 1$ is that *the inversion of the integral now leads to an elliptic function*, that is, to a (single-valued) function of rational character having not just one but *two* independent complex periods; this is the fundamental discovery of Gauss [1799] and Abel [1827]. Indeed, by May 1799, Gauss already knew that the incomplete lemniscatic integral $\int (1-y^4)^{-1/2} dy$ has such an inverse function; it is the original elliptic function.[8]

Discussion. The direct map indicated in Fig. 2.6 is modified by detouring about

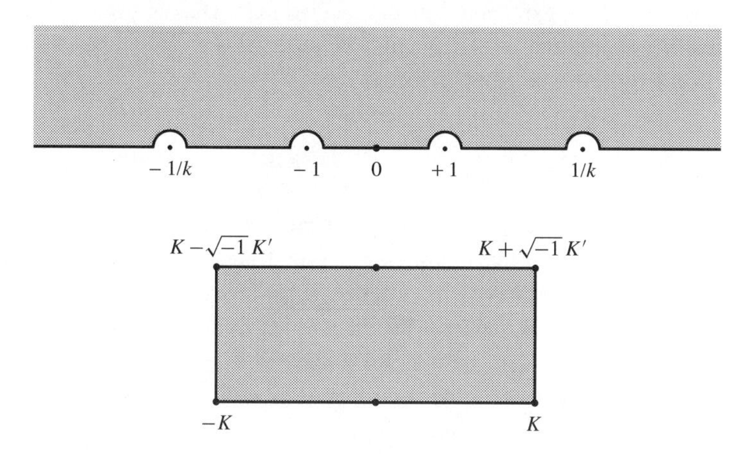

Figure 2.6. Mapping of sin amp. The direct map.

the branch points ± 1 and $\pm 1/k$ in different ways; for example, if you pass under them via infinitesimal semicircles in the lower half-plane, you will obtain the congruent rectangle immediately below the original, double-hatched rectangle of the figure, and the lower half-plane will be mapped 1:1 onto the inside. The recipe can be varied by winding about the branch points by any integral multiple of $180°$. This produces adjacent rectangles forming a **paving** or **tessellation**

[8] Gauss [1870–1929 (10, Abh. 2): 62–84].

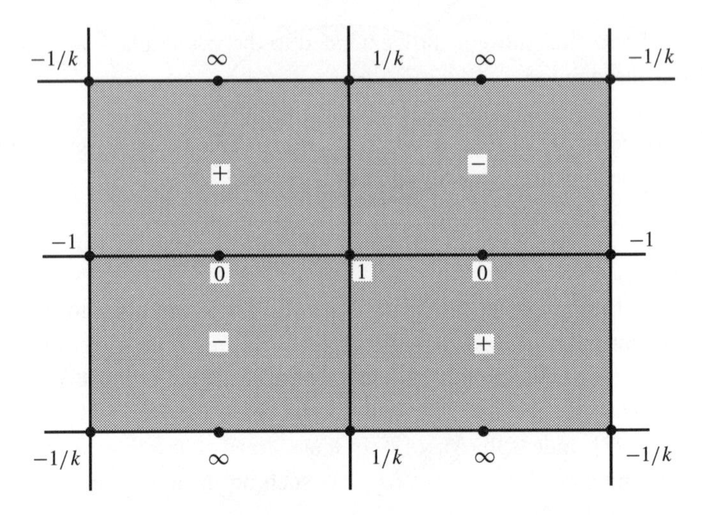

Figure 2.7. Mapping of sin amp. The four rectangles.

of the whole plane. Now consider the inverse map. Jacobi [1829] called it the **sinus amplitudinus**; it is denoted, variously, by sn$x$ or by sn(x, k) if the modulus k requires to be emphasized:

$$x = \int_0^{\operatorname{sn}(x,k)} \left[(1-y^2)(1-k^2y^2)\right]^{-1/2} dy \quad \text{up to periods.}$$

A fourfold copy of the original rectangle of Fig. 2.6 is shaded in Fig. 2.7; the labels $0, \pm 1, \pm 1/k, \infty$ refer to preimages. The function sin amp maps it 2:1 onto $\mathbb{C}+\infty$, alias the projective line $\mathbb{P}^1$; elsewhere, it repeats itself in congruent blocks of four rectangles and so is invariant under translation by $\omega_1 = 4K(k)$ and by $\omega_2 = 2\sqrt{-1}K'(k)$. Here K and K' are the complete elliptic integrals of the first kind of Section 1. This is the same as saying that *sin amp has two independent complex periods* ω_1 *and* ω_2 *and so can be viewed as a function of rational character on the torus* $\mathbf{X} = \mathbb{C}/\mathbb{L}$ *obtained as the quotient of* $\mathbb{C}$ *by the period lattice* $\mathbb{L} = \mathbb{Z}\omega_1 \oplus \mathbb{Z}\omega_2$; compare Fig. 2.8 and also Fig. 2.9 in which $\mathbf{X}$ appears as a two-sheeted cover of the projective line $\mathbb{P}^1$, ramified over the four points $\pm 1, \pm 1/k$ of the latter where the radical $\sqrt{(1-x^2)(1-k^2x^2)}$ vanishes.

Note that there is a general principle here: A function of rational character on the complex torus $\mathbf{X} = \mathbb{C}/\mathbb{L}$ can be viewed as a doubly periodic function on the universal cover $\mathbb{C}$ of $\mathbf{X}$, and vice versa. $\mathbf{X}$ is a so-called **elliptic curve**; in fact, it may be identified with the nonsingular projective quartic $\mathbf{y}^2 = (1-\mathbf{x}^2)(1-k^2\mathbf{x}^2)$ via the substitution sn$x \mapsto \mathbf{x}$, sn$'x \mapsto \mathbf{y}$, as will be further explained in Section 11.

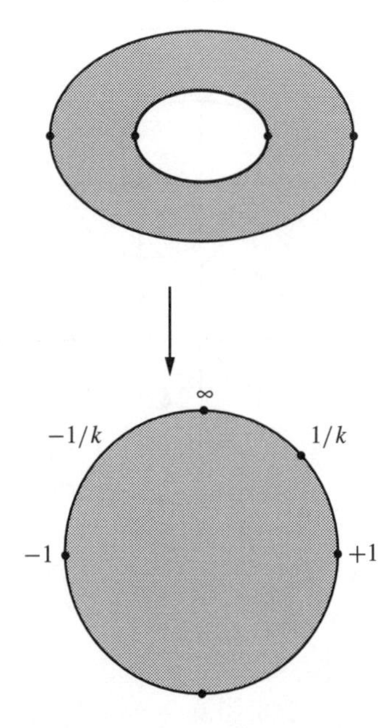

Figure 2.8. **X** as a ramified cover of $\mathbb{P}^1$.

The striking geometric fact, that *the quartic, in turn, may be identified with a torus*, has also a geometric and an algebraic proof; compare Section 16.

Exercise 3. The direct map $x \mapsto \int_0^x \left[(1-y^2)(1-k^2y^2)\right]^{-1/2} dy$ is of rational character, including at ∞, except perhaps at the branch points $\pm 1, \pm 1/k$ where the radical vanishes, so the inverse map $x \mapsto \mathrm{sn}x$ is automatically of rational character except perhaps at their images $\pm K + \mathbb{L}$ and $\pm K + \sqrt{-1}K' + \mathbb{L}$. Check that $\mathrm{sn}x$ is perfectly well behaved at these places, too.

Exercise 4. The complete integral of $[(1-x^2)(1-k^2x^2)]^{-1/2}$ taken between any two of the branch points $\pm 1, \pm 1/k$ is a half-period of sin amp. Prove that the integral of $[(x-e_1)\cdots(x-e_n)]^{-1/2}$ taken between any pair of distinct branch points $e_i (i \le n)$ is a half-period of the inverse function of the incomplete integral

$$\int_0^x [(y-e_1)\cdots(y-e_n)]^{-1/2}\, dy.$$

Deduce that the inverse of such an integral cannot be single-valued if $n \ge 5$. *Hint*: The inverse function satisfies $(f')^2 = (f-e_1)\cdots(f-e_n)$. What kind of poles could it have? How big is it at ∞? What then?

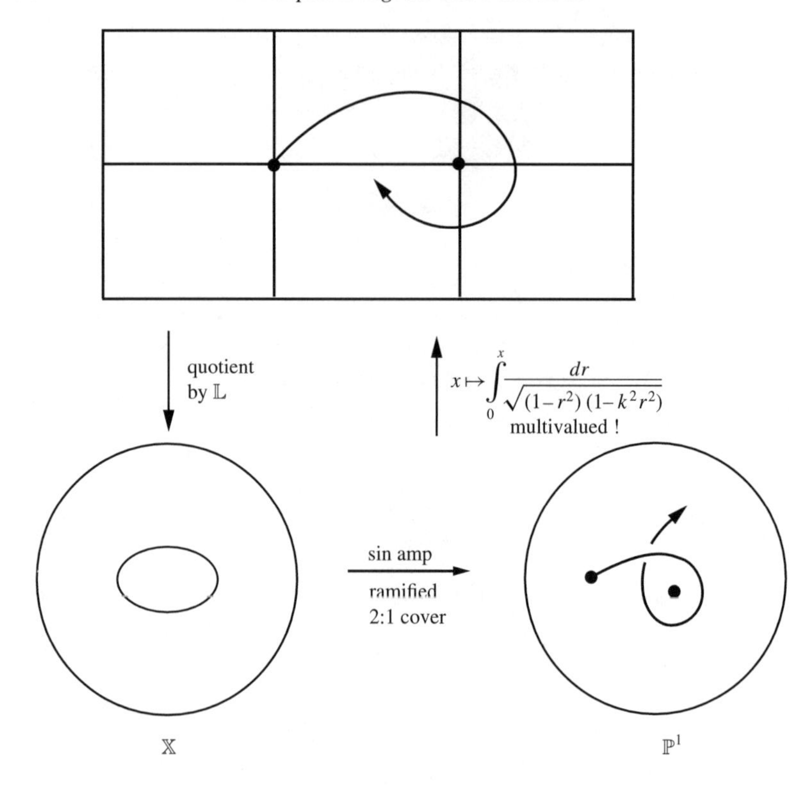

Figure 2.9. Mapping of sin amp.

Exercise 5. Redo ex. 2 for sin amp, that is, think of it as the Riemann map of the double-hatched rectangle of Fig. 2.6 to the upper half-plane $\mathbb{H}$, standardized by the values 0, 1, ∞ at 0, $K(k)$, and $\sqrt{-1}K'(k)$ and extended to the whole plane by reflection.

Exercise 6. Use $(\mathrm{sn}'x)^2 = (1 - \mathrm{sn}^2 x)(1 - k^2\mathrm{sn}^2 x)$ to integrate the equation of motion of the simple pendulum of example 1.4. *Answer*: $\sin(\theta/2) = (k/\sqrt{2})\mathrm{sn}(It/\sqrt{2}, k)$ with $k^2 = 1 - I$. What happens at $k = 0^+$? At $k = 1^-$? Draw the phase diagrams of the motions and compare.

Mechanical aside. The equation of motion of the simple pendulum is in Newton's form: force = mass $\times$ acceleration. Now the acceleration is the second time derivative of position, so the substitution $t \mapsto \sqrt{-1}t$ is equivalent to reversing the gravitational force and leads, on self-evident mechanical grounds, to a complementary periodic motion of the pendulum. The moral is that the solution of the problem with downward-directed gravity has not only its obvious *real* temporal period, but a *purely imaginary* temporal period as well. This remark constitutes a mechanical proof of the double-periodicity of Jacobi's sinus amplitudinus, originating with Appell (1870); see Whittaker [1937:73].[9]

Exercise 7. The conduction of the nervous impulse in, for example, the optic nerve of a squid is caricatured by the equation of Nagumo, Arimoto, and Yoshizawa [1962]: $\partial e/\partial t = \partial^2 e/\partial x^2 + e(1-e)(e-a), 0 < a < 1/2$, in which $e(t, x)$ is the voltage viewed as a function of the time t and the distance x along the nerve. Hodgkin [1971] discusses a more realistic model and its physiological background; see also McKean [1970] for pictures. A traveling wave is a solution of the form $e = e(x - ct)$. Describe the waves of speed 0. *Hint*: $e'' = e(e-1)(e-a)$ implies $(e')^2 = (1/2)e^4 - (2/3)(1+a)e^3 + ae^2 - b = P_4(e)$ with a constant of integration b. Check that the roots of $P_4(e)$ satisfy $e_1 < 0 < e_2 < a < e_3 < 1 < e_4$ if $a^3(2-a) > 6b > 0$. Then express the solution in the form

$$\frac{e_2 - e_1}{e_3 - e_2} \cdot \frac{e_3 - e(x)}{e(x) - e_1} = \frac{2}{1+k} \frac{kf(x) - 1}{f(x) - 1}$$

with $f(x) = \operatorname{sn}\left[2^{-3/2}\sqrt{k}\sqrt{(e_1 - e_2)(e_3 - e_4)}x, k\right]$ and $4k(1+k)^{-2}$ = the cross ratio of e_1, e_2, e_3, e_4. *Hint*: See step 2 of Section 2 for help.

2.6 Periods in General

Jacobi's sinus amplitudinus illustrates the possibility that a function of rational character on the plane can have two (or more) complex periods. Let f be such a (nonconstant, single-valued) function with periods ω: $f(x+\omega) = f(x)$. These form a **module** $\mathbb{L}$ over $\mathbb{Z}$, that is, any integer combination of periods $n_1\omega_1 + n_2\omega_2$ is also a period. $\mathbb{L}$ can be trivial ($\equiv 0$) as for most functions, or of rank 1 as for sin ($\mathbb{L} = 2\pi\mathbb{Z}$), or of rank 2 as for sin amp ($\mathbb{L} = 4K\mathbb{Z} \oplus 2\sqrt{-1}K'\mathbb{Z}$).

Period Lattices. Jacobi [1829] proved that there are no further possibilities: Either $\mathbb{L} = 0$ or else it is of rank $r = 1$ or 2 over $\mathbb{Z}$; in the third case, $\mathbb{L}$ is a **period lattice** and f is an **elliptic function**.

Proof. $\mathbb{L}$ cannot contain infinitesimal (= arbitrarily small) periods since f is not constant, and as $\mathbb{L}$ is obviously closed, it contains a nonvanishing period ω_1 closest to the origin. $\mathbb{L} \supseteq \omega_1\mathbb{Z}$ and $r = 1$ if $\mathbb{L} = \omega_1\mathbb{Z}$. In any case, there are no other periods on the line $\omega_1\mathbb{R}$: Such a period, reduced mod ω_1, would be too

[9] I owe my own appreciation of this pretty device to M. Kac. [H. McKean]

near the origin. Now if $\mathbb{L}$ is more extensive than $\omega_1\mathbb{Z}$, then any new period ω_2 can be brought into the shaded strip of Fig. 2.10 by reduction mod ω_1. The fact

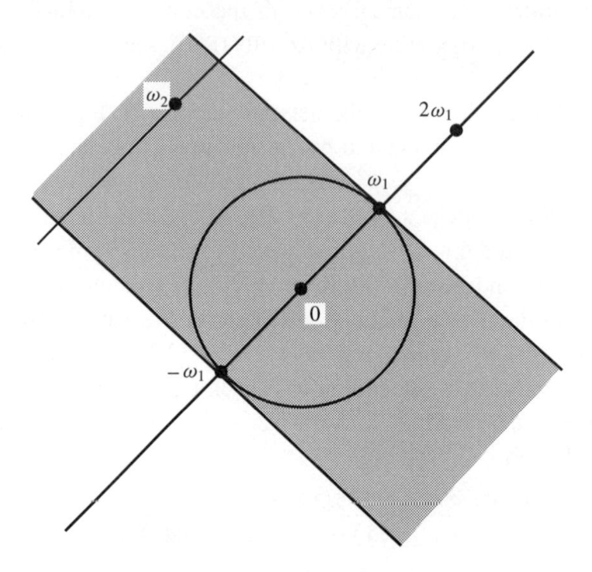

Figure 2.10. The shaded strip.

that $\mathbb{L}$ is closed is invoked a second time to make the distance between $\omega_1\mathbb{R}$ and $\omega_1\mathbb{R}+\omega_2$ as small as possible. The lattice $\mathbb{L}' = \omega_1\mathbb{Z} \oplus \omega_2\mathbb{Z} \subset \mathbb{L}$ is depicted in Fig. 2.11. It is the whole of $\mathbb{L}$ since any other period, reduced mod $\mathbb{L}'$, would have the form $\alpha\omega_1 + \beta\omega_2$ with $0 \le \alpha, \beta < 1$, and either $\beta > 0$, violating the choice of ω_2, or $\beta = 0$ and $\alpha > 0$, violating the choice of ω_1. The proof is finished.

The Modular Group. The lattice $\mathbb{L} = \omega_1\mathbb{Z} \oplus \omega_2\mathbb{Z}$ can be provided with many pairs of **primitive periods**; for example, ω_1 and $5\omega_1 + \omega_2$ will do just as well as ω_1 and ω_2. The question is: *What are all the possible pairs of primitive periods for a fixed lattice* $\mathbb{L}$? Let ω_1' and ω_2' be such a pair. Then

$$\omega_2' = i\omega_2 + j\omega_1,$$
$$\omega_1' = k\omega_2 + l\omega_1$$

with $i, j, k, l \in \mathbb{Z}$ and vice versa:

$$\omega_2 = i'\omega_2' + j'\omega_1',$$
$$\omega_1 = k'\omega_2' + l'\omega_1'$$

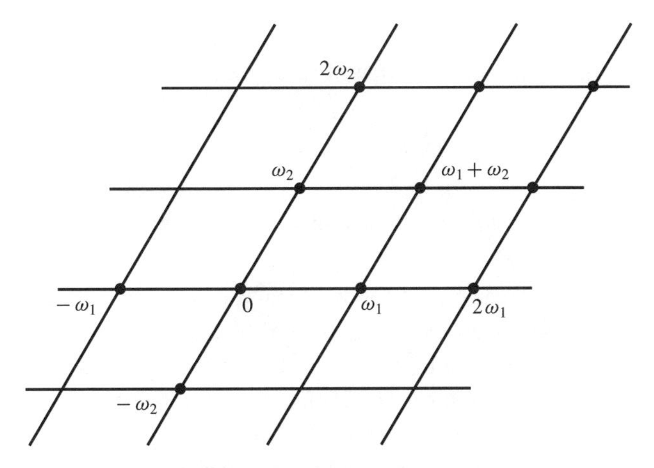

Figure 2.11. The lattice.

with $i', j', k', l' \in \mathbb{Z}$. Now any pair of primitive periods represents two independent directions in the plane. It follows that the 2×2 integral matrices $[ij/kl]$ and $[i'j'/k'l']$ are inverse to each other; in particular, the product of their (integral) determinants is unity and so must have the common value $+1$ or -1. The value $+1$ is enforced by the rule, imposed from now on without further comment, that the **period ratio** $\omega = \omega_2/\omega_1$ have positive imaginary part. The moral is that *two pairs of primitive periods are related by a linear substitution* $[ij/kl]$ *from the special linear group* $SL(2, \mathbb{R})$ *of* 2×2 *integral matrices of determinant* $+1$, *and conversely, any such substitution produces new primitive periods from old.* The corresponding period ratios $\omega = \omega_2/\omega_1$ are related by the associated *fractional* linear substitution $\omega \mapsto (i\omega + j)(j\omega + l)^{-1}$ from the projective group $PSL(2, \mathbb{Z}) = SL(2, \mathbb{Z})/(\pm 1)$. The latter is the **modular group of first level** Γ_1; its elements are **modular substitutions**; see Chapter 4 for more on this subject.

Torus with Complex Structure. The quotient $\mathbf{X} = \mathbb{C}/\mathbb{L}$ of the plane by a lattice $\mathbb{L} = \omega_1\mathbb{Z} \oplus \omega_2\mathbb{Z}$ is a complex manifold with complex structure inherited from $\mathbb{C}$. These are all the complex tori, as noted in Section 1.14. Now the question before you is: *When are two such tori conformally equivalent, that is, when is there a* 1:1 *analytic map between them*? Let $\mathfrak{F}$ be the **fundamental cell** $\{x = \alpha\omega_1 + \beta\omega_2 : 0 \le \alpha, \beta < 1\}$ of the complex torus $\mathbf{X} = \mathbb{C}/\mathrm{L}$ seen in Fig. 2.12; the top and the right sides are omitted since they are identified with the bottom and the left sides when the cell is folded up to make the torus. Now any 1:1 conformal map f of $\mathbf{X}$ to a second torus $\mathbf{X}'$ may be lifted to a map f of their universal covering planes $\mathbb{C}$ as in Section 1.11: With a preliminary translation

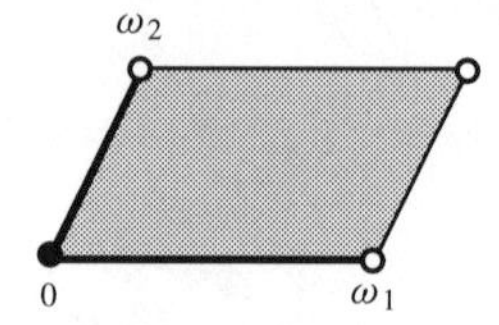

Figure 2.12. Fundamental cell of the torus.

$x \mapsto x + c$ of $\mathbf{X}$, f can be viewed as a map of fundamental cells $\mathfrak{F} \mapsto \mathfrak{F}'$ fixing the origin and then extended, periodically, to a map of tessellations $\mathbb{C} = \mathfrak{F} \oplus \mathbb{L} \mapsto \mathbb{C} = \mathfrak{F}' \oplus \mathbb{L}'$ preserving the period lattices $\mathbb{L}$ and $\mathbb{L}'$. The lifted map is an automorphism of $\mathbb{C}$, of the form $x \mapsto ax + b$, $f(0) = 0$ kills b, and you have $a\mathbb{L} = \mathbb{L}'$, from which you learn that *two complex tori are conformally equivalent if (and only if) their period ratios are related by a modular substitution* $\omega' = (i\omega + j)(k\omega + l)^{-1}$.

Moral. The family of orbits of the modular group Γ_1, acting on the (open) upper half-plane $\mathbb{H}$, can be viewed as *a list of all the possible complex structures on the topological torus*; compare Section 4.3 where this list is identified with the fundamental region of the modular group Γ_1; compare also Section 1.14 on the conformal structure of annuli.

Complex Multiplication. Let f (not the identity) be a conformal self-map of the torus $\mathbb{C} = \mathbf{X}/\mathbb{L}$ of any degree, that is, not necessarily 1:1. After a preliminary adjustment, it can be lifted, as before, to an automorphism $x \mapsto cx$ of the cover $\mathbb{C}$. Then $c\mathbb{L} \subset \mathbb{L}$ and the inclusion is (mostly) proper; for example, the self-map $x \mapsto nx$ is always available. $\mathbf{X}$ is said to admit **complex multiplication** if it has such a self-map $x \mapsto cx$ with $c \notin \mathbb{Z}$, in which case

$$c\omega_2 = i\omega_2 + j\omega_1,$$

$$c\omega_1 = k\omega_2 + l\omega_1$$

with $i, j, k, l \in \mathbb{Z}$, and the period ratio ω is a fixed point of the substitution $[ij/kl]$; as such, it is, as you will check, a quadratic irrationality from the field $\mathbb{Q}\left[\sqrt{(i+l)^2 - 4(il - jk)}\right]$.

Exercise 1. The matter simplifies if the self-map is 1:1. Then $\mathbb{L} = c\mathbb{L}$ and c is of modulus 1 (why?) and also a root of unity (why is this?). The possibilities are $c = 1, \sqrt{-1}$, and $e^{\pm 2\pi\sqrt{-1}/3}$, up to sign, so the only interesting lattices of this type are square ($\omega = \sqrt{-1}$), as in Fig. 2.13, or triangular ($\omega = e^{2\pi\sqrt{-1}/3} = (1 + \sqrt{-3})/2$), as in Fig. 2.14. Check it.

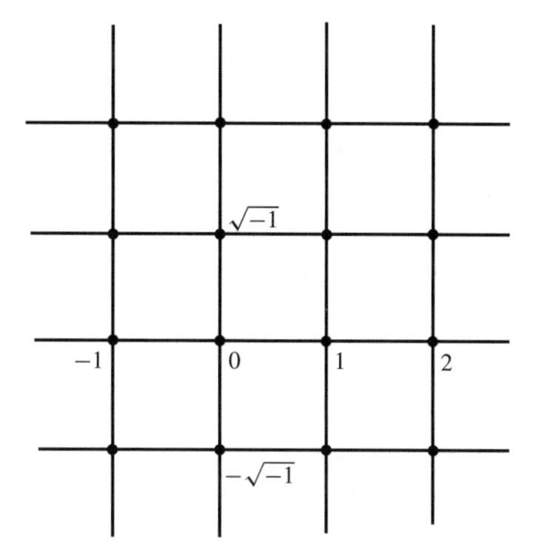

Figure 2.13. The square lattice $\omega = \sqrt{-1}$.

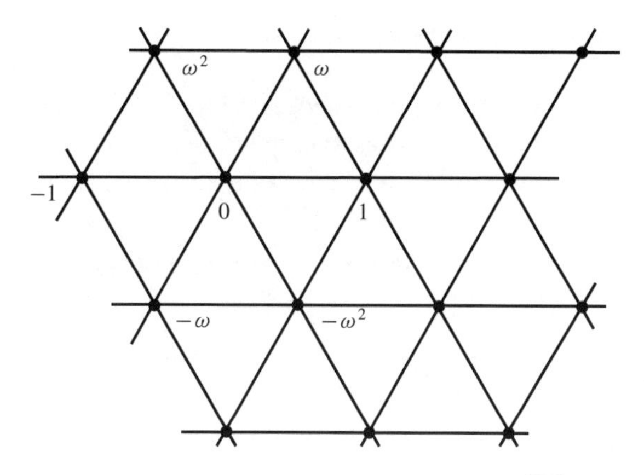

Figure 2.14. The triangular lattice $\omega = (1 + \sqrt{-3})/2$.

Exercise 2. $\mathbf{X} = \mathbb{C}/\mathbb{L}$ always admits, as automorphisms, translations $x \mapsto x+c$ and reflection $x \mapsto -x$. Prove that addition of half-periods and reflection are the only involutions of $\mathbf{X}$.

2.7 Elliptic Functions in General

An **elliptic function field** is the class $\mathbf{K} = \mathbf{K}(\mathbf{X})$ of functions of rational character on a complex torus $\mathbf{X}$. The identification of $\mathbf{X}$ as the quotient $\mathbb{C}/\mathbb{L}$ of $\mathbb{C}$ by the (period) lattice $\mathbb{L} = \omega_1\mathbb{Z} \oplus \omega_2\mathbb{Z}$ permits the identification of $\mathbf{K}$ with doubly periodic functions on the (universal) cover $\mathbb{C}$ of $\mathbf{X}$. The **primitive**

periods ω_1 and ω_2 are taken as before, so that the **period ratio** $\omega = \omega_2/\omega_1$ has positive imaginary part. The parallelogram they span is the **fundamental cell** $\mathfrak{F}$, the top and right sides being omitted, as before. Note that **K** is a *differential field*: The global coordinate x of the cover $\mathbb{C}$ gives $f' = df/dx$ an unambiguous meaning and f' belongs to **K** if f does.

Now the source of all good things in this subject is the

Fundamental Lemma. The integral of any function $f \in \mathbf{K}$, taken about the perimeter of the fundamental cell, vanishes.

Amplification. $f \in \mathbf{K}$ is a function of the point $\mathfrak{p} \in \mathbf{X}$; it is also a (periodic) function of x on the universal cover $\mathbb{C}$. The map of $x \in \mathbb{C}$ to $\mathfrak{F}$ by reduction modulo $\mathbb{L}$ is the covering map. $dx(\mathfrak{p})$ is *the* differential of the first kind on **X**, and $f(\mathfrak{p})dx(\mathfrak{p})$ is to be integrated; see ex. 9.1.

Exercise 1. Think it over; compare ex. 1.3.6.

Proof. The path of integration is indented, as in Fig. 2.15, to avoid any poles of f on the perimeter, in such a way as to keep just one copy of each pole inside the modified path. These are finite in number (why?) so there is no difficulty.

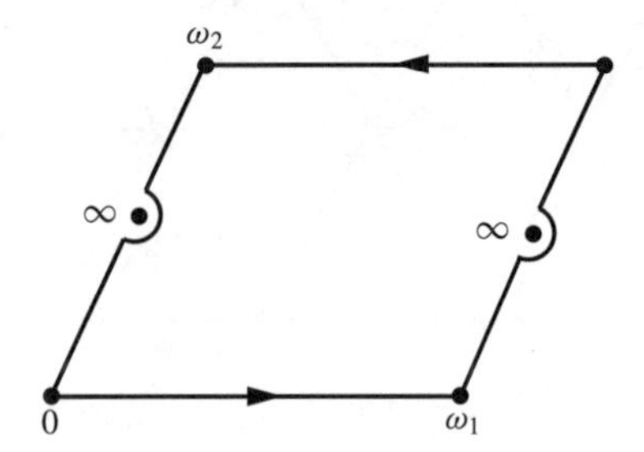

Figure 2.15. The indented path of integration.

Now the function repeats its values on the top and bottom of the cell, and as these are tranversed in opposite directions, so their contributions to the integral cancel, and likewise for the sides.

This simple lemma has very important consequences. The next few items illustrate its content.

Item 1. A function $f \in \mathbf{K}$ has a finite number of poles per cell. This number is its **degree** d. It has the same number of roots per cell, as you will check by integrating $(2\pi\sqrt{-1})^{-1} f'/f \in \mathbf{K}$ about the perimeter. More generally, it takes on every complex value d times *per* cell, just like a function f in $\mathbf{K}(\mathbb{P}^1)$: $f - c$ has the same number of poles as f and so, also, the same number of roots; compare Sections 1.3 and 1.16.

Item 2. The only functions $f \in \mathbf{K}$ of degree $d = 0$ are the constants; indeed, such a function is bounded in the cell and so, also, in the whole of $\mathbb{C}$, by periodicity.

Item 3. $\mathbf{K}$ has no functions of degree 1; such a function would have a single pole with residue $(2\pi\sqrt{-1})^{-1} \oint f\,dx = 0$, by the lemma. Here you see the simplest function-theoretic distinction between $\mathbf{X}$ and the projective line $\mathbb{P}^1$: $\mathbf{K}(\mathbb{P}^1)$ contains functions of any degree you like.

Item 4. The roots and poles of a function $f \in \mathbf{K}(\mathbb{P}^1)$ can be placed on the sphere as you will; see ex. 1.3.3. This is *not* so for $\mathbf{K} = \mathbf{K}(\mathbf{X})$! Let $\mathfrak{p}_1, \ldots, \mathfrak{p}_d$ be the poles and $\mathfrak{q}_1, \ldots, \mathfrak{q}_d$ the roots of $f \in \mathbf{K}$ and integrate $2\pi\sqrt{-1}x(\mathfrak{p}) \times d\log f(\mathfrak{p})$ around the perimeter of the cell. The integrand xf'/f is *not* a member of $\mathbf{K}$ so the fundamental lemma does not apply, but the same reasoning produces

$$
\begin{aligned}
\sum_{\text{poles}} x(\mathfrak{p}) - \sum_{\text{roots}} x(\mathfrak{q}) &= \frac{1}{2\pi\sqrt{-1}} \oint x d \log f \\
&= \frac{1}{2\pi\sqrt{-1}} \int_{\text{bottom}} [x - (x + \omega_2)]\, d \log f \\
&\quad + \frac{1}{2\pi\sqrt{-1}} \int_{\text{right side}} [x - (x - \omega_1)]\, d \log f \\
&= -\omega_2 \times \frac{1}{2\pi\sqrt{-1}} \int_{\text{bottom}} d \log f \\
&\quad - \omega_1 \times \frac{1}{2\pi\sqrt{-1}} \int_{\text{right side}} d \log f \\
&= n_1\omega_1 + n_2\omega_2,
\end{aligned}
$$

in which the first line is the evaluation of the second by residues, and the last line reflects the fact that f takes the *same* value at all the four corners of the cell. In short, pole sum − root sum = a period. Abel [1827] proved that this necessary condition, that the $\mathfrak{p}$s be the poles and the $\mathfrak{q}$s the roots of a function $f \in \mathbf{K}$, is also sufficient; see Section 15 for the proof.

Exercise 2. The function $f \in \mathbf{K}$ is viewed as a projection of $\mathbf{X}$ onto $\mathbb{P}^1$, so $\mathbf{X}$ appears as a ramified cover of $\mathbb{P}^1$ to which the Riemann–Hurwitz formula $r = 2(d + g - 1)$ of Section 1.12 applies: d is the sheet number, alias the degree of f, and $g = 1$ is the genus of $\mathbf{X}$ so $r = 2d$. The ramifications of the cover of index m are the points of $\mathbf{X}$ where the function takes its value with multiplicity $m + 1 \geq 2$. Confirm the evaluation $r = 2d$ by the fundamental lemma. *Hint*: $d = \sum_{f=0}(m - 1) = \sum_{f=\infty}(m + 1)$. Now evaluate the degree of f' as $\sum_{f=\infty}(m + 1) + \sum_{f\neq\infty} m$.

2.8 The $\wp$-Function

The degree d of a nonconstant function $f \in \mathbf{K}$ is at least 2 (see item 7.3), so the simplest possibilities are: two simple poles or one double pole. Jacobi's function sinus amplitudinus exemplifies the first possibility for tori with period ratio $\sqrt{-1}K'/2K$, but at this point it is not clear that every period ratio can be so replicated by choice of the modulus $k^2 \neq 0$ or 1. The fact is that it *can* (see Section 16), but for now it is simpler to follow Weierstrass (1850) in constructing a function with a single *double* pole. This is the celebrated $\wp$-function. The idea is simple: The pole is placed at the origin so it appears as x^{-2}, plain, and the function is made periodic by summing its translates over the period lattice. The attempt is naive, the sum being divergent like $\sum_{\mathbb{Z}^2}(n^2 + m^2)^{-1} \sim 2\pi \int_1^\infty r^{-2} \times r dr$, but the idea is salvaged by taking

$$\wp(x) = x^{-2} + \sum_{\omega\in\mathbb{L}^2-0} \left[(x - \omega)^{-2} - \omega^{-2}\right].$$

Now the general summand is comparable to ω^{-3} and $\sum_{\mathbb{L}-0} |\omega|^{-3}$ converges like $\int_1^\infty r^{-3} \times r dr$, so the divergence is cured at the price of obscuring the periodicity of the function. This will be dealt with in a moment. Note first that $\wp$ is an even function of x in view of $\mathbb{L} = -\mathbb{L}$. Now the derivative $\wp' = -2\sum_{\mathbb{L}}(x - \omega)^{-3}$ is kosher: The sum converges fine. It is also periodic, by inspection, so $\wp(x+\omega) - \wp(x)$ is constant in x for any $\omega \in \mathbb{L}$, and the constant is seen to vanish by evaluation at the half-period $-\omega/2$: $\wp(\omega/2) - \wp(-\omega/2) = 0$, by the symmetry of $\wp$.

Moral. $\wp \in \mathbf{K}(\mathbf{X})$ *is an even function of degree* 2 *with a double pole at* $x = 0$ *and no others in the cell.*

Exercise 1. Show that $\wp(cx|c\mathbb{L}) = c^{-2}\wp(x|\mathbb{L})$ for any complex number $c \neq 0$.

Now look at the derivative $\wp'(x)$: In the cell, it has just one (triple) pole at $x = 0$, so it is of degree 3 and, as such, has three (and only three) roots per cell. It is also odd and so vanishes at half-periods: $\wp'(\omega/2) = \wp'(\omega/2 - \omega) = \wp'(-\omega/2) = -\wp'(\omega/2)$. There are three such (inequivalent) half-periods per cell, so $\wp'(x) = 0$ has no further roots and these are simple. The corresponding values $e_1 = \wp(\omega_1/2)$, $e_2 = \wp(\omega_1/2 + \omega_2/2)$, $e_3 = \wp(\omega_2/2)$ of $\wp(x)$ play an important role in the sequel; see Fig. 2.16.

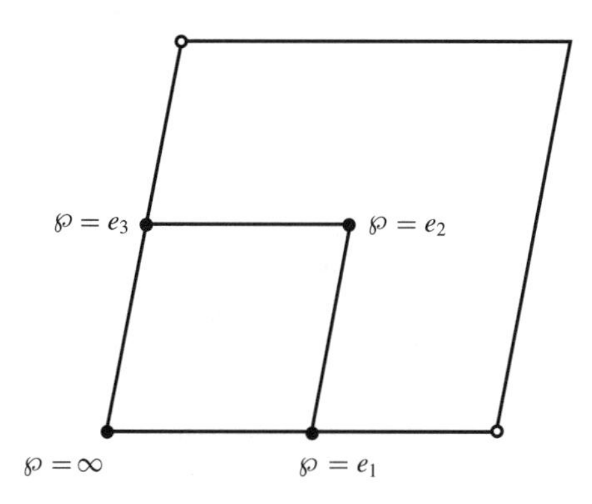

Figure 2.16. The values at half-periods.

Exercise 2. The numbers e_1, e_2, e_3 are distinct. Why?

Exercise 3. The map $\wp: \mathbf{X} \to \mathbb{P}^1$ exhibits the torus as a 2:1 cover of the projective line, ramified over the four distinct points e_1, e_2, e_3, ∞, as in Fig. 2.17. The ramification indices are all 1 in agreement with the Riemann–Hurwitz formula:

$$4 = \text{ramification index} = 2 \times (\text{sheets} + \text{genus} - 1) = 2 \times (2 + 1 - 1).$$

Think it over.

Exercise 4. What happens if one or both of the primitive periods of $\wp$ is taken to ∞? *Answer*: $\wp(x) \to (\pi/\omega_1)^2 \times \left[\sin^{-2}(\pi x/\omega_1) - 1/3\right]$ for $\omega_2 = \infty$ and $\wp(x) \to 1/x^2$ if ω_1 is taken to ∞ as well.

Exercise 5. The map $j: x \mapsto -x$ of ex. 6.2 is the only interesting involution of the torus. Identify the quotient $\mathbf{X}/j$ as the projective line and the quotient map $\mathbf{X} \mapsto \mathbf{X}/j$ as the $\wp$-function.

The Differential Equation. The subject of elliptic functions is full of extraordinary identities proved by a clever stroke or two. This provides much of its

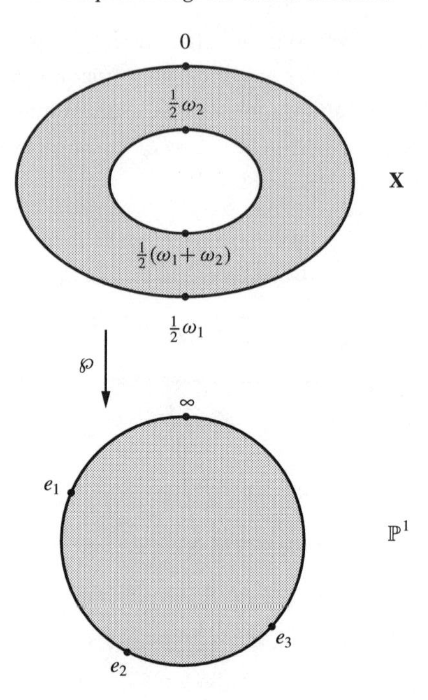

Figure 2.17. The torus as a ramified cover of $\mathbb{P}^1$.

charm. The simplest example is the differential equation for the $\wp$-function:

$$(\wp')^2 = 4(\wp - e_1)(\wp - e_2)(\wp - e_3).$$

Exercise 6. Try (for a few minutes) to check this directly from the sums.

Now try this: $\wp - e_1$ has a double root at the half-period $x = \omega_1/2$, as does $(\wp')^2$, and likewise at the other two half-periods; also, $(\wp - e_1)(\wp - e_2)(\wp - e_3)$ has a pole of degree 6 at $x = 0$. $(\wp')^2$ imitates this feature, too, so the ratio $4(\wp - e_1)(\wp - e_2)(\wp - e_3)/(\wp')^2 \in \mathbf{K}$ is constant ($= 1$) as it is pole-free in the cell and looks like $4(x^{-2})^3(-2x^{-3})^{-2} = 1$ at $x = 0$.

A second form of the differential equation is often preferred: In the vicinity of $x = 0$, $\wp(x) = x^{-2} + 0 \cdot x^0 + ax^2 + bx^4 + \cdots$ with[10] $a = 3\sum'\omega^{-4}$ and $b = 5\sum'\omega^{-6}$, by inspection of the sum for $\wp$, so $\wp'(x) = -2x^{-3} + 2ax +$

[10] The *prime* signifies that only the nonvanishing periods enter into the sum.

$4bx^3 + \cdots$ and

$$\frac{1}{4}\left[\wp'(x)\right]^2 = x^{-6} - 2ax^{-2} - 4b + \cdots$$
$$= [\wp(x) - e_1][\wp(x) - e_2][\wp(x) - e_3]$$
$$= \wp^3(x) - \sigma_1\wp^2(x) + \sigma_2\wp(x) - \sigma_3$$
$$= x^{-6} + 3ax^{-2} + 3b - \sigma_1(x^{-4} + 2a) + \sigma_2 x^{-2} - \sigma_3 + \cdots,$$

with

$$\sigma_1 = e_1 + e_2 + e_3 = 0,$$
$$\sigma_2 = e_1e_2 + e_2e_3 + e_3e_1 = -5a = -15\sum{}'\omega^{-4},$$
$$\sigma_3 = e_1e_2e_3 = 7b = 35\sum{}'\omega^{-6}.$$

The differential equation is now written $(\wp')^2 = 4\wp^3 - g_2\wp - g_3$ with

$$g_2 = 60\sum{}'\omega^{-4} \text{ and } g_3 = 140\sum{}'\omega^{-6}.$$

The numbers g_2 and g_3 are the **invariants of the cubic**; they depend on the lattice in an interesting way to be studied in Chapter 4.

Exercise 7. It is to be proved that, in the full expansion $\wp(x) = x^{-2} + ax^2 + bx^4 + \cdots$, the higher coefficients are (universal) polynomials in a and b. Write out a few of them. *Hint*: $2x^4\wp''(x) + g_2x^4 = 3[x^2\wp'(x)]^2$.

Exercise 8. Reduce the equation of motion of the simple pendulum of example 1.4 from its Jacobian form $(\dot{x})^2 = (1 - x^2)(1 - k^2x^2)$ to its (more or less) Weierstrassian form $(\dot{y})^2 = \frac{1}{2}(1 - k^2)y^3 + (1 + k^2)y^2 + \frac{1}{2}y$ by the substitution $y = (1 - x)/(1 + x)$. What are the numbers e_1, e_2, e_3? Express the motion explicitly in terms of the $\wp$-function.

2.9 Elliptic Integrals, Complete and Incomplete

The differential equation states that $dx = d\wp/\wp' = \frac{1}{2}\,[(\wp - e_1)(\wp - e_2)(\wp - e_3)]^{-1/2}\,d\wp$, so $\wp$ inverts the *incomplete integral*

$$\frac{1}{2}\int_\infty^\wp [(y - e_1)(y - e_2)(y - e_3)]^{-1/2}\,dy.$$

More precisely,

$$x = \frac{1}{2}\int_\infty^{\wp(x)} [(y - e_1)(y - e_2)(y - e_3)]^{-1/2}\,dy \text{ up to periods;}$$

in particular, the primitive periods of $\mathbb{L}$ are expressed by the complete integrals

$$\omega_1 = \int_{\infty}^{e_1} [(y-e_1)(y-e_2)(y-e_3)]^{-1/2}\,dy,$$

$$\omega_2 = \int_{e_1}^{e_2} [(y-e_1)(y-e_2)(y-e_3)]^{-1/2}\,dy$$

with the proper determination of the radicals; compare example 1.1 and Section 5.

Exercise 1. Check that, up to constant multiples, $dx = d\wp/\wp'$ is the only differential of the first kind on $\mathbf{X}$; see Section 1.3 for terminology and ex. 1.3.6 for the fact that the projective line does not carry such an object.

The next examples and exercises relate to specific lattices and cubics.

Example 1. The cubic $\mathbf{X}_1\colon y^2 = x^3 - x$ has its roots at $-1, 0, 1$; it corresponds to a **square lattice** in view of

$$\omega_1 = \int_0^1 (x-x^3)^{-1/2}dx = \int_1^{\infty} (x^3-x)^{-1/2}dx = \omega_2/\sqrt{-1},$$

which you will check by the substitution $x \mapsto 1/x$.

Exercise 2. $\mathbf{X}_1$ admits the involution $(x, y) \mapsto (-1/x, y/x^2)$. It is not the involution $j\colon (x, y) \mapsto (x, -y)$ so it comes from addition of some half-period; see ex. 6.2. What is that half-period?

Exercise 3. Prove that e_1, e_2, e_3 are real if and only if ω_1 is real and ω_2 is purely imaginary. Check that in this case $e_1 > e_2 > e_3$ and that, in the fundamental cell, $\wp(x = a + \sqrt{-1}b)$ is real only on the lines $a = 0$ or $\omega_1/2$ and $b = 0$ or $\omega_2/2$.

Exercise 4. Check that if $e_1 > e_2 > e_3$ are real as in example 1, then $\frac{1}{2}\int_{\infty}^{x} [(y-e_1)(y-e_2)(y-e_3)]^{-1/2}\,dy$ maps the upper half-plane 1:1 onto the fundamental cell, as in ex. 3.

Example 2. This has to do with the triangular lattice of period ratio $\omega = e^{2\pi\sqrt{-1}/3}$ of ex. 6.1. It is taken from Baker [1911: 321]. The lattice is highly symmetrical in that it admits complex multiplication (by ω). The $\wp$-function obeys the rule $\wp(\omega x) = \omega^{-2}\wp(x)$, so the numbers e_1, e_2, e_3 stand in the proportion $1, \omega^2, \omega$; also $g_2 = 0$, $g_3 = 4e_1^3 > 0$, $\omega_1 = \int_1^{\infty}(x^3-1)^{-1/2}dx$ if $e_1 = 1$;

and with this choice of e_1, the curve is $y^2 = 4(x^3 - 1)$. Figure 2.18 displays the locus of real and imaginary values of $\wp'$ by dark and light lines, respectively.

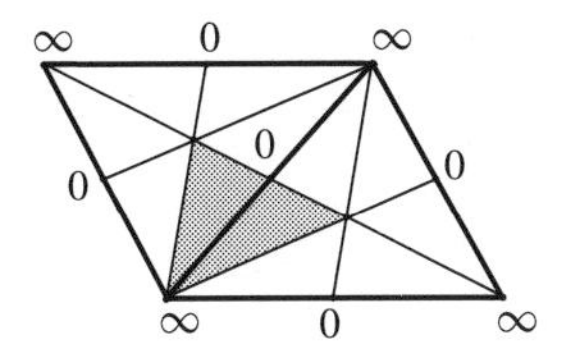

Figure 2.18. Real and imaginary values.

Exercise 5. Check Fig. 2.18 with the help of $\wp'(\omega^{1/2}x) = -\wp'(x)$ and the fact that $\wp'$ is of degree 3. Conclude that $\wp'$ maps the shaded triangle onto the right half-plane $x = a + \sqrt{-1}b$, $a > 0$, and identify this map with the inversion of the incomplete integral $\int_0^x 2^{1/3}3^{-1}(x^2 + 4)^{-2/3}dx$.

Exercise 6. (Trinity College 1905) Prove that, with $e_1 = -a$, $e_2 = \frac{1}{2}(a + b)$, and $e_3 = \frac{1}{2}(a - b)$, $\wp'/2(\wp + a)$ inverts the incomplete integral $\int_\infty^x (x^4 + 6ax^2 + b^2)^{-1/2}dx$.

2.10 Two Mechanical Applications

Wonderful examples from classical mechanics abound. C. Neumann's constrained harmonic oscillators [1859] and the long-wave shallow-water equation of Korteweg–de Vries [1895] are treated here; see also Section 15 for a third illustration; others can be found in Jacobi [1866] and Whittaker [1937].

Example 1. C. Neumann [1859] studied the motion of a pair of uncoupled oscillators, $\ddot{x}_1 + \omega_1^2 x_1 = 0$ and $\ddot{x}_2 + \omega_2^2 x_2 = 0$, with distinct frequencies $0 < \omega_1 < \omega_2$, the joint motion $x = (x_1, x_2)$ being constrained to move on the circle $x_1^2 + x_2^2 = 1$ by the imposition of an external force $-f(x)$. A simple calculation shows $f = 2H - 2(\omega_1^2 x_1^2 + \omega_2^2 x_2^2)$, H being the total energy $(1/2)\left[(\dot{x}_1)^2 + \omega_1^2 x_1^2 + (\dot{x}_2)^2 + \omega_2^2 x_2^2\right]$ of the unconstrained oscillators, and it is easy to see that H is still a constant of motion of the constrained system

$$\ddot{x}_1 + \omega_1^2 x_1 = -f x_1, \quad \ddot{x}_2 + \omega_2^2 x_2 = -f x_2.$$

Exercise 1. Check it.

The integration of the constrained system is effected by means of ellipsoidal coordinates, Jacobi's trick [1866: 198–206]. Fix (x_1, x_2) and introduce the function

$$R(a) = \frac{x_1^2}{\omega_1^2 - a} + \frac{x_2^2}{\omega_2^2 - a} = \frac{b - a}{(\omega_1^2 - a)(\omega_2^2 - a)}$$

seen in Fig. 2.19. R is a rational function of degree 2 with one root at $b =$

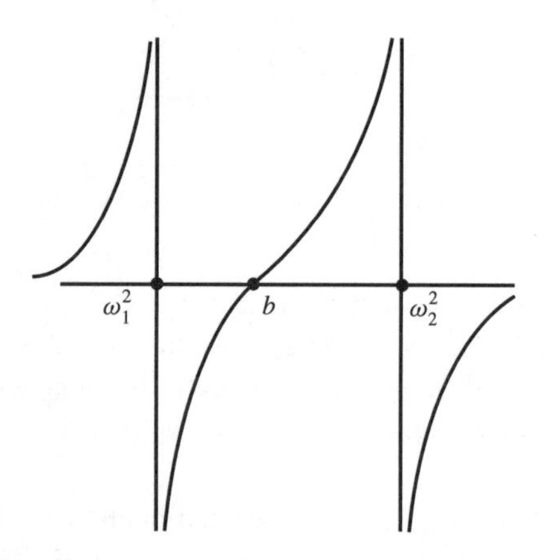

Figure 2.19. The rational function R.

$x_1^2\omega_2^2 + x_2^2\omega_1^2$, located between ω_1^2 and ω_2^2, and one root at ∞; also,

$$x_1 = \sqrt{\frac{b - \omega_1^2}{\omega_2^2 - \omega_1^2}}, \qquad x_2 = \sqrt{\frac{\omega_2^2 - b}{\omega_2^2 - \omega_1^2}}$$

with the proper determination of the radicals. Now compute $\dot{x}_1$ and $\dot{x}_2$ in terms of $\dot{b}$ and substitute into H to obtain

$$(\dot{b})^2 = 4(\omega_2^2 - b)(b - \omega_1^2)\left[b - (\omega_1^2 + \omega_2^2 - 2H)\right].$$

This permits a further integration in terms of a suitable $\wp$-function, b being expressed as $\wp(\sqrt{-1}t + t_0)$, assuming the content of Section 12 to follow that every cubic $y^2 = 4(x - e_1)(x - e_2)(x - e_3)$ is realized by the $\wp$-function of some lattice.

Exercise 2. Supply the details; especially find the value of the shift t_0.

Actually, Neumann treated three oscillators. This leads to the inversion of (sums of) **hyperelliptic integrals** of the form $\int_\infty^x [(x - e_1)\cdots(x - e_5)]^{-1/2}\, dx$; compare Section 18 under the **addition theorem**. The same can be done for any number of oscillators; see Moser [1980] for this and connections with other beautiful mechanical problems such as Jacobi's integration of the geodesic flow on the ellipsoid [1866: 212–14], Kovalevskaya's integration of the top [1889], and more modern examples.

Example 2. Korteweg and de Vries [1895] described the motion of long water waves in a shallow canal by means of the equation

$$\text{(KdV)} \qquad \frac{\partial Q}{\partial t} = 3Q\frac{\partial Q}{\partial x} - \frac{1}{2}\frac{\partial^3 Q}{\partial x^3}.$$

Exercise 3. Check that, if ω_1 is real and positive and ω_2 is purely imaginary, then $Q = 2k^2\wp\,[k(x-ct)+(1/2)\omega_2] - c/3$ solves KdV for any choice of k and c; it is real and smooth if k and c are real. What is the most general such (real, smooth) traveling wave for KdV? What happens if you take ω_1 to $+\infty$ and adjust the wave speed so that Q dies out at $x = \pm\infty$? *Hint*: See ex. 8.4 and ex. 9.3. *Answer*: $Q = -c \times \cosh^{-2}\left[\sqrt{c/2}(x-ct)\right]$ with speed $c > 0$. This is the celebrated **soliton**; see Drazin and Johnson [1989] for details.

Airault, McKean, and Moser [1977] proved that if $Q(t,x)$ belongs, for each value of t, to one and the same elliptic function field $\mathbf{K}(\mathbf{X})$, then it must have the form $Q = +2\sum_1^n \wp(x - x_i(t)) + c$ with moving poles and constant c. Let the primitive periods of $\mathbf{X} = \mathbb{C}/\mathbb{L}$ be one real ($\omega_1 = 1$, say) and one purely imaginary, for reasons of reality (compare ex. 9.3), and let $Q(t,x) = a(x-x_0)^{-m} + b(x-x_0)^{-m+1} + \cdots$, with time dependent $a \neq 0$, b, and x_0, be the expression at an isolated (moving) pole of Q, persisting in some small interval of time. To balance $(\partial Q/\partial t) = 3QQ' - \frac{1}{2}Q'''$ in the vicinity of $x = x_0(t)$, you need

$$\frac{3ma^2}{(x-x_0)^{2m+1}} + \frac{3ab(2m+1)}{(x-x_0)^{2m}} - \frac{1}{2}\frac{am(m+1)(m+2)}{(x-x_0)^{m+3}} - \frac{1}{2}\frac{b(m-1)m(m+1)}{(x-x_0)^{m+2}}$$

to vanish, and that is not possible unless (1) $2m+1 = m+3$, which is to say $m = 2$, (2) $12a = 6a^2$, which is to say $a = 2$, and (3) $b = 0$. Now let $x_i(t)(i \le n)$ be the poles of $Q(t,\bullet)$. Then $c \equiv Q(t,x) - 2\sum_1^n \wp[x - x_i(t)]$ is pole-free and so constant (in x). But $\int_0^1 Q\,dx$ is also a constant of motion, as you will check, and $\int_0^1 \wp(x-x_0)dx$ is independent of x_0 (why?), the upshot being that c is constant, not only in x but in t as well. The poles must be $n = (1/2)m(m+1)(= 1, 3, 6, 10, \text{etc.})$ in number for deeper reasons and move in intricate ways; see Krichever [1980] and Treibich and Verdier [1990] for more (geometric) information about that. The fact is that the most general solution $Q(t,x)$, of spatial period 1, say, comes from a (nonclassical) curve $y^2 = \prod_0^{2g}(x - e_n)$, mostly of genus $g = \infty$; see Novikoff et al. [1984] and, for an overview, McKean [1978].

2.11 The Projective Cubic

The differential equation $(\wp')^2 = 4(\wp - e_1)(\wp - e_2)(\wp - e_3)$ has a geometrically striking interpretation: *The torus* $\mathbf{X} = \mathbb{C}/\mathbb{L}$ *is the same, complex structure and all, as the projective cubic* $y^2 = 4(x - e_1)(x - e_2)(x - e_3)$, *and conversely, every smooth projective cubic arises in this way.* The second statement embodies Abel's **problem of inversion** [1827]; it occupies the next section.

Cubics. A preliminary review of projective cubics is required; see Section 1.10 for background. For the general cubic, the defining relation is

$$\begin{aligned} P(x, y, z) &= 4(x - e_1 z)(x - e_2 z)(x - e_3 z) - y^2 z \\ &= 4x^3 - g_2 x z^2 - g_3 z^3 - y^2 z = 0 \end{aligned}$$

under the harmless assumption that $e_1 + e_2 + e_3 = 0$. The original curve is seen in the patch $\mathbb{C}^2 \times 1$, and there is the single point $(0, 1, 0)$ at infinity. The curve is smooth, that is, it fits into $\mathbb{P}^2$ without singularities, which is to say that the gradient of the defining relation $P = 0$ does not vanish any place on the curve. That is the content of the next two exercises.

Exercise 1. Check that $g_2^3 - 27g_3^2$ is $16\times$ the discriminant $(e_1 - e_2)^2(e_3 - e_2)^2(e_3 - e_1)^2$ of the cubic $x^3 - (g_2/4)x - (g_3/4)$.

Exercise 2. Show that the only point on the curve that might be singular is $(\sqrt{g_2}, 0, 2\sqrt{3})$, and then only if $g_2^3 - 27g_3^2$ could vanish.

The Correspondence. If $x \neq 0$, then the differential equation states that $(\mathbf{x} = \wp(x), \mathbf{y} = \wp'(x), \mathbf{z} = 1) \in \mathbb{C}^2 \times 1$ is a point of the cubic. *This accounts for all such points in a* 1:1 *manner.* Indeed, $\wp(x)$ takes every value $\mathbf{x} \in \mathbb{C} - (e_1, e_2, e_3)$ twice, at x and $-x$, and the two values $\mathbf{y} = \pm\wp'(x)$ discriminate between them, as the two copies of $\mathbb{P}^1$ seen in Fig. 1.18 are meant to suggest; as to $\mathbf{x} = e_1, e_2, e_3$, $\mathbf{y}$ is not needed, the value itself being distinctive already. The correspondence is completed by associating $x = 0$ to the point at infinity $(0, 1, 0)$ of the cubic, in accord with the fact that $\mathbf{x}/\mathbf{y} = \wp(x)/\wp'(x)$ vanishes there. Now the projective cubic has its own complex structure, as expounded in Section 1.10. What it amounts to is that $\mathbf{x}$ serves as a local parameter at most places on the curve, the exceptions being $(\mathbf{x} = e, \mathbf{y} = 0, \mathbf{z} = 1)$ with $e = e_1, e_2, e_3$, where you must use $1/\sqrt{\mathbf{x} - e}$ instead, and $(\mathbf{x} = \infty, \mathbf{y} = \infty, \mathbf{z} = 1)$ or, more correctly, the infinite point $(0, 1, 0)$, where it is $1/\sqrt{\mathbf{x}}$ instead. As to the torus, the covering coordinate $x(\mathfrak{p})$ serves as local parameter everywhere and now you have only to match these parameters up. In detail,

- at most places, $\wp'(x) \neq 0$, so $\mathbf{x} = \wp(x)$ is an invertible function of x, in the small;
- in the vicinity of $e = e_1, e_2$, or e_3, $\wp(x) = e + \frac{1}{2}\wp''(e)(x-e)^2 + \cdots$ and $\wp''(e) \neq 0$, $\wp$ being of degree 2, so $\sqrt{\mathbf{x} - e} = \sqrt{\wp''(e)/2} \times (x - e)$ is an invertible function of x;
- in the vicinity of $x = 0$, $\wp(x) = x^{-2} + \cdots$, so $1/\sqrt{\mathbf{x}} = x + \cdots$ is likewise invertible.

Moral. $\mathbf{X} = \mathbb{C}/\mathbb{L}$ *and the projective cubic are one and the same, complex structure and all.*

Exercise 3. In the terminology of Section 1.15, the map $(\wp(x), \wp'(x)) \mapsto (\mathbf{x}, \mathbf{y}, \mathbf{z})$ **uniformizes** the cubic by means of elliptic functions; this jibes with the standard nomenclature *elliptic curve.* Carry out the following informal proof that, unlike the conics of Section 1.10, *the cubic is not a rational curve*, that is, it cannot be uniformized by functions from $\mathbf{K}(\mathbb{P}^1)$. If it were, then $\mathbb{P}^1$ would be a (ramified) cover of the torus (and more besides), and that is already out of the question for topological reasons; compare Clemens [1980: 75].

Exercise 4. The standard cubic $y^2 = 4(x - e_1)(x - e_2)(x - e_3)$ looks more special than

$$\mathbf{X}\colon c_{30}x^3 + c_{21}x^2y + c_{12}xy^2 + c_{03}y^3 + c_{20}x^2 + \cdots + c_{00} = 0.$$

Prove that this is only apparent: More precisely, check that $\mathbf{X}$ is projectively the same as the standard cubic except if it degenerates into one, two, or three copies of the projective line. *Samples*: $x^3 = y^3$ represents three copies of the projective line; $x^3 + y^3 = 1$ reduces to $y^2 = 4x^3 - 1/27$ via the substitutions $x \mapsto (1 + 3y)/(6x)$, $y \mapsto (1 - 3y)/(6x)$. What happens if the degree is raised from 3 to 4? *Hint*: What is the handle number of $y^4 = (x - e_1)(x - e_2)$ or $y^4 = (x - e_1)(x - e_2)(x - e_3)$? Help with this will be found in Section 1.12; compare Nagell [1964] and Chapter 7 for number-theoretic connections.

Exercise 5. Check that the quartic $y^2 = x^4 - 1$ is equivalent to the cubic $y^2 = x^3 + 4x$ via the substitution $x \mapsto 1 + 4(x-2)^{-1}$, $y \mapsto -4y(x-2)^{-2}$.

2.12 The Problem of Inversion

The identification of $\mathbb{C}/\mathbb{L}$ with the projective $y^2 = 4x^3 - g_2x - g_3$ raises the **problem of inversion** of Abel [1827]: *Does every such cubic arise in this way?*

That is, does every pair of numbers $(g_2, g_3) \in \mathbb{C}^2 - 0$ with $\Delta = g_2^3 - 27g_3^2 \neq 0$ arise from some lattice $\mathbb{L}$ via the formulas of Section 8:

$$g_2 = 60 \sum{}' \omega^{-4}, \quad g_3 = 140 \sum{}' \omega^{-6}?$$

The point may be settled affirmatively in two very different ways.

Method 1. The functions $\mathbf{x}/\mathbf{z}$ and $\mathbf{y}/\mathbf{z}$ introduced in Section 11, make projective sense on the cubic, as does the differential of the first kind $d\mathbf{x}/\mathbf{y}$, and you may define a covering parameter x by integrating this differential along paths on the cubic that start at ∞, avoiding the three special points $(e, 0, 1)$ at which $\mathbf{y}$ vanishes. Then you can show, much as in Section 9, that $\mathbf{x}/\mathbf{z} = \wp(x)$ and $\mathbf{y}/\mathbf{z} = \wp'(x)$ are doubly periodic functions of $x \in \mathbb{C}$ with primitive periods

$$\omega_1 = \int_\infty^{e_1} \mathbf{y}^{-1} d\mathbf{x} \quad \text{and} \quad \omega_2 = \int_{e_1}^{e_2} \mathbf{y}^{-1} d\mathbf{x}$$

and identify the cubic with the quotient of $\mathbb{C}$ by the corresponding lattice $\mathbb{L}$. This would be the preferred geometric method. Notice that $\mathbf{x}/\mathbf{z}$ and $\mathbf{y}/\mathbf{z}$ are introduced purely algebraically – it is only their relation to the covering coordinate x that is transcendental; compare Section 1.15.

Method 2. The numbers g_2 and g_3 cannot both vanish since $\Delta = g_2^3 - 27g_3^2 \neq 0$; also, the multiplication $\mathbb{L} \mapsto c\mathbb{L}$ changes g_2 into $c^{-4}g_2$ and g_3 into $c^{-6}g_3$. Now suppose that the quantity $j = g_2^3/\Delta$ assumes all possible complex values as $\mathbb{L}$ runs through the family of possible period lattices. The value 0 requires the vanishing of g_2. Then $g_3 \neq 0$ can be adjusted at will by multiplication of the lattice, so any pair $(0, g_3)$ with $g_3 \neq 0$ can be obtained. To obtain the other pairs $(g_2, g_3) \in \mathbb{C}^2 - 0$, pick

$$a = \frac{g_3^2}{g_2^3} = \frac{1}{27}\left(1 - \frac{1}{j}\right) \neq \frac{1}{27}$$

at will and multiply the lattice by c to produce

$$g_2' = c^{-4}g_2 = b \quad \text{and} \quad g_3' = c^{-6}g_3 = a^{1/2}b^{3/2}.$$

The number $b \neq 0$ is arbitrary and the restriction $a \neq 1/27$ is equivalent to the nonvanishing of $c^{-12}\Delta = b^3(1 - 27a)$. In this way, every pair $(g_2, g_3) \in \mathbb{C}^2 - 0$ with $\Delta \neq 0$ is obtained, up to the sign of g_3, and that can be adjusted by multiplication of the lattice by $\sqrt{-1}$. The problem of inversion is now reduced to proving that the so-called **absolute invariant** $j = g_2^3/\Delta$ takes all possible values; see Section 4.9 in which this version of the question is resolved.

Exercise 1. The absolute invariant is a function only of the period ratio $\omega = \omega_2/\omega_1$, as you will see by inspection of the sums. Prove that it is unchanged by the action of the modular group $\Gamma_1 = PSL(2, \mathbb{Z})$ of Section 6. This will explain the name *invariant*. The adjective *absolute* receives its full explanation in Sections 4.9 and 4.16: Any other suitably restricted invariant function of Γ_1 is a rational function of j; compare Section 1.7 on the Platonic solids.

Exercise 2. For cubics with real roots, the problem of inversion is to show that any $e_1 > 0$ and any real e_2 between e_1 and $-e_1/2$ can be obtained by adjusting the positive numbers

$$\omega_1 = \int_{e_1}^{\infty} [(x - e_1)(x - e_2)(x + e_1 + e_2)]^{-1/2} dx,$$

$$\frac{\omega_2}{\sqrt{-1}} = \int_{e_2}^{e_3} [(e_1 - x)(x - e_2)(x + e_1 + e_2)]^{-1/2} dx.$$

Prove that this is so. *Hint*: The correspondence

$$a' = \int_1^{\infty} [x(x-1)(ax+b)]^{-1/2} dx, \quad b' = \int_0^1 [x(x-1)(ax+b)]^{-1/2} dx$$

effects a 1:1 self-map of the quadrant $a > 0, b > 0$. Why? Now identify a as $e_1 - e_2$ and b as $e_1 + 2e_2$.

Exercise 3. Let $k^2 = (e_2 - e_3)/(e_1 - e_3)$; it is different from 0, 1, ∞, the numbers e being distinct. Check that $j = g_2^3/\Delta = (4/27)(k^4 - k^2 + 1)^3/[k^4(1 - k^2)^2]$; compare ex. 1.7.4, where this rational function with x in place of k^2 appears as the absolute invariant of the group of anharmonic ratios. The functions j and k^2 have leading roles to play in much of Chapters 4 and 5. k^2 is nothing but Jacobi's modulus, considered as a function of the period ratio, in an easily penetrable Weierstrassian disguise; see Section 17.

2.13 The Function Field

The algebraic structure of the (elliptic) function field $\mathbf{K}(\mathbf{X})$ is easily described:

$$\mathbf{K} = \mathbb{C}(\wp)[\wp'] = \mathbb{C}(\wp)\left[\sqrt{(\wp - e_1)(\wp - e_2)(\wp - e_3)}\right],$$

that is, $\mathbf{K}$ is *the field of rational functions of* $\wp$ *with the quadratic irrationality* $\wp' = 2\sqrt{(\wp - e_1)(\wp - e_2)(\wp - e_3)}$ *adjoined*, for an extension of degree 2. Notice that the subfield $\mathbf{K}_0 = \mathbb{C}(\wp)$ of even functions *is not* isomorphic to

$\mathbf{K}(\mathbf{X})$, in contrast to the rule for the field $\mathbf{K}(\mathbb{P}^1)$; see Section 1.4 on Luroth's theorem.

Proof. $\wp' f \in \mathbf{K}$ is even if $f \in \mathbf{K}$ is odd, so it suffices to deal with even $f \in \mathbf{K}$. Now the product of f with an appropriate factor $\wp - \wp(x_0)$ lowers the degree of any pole of f not at $x_0 = 0$, so the product of f and an appropriate polynomial $a \in \mathbb{C}[\wp]$ is a function of $\mathbf{K}$ with poles only at $x = 0$. The product is even, f being such, and now subtraction of a second polynomial $b \in \mathbb{C}[\wp]$ kills the pole. Then $a(\wp)f - b(\wp) \in \mathbf{K}$ is pole-free and so constant $(= c)$, and $f = [b(\wp) + c]/a(\wp)$ belongs to $\mathbb{C}(\wp)$. The proof is finished.

Algebraic Relations. It is an important general principle that *any two functions from the same elliptic function field* $\mathbf{K}(\mathbf{X})$ *satisfy an irreducible polynomial in two variables, that is, the first is an algebraic function of the second (and vice versa).* It is a special case that any function $f \in \mathbf{K}$ satisfies an (algebraic) differential equation, that is, $f' \in \mathbf{K}$ is algebraic over $\mathbb{C}(f)$. The cubic $y^2 = 4(x - e_1)(x - e_2)(x - e_3)$ relating $y = \wp'$ to $x = \wp$ is the simplest illustration.

Proof. $f \in \mathbf{K}$ is of the form $[a(\wp) + b(\wp)\wp']/c(\wp)$ with $a, b, c \in \mathbb{C}[\wp]$, so

$$[f \times c(\mathbf{x}) - a(\mathbf{x})]^2 - b^2(\mathbf{x}) \times 4(\mathbf{x} - e_1)(\mathbf{x} - e_2)(\mathbf{x} - e_3) \in \mathbb{C}[\mathbf{x}, f]$$

has the root $\mathbf{x} = \wp$; in particular, $\wp$ is algebraic over $\mathbb{C}(f)$, as is $\wp'$, and $[\mathbf{K}: \mathbb{C}(f)] < \infty$. The general principle follows from that.

Isomorphism of Fields. It is plain that conformally equivalent curves have algebraically isomorphic function fields. It is a deeper general principle of geometry that the converse is also valid: *An isomorphism of the function fields fixing the constant field* $\mathbb{C}$ *requires conformality of the underlying curves*; see Shafarevich [1977] for more information on such matters

Proof for tori. Let $\mathbf{K}$ and $\mathbf{K}^*$ be elliptic function fields with an isomorphism I between them. The key to the proof is the fact that the **topological degree** d of a function $f \in \mathbf{K}$, considered as a map of $\mathbf{X} = \mathbb{C}/\mathbb{L}$ onto the projective line $\mathbb{P}^1$, is also the **algebraic degree** of $\mathbf{K}$ over its subfield $\mathbf{K}_0 = \mathbb{C}(f)$. This may be seen much as in Section 1.11: The lower extension of Fig. 2.20 has degree m, while the upper has degree $n = 1$ or 2. Now $\wp$ is of degree 2, so the map $\wp\colon \mathbf{X} = \mathbb{C}/\mathbb{L} \mapsto \mathbb{P}^1$ is mostly 2:1, and the two points $[\pm x]$ of $\mathbf{X}$ lying over a fixed value of $\wp(x)$ may be distinguished by the sign of the radical $\wp'(x) = 2\sqrt{(\wp - e_1)(\wp - e_2)(\wp - e_3)}$; in short, the pair $\mathfrak{p} = (\wp, \wp')$ represents an unequivocal point of $\mathbf{X}$, as already noted in Section 11. It follows

$$\mathbf{K}_0(\wp)[\sqrt{(\wp - e_1)(\wp - e_2)(\wp - e_3)}] = \mathbf{K}$$

degree = 1 or 2

$$\mathbf{K}_0(\wp)$$

degree = m

$$\mathbf{K}_0 = \mathbb{C}(f)$$

Figure 2.20. Comparing degrees.

from the meaning of degree that $d = mn = [\mathbf{K}\colon \mathbf{K}_0]$; indeed, $\mathbf{x} = \wp$ is of algebraic degree m over $\mathbf{f} = f$, so over most *numerical values* f of $\mathbf{f}$ lie m distinct values of $\mathbf{x} = \wp$. Similarly, $\mathbf{y} = \wp'$ is of algebraic degree n over $\mathbf{K}_0(\mathbf{x}) = \mathbb{C}(\mathbf{f}, \mathbf{x})$ and *either* $n = 1$ and the value of $\wp'$ is fixed by knowledge of $\mathbf{f} = f$ and $\wp$, *or else* $n = 2$ and $\wp' = 2\sqrt{(\wp - e_1)(\wp - e_2)(\wp - e_3)}$ takes (mostly) two distinct values. The upshot is that over most numerical values of $\mathbf{f}$ lie mn distinct numerical pairs $(\wp, \wp')$. This number is just the degree $d = mn = [\mathbf{K}\colon \mathbf{K}_0]$.

The rest of the proof is easy. The image $I[\wp]$ of $\wp \in \mathbf{K}$ is of topological degree 2, this being the *algebraic degree* $[\mathbf{K}^*\colon \mathbb{C}(\wp^*] = [\mathbf{K}\colon \mathbb{C}(\wp)]$, and has one double pole rather than two simple poles since the image of $(\wp')^2 = 4(\wp - e_1)(\wp - e_2)(\wp - e_3)$ is a square; in short, $I[\wp] = a\wp^*(x - b) + c$, $\wp^*$ being the $\wp$-function of $\mathbf{K}^*$.

Exercise 1. Check it.

Now the scaling $\wp(c^{-1}x|c^{-1}\mathbb{L}) = c^2\wp(x|\mathbb{L})$ of ex. 8.1 shows that $I[\wp]$ may be identified with $\wp^*$ itself after a preliminary (conformal) translation and magnification of the second torus. In detail, $I[\wp] = a\wp^*(x - b) + c$ is reduced to $\wp^* + c$ by the substitution $x \mapsto a^{-1/2}x + b$, and c is seen to vanish by inspection of

$$I[\wp']^2 = 4(\wp^* + c - e_1)(\wp^* + c - e_2)(\wp^* + c - e_3);$$

in detail, the presence of the left-hand square permits the identification of $e_1 - c$, $e_2 - c$, $e_3 - c$ with e_1^*, e_2^*, e_3^* in some order, and the vanishing of $e_1 + e_2 + e_3$ and $e_1^* + e_2^* + e_3^*$ does the rest. The identification of lattices $\mathbb{L} = \mathbb{L}^*$ follows from

$$\frac{\omega_1}{2} = \int_{\infty}^{e_1} [(x - e_1)(x - e_2)(x - e_3)]^{-1/2} dx,$$

$$\frac{\omega_2}{2} = \int_{e_1}^{e_2} [(x - e_1)(x - e_2)(x - e_3)]^{-1/2} dx.$$

The proof is finished.

Exercise 2. Check the identification of e_1^*, e_2^*, e_3^*.

2.14 Addition on the Cubic

To the beginner, it will come as something unforeseen that the cubic $\mathbf{X}$: $y^2 = 4x^3 - g_2x - g_3$ has a (commutative) **law of addition**, but the identification of $\mathbf{X}$ with the torus $\mathbb{C}/\mathbb{L}$ says it must be so. Write x for a point of the cover $\mathbb{C}$ and $\mathfrak{p} = (\mathbf{x}, \mathbf{y})$ for the corresponding point of the cubic determined by $\mathbf{x} = \wp(x)$ and $\mathbf{y} = \wp'(x)$. Then, with a self-evident notation, $\mathfrak{p}_1$ and $\mathfrak{p}_2$ determine x_1 and x_2, separately, and also their sum $x_3 = x_1 + x_2 \in \mathbb{C}/\mathbb{L}$, which in turn, determines a third point $\mathfrak{p}_0 = (\mathbf{x}_0, \mathbf{y}_0) =$ the sum $\mathfrak{p}_1 + \mathfrak{p}_2$. That is **the law of addition** of the cubic; it is even *rationally expressible* in that $\mathbf{x}_0$ and $\mathbf{y}_0$ belong to the field $\mathbb{C}(\mathbf{x}_1, \mathbf{y}_1, \mathbf{x}_2, \mathbf{y}_2)$. The explicit expression of $\mathbf{x}_0$ and $\mathbf{y}_0$ as rational functions in these four variables is postponed in favor of the very pretty *geometric* interpretation of the addition. To fix ideas, take ω_1 real and ω_2 purely imaginary so that $e_1 > e_2 > e_3$ are real; see ex. 9.3. Then $\mathbf{X}$ has two *real ovals*, seen in Fig. 2.21, one over the interval $e_3 \le \mathbf{x} \le e_2$, the other over the interval $e_1 \le \mathbf{x} \le \infty$. Let $\mathfrak{p}_1$ and $\mathfrak{p}_2$ lie on the first oval, extend the line joining them until it cuts the second one at $\mathfrak{p}_3 = (\mathbf{x}_3, \mathbf{y}_3)$, and then drop straight down to the involute $(\mathbf{x}_3, -\mathbf{y}_3)$; that is the sum $\mathfrak{p}_0 = (\mathbf{x}_0, \mathbf{y}_0)$ of $\mathfrak{p}_1$ and $\mathfrak{p}_2$! Abel's [1827] proof is followed as it has the preferred mixture of geometry and analysis, but see Shafarevich [1977: 148–50] for a purely algebraic proof avoiding the (transcendental) parameter x of the universal cover.

Abel's Proof. This has to do with the movement of the three intersections of the line $\mathbf{y} = a\mathbf{x} + b$ and the cubic $\mathbf{y}^2 = 4\mathbf{x}^3 - g_2\mathbf{x} - g_3$ in response to infinitesimal changes $\dot{a}$ and $\dot{b}$ of a and b. $\mathfrak{p}$ is a point of intersection if and only if $F(\mathbf{x}) = 4\mathbf{x}^3 - g_2\mathbf{x} - g_3 - (a\mathbf{x} + b)^2$ vanishes, as it must threefold being of degree 3; it is assumed, for the moment, that these three roots are distinct, as for most values of a and b. Then $F'(\mathbf{x})\dot{\mathbf{x}} - 2(a\mathbf{x} + b)(\dot{a}\mathbf{x} + \dot{b}) = 0$ implies

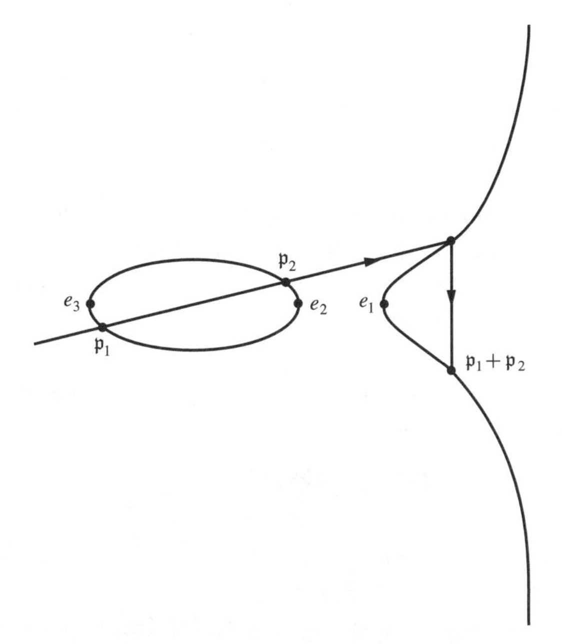

Figure 2.21. Addition on the cubic.

$\dot{\mathbf{x}}/\mathbf{y} = 2(\dot{a}\mathbf{x} + \dot{b})/F'(\mathbf{x})$ at each of the three intersections; in particular,

$$\frac{\dot{\mathbf{x}}_1}{\mathbf{y}_1} + \frac{\dot{\mathbf{x}}_2}{\mathbf{y}_2} + \frac{\dot{\mathbf{x}}_3}{\mathbf{y}_3} = \frac{1}{2\pi\sqrt{-1}} \oint \frac{\dot{a}\mathbf{x} + \dot{b}}{F(\mathbf{x})} d\mathbf{x}.$$

The integral is taken about a big circle enclosing the roots of $F(\mathbf{x}) = 0$; it vanishes in view of $F(\mathbf{x}) \simeq 4\mathbf{x}^3$ at ∞. Now let $x = x_1, x_2, x_3 \in \mathbb{C}$ cover the three intersections. Then $\mathbf{x} = \wp(x)$ implies $\dot{\mathbf{x}}/\mathbf{y} = \dot{x}$ at each of these places and $\dot{x}_1 + \dot{x}_2 + \dot{x}_3 = 0$, at which point you should notice that the outcome does not really depend on the simplicity of the roots of $F(\mathbf{x}) = 0$. But now you see that $x_1 + x_2 + x_3$ does not in fact require the line at all and so may be evaluated (as a lattice point) by moving the line to ∞; for example, if $a = 0$ and b is large, then the roots $\mathbf{x} = \wp(x)$ of $F(\mathbf{x}) = 0$ are also large and that happens only if x is close to 0 mod $\mathbb{L}$. The upshot is that $x_1 + x_2 = -x_3$ in $\mathbb{C}/\mathbb{L}$, which is to say that the sum $\mathfrak{p}_0$ of $\mathfrak{p}_1$ and $\mathfrak{p}_2$ *is the involute of the third point of intersection of the cubic with the line that passes through them.* Naturally, if $\mathfrak{p}_1 = \mathfrak{p}_2$, it is the line tangent to $\mathbf{X}$ at that place which is to be employed. Note also that the point $\mathfrak{p}_\infty$ corresponding to $x = 0$, which was ignored for reasons of simplicity, acts as the additive identity: The sum of $\mathfrak{p}_\infty$ and any other $\mathfrak{p} \in \mathbf{X}$ is $\mathfrak{p}$ itself.

Exercise 1. To be candid, the proof is skimpy. To do it right is to do it *projectively*. Check first that *any* projective line $a\mathbf{x}+b\mathbf{y}+c\mathbf{z}=0$ with $a, b, c \in \mathbb{C}^3-0$ cuts the projective cubic $\mathbf{y}^2\mathbf{z} - 4\mathbf{x}^3 + g_2\mathbf{x}\mathbf{z}^2 + g_3\mathbf{z}^3 = 0$ three times, counting multiplicities. Carry out the rest of the proof in this projective style.

The result of the addition will now be spelled out in three different ways, expressing the output $\mathfrak{p}_0 = (\mathbf{x}_0, \mathbf{y}_0)$ rationally in terms of the input $\mathbf{x}_1, \mathbf{y}_1, \mathbf{x}_2, \mathbf{y}_2$. The subsequent exercises provide a sample of the bewildering number of identities that the $\wp$-function satisfies.

Version 1. $\mathbf{x}_1 + \mathbf{x}_2 + \mathbf{x}_3 = a^2/4$ by the form of $F(\mathbf{x})$, and a is the slope of the line, so for distinct $\mathbf{x}_1$, $\mathbf{x}_2$, and $\mathbf{x}_3$ you have

$$a = \frac{\mathbf{y}_2 - \mathbf{y}_1}{\mathbf{x}_2 - \mathbf{x}_1} = \frac{\mathbf{y}_3 - \mathbf{y}_2}{\mathbf{x}_3 - \mathbf{x}_2} = \frac{\mathbf{y}_1 - \mathbf{y}_3}{\mathbf{x}_1 - \mathbf{x}_3},$$

and, by a little rearrangement,

$$\mathbf{x}(\mathfrak{p}_1 + \mathfrak{p}_2) = \mathbf{x}_3 = -\mathbf{x}_1 - \mathbf{x}_2 + \frac{1}{4}\left(\frac{\mathbf{y}_2 - \mathbf{y}_1}{\mathbf{x}_2 - \mathbf{x}_1}\right)^2,$$

$$\mathbf{y}(\mathfrak{p}_1 + \mathfrak{p}_2) = \mathbf{y}_3 = -\frac{\mathbf{y}_2(\mathbf{x}_1 - \mathbf{x}_3) + \mathbf{y}_1(\mathbf{x}_3 - \mathbf{x}_2)}{\mathbf{x}_2 - \mathbf{x}_1}.$$

This is the *rational expression* of the law of addition. The supplementary rule

$$\mathbf{x}(2\mathfrak{p}) = -2\mathbf{x}(\mathfrak{p}) + \mathbf{y}^{-2}(\mathfrak{p})\left[3\mathbf{x}^2(\mathfrak{p}) + e_1e_2 + e_2e_3 + e_3e_1\right]^2$$

$$\mathbf{y}(2\mathfrak{p}) = -\mathbf{y}(\mathfrak{p}) + 2\mathbf{y}^{-1}(\mathfrak{p})\left[\mathbf{x}(2\mathfrak{p}) - x(\mathfrak{p})\right] \times \left[3\mathbf{x}^2(\mathfrak{p}) + e_1e_2 + e_2e_3 + e_3e_1\right]$$

covers the exceptional case $\mathfrak{p}_1 = \mathfrak{p}_2$; see ex. 4 to follow.

Version 2. The formula for $\mathbf{x}_3$ leads to an irreducible polynomial in three variables satisfied by $\mathbf{x}_1$, $\mathbf{x}_2$, and $\mathbf{x}_3$ after elimination of $\mathbf{y} = \wp'(x)$ by means of the cubic. This is the *addition theorem* of the $\wp$-function; it shows that $\mathbf{x}_3$ is of degree 2 over the field $\mathbb{C}(\mathbf{x}_1, \mathbf{x}_2)$.

Version 3. $\mathbf{y} = a\mathbf{x} + b$ at the three intersections, so the determinant vanishes.

$$\begin{vmatrix} 1 & \mathbf{x}_1 & \mathbf{y}_1 \\ 1 & \mathbf{x}_2 & \mathbf{y}_2 \\ 1 & \mathbf{x}_3 & \mathbf{y}_3 \end{vmatrix} = 2 \times \begin{vmatrix} 1 & \mathbf{x}_1 & \sqrt{(\mathbf{x}_1 - e_1)(\mathbf{x}_1 - e_2)(\mathbf{x}_1 - e_3)} \\ 1 & \mathbf{x}_2 & \sqrt{(\mathbf{x}_2 - e_1)(\mathbf{x}_2 - e_2)(\mathbf{x}_2 - e_3)} \\ 1 & \mathbf{x}_3 & \sqrt{(\mathbf{x}_3 - e_1)(\mathbf{x}_3 - e_2)(\mathbf{x}_3 - e_3)} \end{vmatrix}$$

Exercise 2. Check that in the presence of the cubic, version 3 and version 1 are the same despite appearances.

Exercise 3. Version 1 is commonly stated in the language of the $\wp$-function:

$$\wp(x_1 + x_2) = -\wp(x_1) - \wp(x_2) + \frac{1}{4}\left[\frac{\wp'(x_2) - \wp'(x_1)}{\wp(x_2) - \wp(x_1)}\right]^2,$$

$$\wp'(x_1 + x_2) = \frac{\wp'(x_1)\,[\wp(x_3) - \wp(x_2)] + \wp'(x_2)\,[\wp(x_1) - \wp(x_3]}{\wp(x_2) - \wp(x_1)}.$$

Check the first line by chasing roots and poles. *Hint*: For variable x_1 and fixed $x_2 \notin \mathbb{L}$, the difference of left and right sides is pole-free, vanishes at $x_1 = 0$, and so must vanish identically.

Exercise 4. The **duplication formula**

$$\wp(2x) = -2\wp(x) + \left[\wp'(x)\right]^{-2}\left[3\wp^2(x) + e_1e_2 + e_2e_3 + e_3e_1\right]$$

follows. Use it to confirm the supplementary rules stated under version 1.

Exercise 5. Use ex. 4 to write the duplication formula as

$$\begin{aligned} 4\wp(2x) = \wp(x) \\ + \frac{(e_1 - e_2)(e_1 - e_3)}{\wp(x) - e_1} + \frac{(e_2 - e_3)(e_2 - e_1)}{\wp(x) - e_2} + \frac{(e_3 - e_1)(e_3 - e_2)}{\wp(x) - e_3}. \end{aligned}$$

Exercise 6. $\wp(x+\omega_1/2) = e_1+(e_1-e_2)(e_1-e_3)\,[\wp(x) - e_1]^{-1}$ comes from the addition theorem. Use it to obtain $\wp(\omega_1/4) = e_1 + \sqrt{(e_1 - e_2)(e_1 - e_3)}$. Find similar expressions for the value of $\wp$ at the other quarter-periods $(\omega_1 + \omega_2)/4$ and $\omega_2/4$.

Exercise 7. Check that $4 \times \wp(2x)$ is the sum of $\wp(x + \omega/2)$ as ω runs through the periods $0, \omega_1, \omega_1 + \omega_2$, and ω_2. What is the corresponding product? See ex. 6 for help. *Answer*: $\left[\wp'(x)\right]^2 \times 16(e_1-e_2)^2(e_2-e_3)^2(e_3-e_1)^2$. The second factor is, of course, the familiar discriminant Δ.

Exercise 8. Check that $\wp'(x_1+x_2) = \frac{3}{2}a\,[\wp(x_1)+\wp(x_2)] - \frac{1}{4}a^3 - \frac{1}{2}\left[\wp'(x_1) + \wp'(x_2)\right]$ with $a = \left[\wp'(x_1)-\wp'(x_2)\right][\wp(x_1)-\wp(x_2)]^{-1}$ as in version 1.

Exercise 9. Prove that $\wp(x_1-x_2)-\wp(x_1+x_2) = [\wp(x_1)-\wp(x_2)]^{-1} \times \wp'(x_1)\,\wp'(x_2)$.

Exercise 10. Prove that every function $f \in \mathbf{K}(\mathbf{X})$ has an addition theorem in the sense of version 2: *If x_1 and x_2 are independent variables on the covering plane $\mathbb{C}$, then $\mathbf{f}_3 = f(x_1+x_2)$ is* **algebraic** *over $\mathbf{f}_1 = f(x_1)$ and $\mathbf{f}_2 = f(x_2)$.* *Hint*: $\mathbf{K} = \mathbb{C}(\wp)[\wp']$.

Exercise 11. Prove the converse:[11] If $\mathbf{f}_3$ is algebraic over $\mathbb{C}(\mathbf{f}_1, \mathbf{f}_2)$, then either f is rational, or it is a rational function of a single exponential such as $\sin x = (2\sqrt{-1})^{-1}[e^{\sqrt{-1}x} - e^{-\sqrt{-1}x}]$, or else it is an elliptic function. *Hints*: The irreducible polynomial satisfied by $\mathbf{f}_1, \mathbf{f}_2, \mathbf{f}_3$ is of degree $d < \infty$ in $\mathbf{f}_3$, so f has a nonvanishing period if it takes on the value $f(x) = c$ at more than d places x. Why? Now let the number of roots of $f(x) = c$ be $\le d$ for every $c \in \mathbb{C}+\infty$. Then f is not rational: In fact, near $x = \infty$ it must omit a little disk of values $f = c$. Why? What then? It remains to check that f is of rational character in the parameter $\exp\left[2\pi\sqrt{-1}x/\omega\right]$ if it has just one primitive period ω. Do it.

An Algebraic Illustration. Tartaglia [1546] expressed the roots of the general quartic $P_4(x) = x^4+ax^2+bx+c \in \mathbb{C}[x]$ in terms of a, b, c, by the extraction of square roots; see Section 5.1 for a refresher on such matters. The present article exhibits them in a different way, based upon the addition theorem: $\mathbf{x}_1, \mathbf{x}_2, \mathbf{x}_3$ are the roots of

$$F(\mathbf{x}) = 4\mathbf{x}^3 - g_2\mathbf{x} - g_3 - (a\mathbf{x}+b)^2,$$

so

$$\begin{aligned} 4(\mathbf{x}_1-\mathbf{x}_2)(\mathbf{x}_1-\mathbf{x}_3) = F'(\mathbf{x}_1) &= 12\mathbf{x}_1^2 - g_2 - 2a(a\mathbf{x}_1+b) \\ &= 12\mathbf{x}_1^2 - g_2 - 2a\mathbf{y}_1, \qquad (1) \end{aligned}$$

$$(\mathbf{x}_2-\mathbf{x}_1)+(\mathbf{x}_3-\mathbf{x}_1) = \mathbf{x}_1+\mathbf{x}_2+\mathbf{x}_3 - 3\mathbf{x}_1 = a^2/4 - 3\mathbf{x}_1, \qquad (2)$$

[11] Weierstrass [1870]. See Schwarz [1893].

and

$$(2)^2 - (1) = (\mathbf{x}_3 - \mathbf{x}_2)^2 = (a^2/4 - 3\mathbf{x}_1)^2 - 12\mathbf{x}_1^2 + g_2 - 2\mathbf{y}_1 a$$
$$= (a/2)^2 - 6\mathbf{x}_1(a/2)^2 + 4\mathbf{y}_1(a/2) - 3\mathbf{x}_1^2 + g_2.$$

This is spelled out with

$$X = \frac{a}{2} = \frac{1}{2}\frac{\mathbf{y}_2 - \mathbf{y}_1}{\mathbf{x}_2 - \mathbf{x}_1} = \frac{1}{2}\frac{\wp'(x_2) - \wp'(x_1)}{\wp(x_2) - \wp(x_1)},$$

$$Y = \mathbf{x}_3 - \mathbf{x}_2 = \wp(x_1 + x_2) - \wp(x_2)$$

as

$$Y^2 = X^4 - 6\wp(x_1)X^2 + 4\wp'(x_1)X - 3\wp^2(x_1) + g_2 \equiv P_4(X),$$

and the following conclusion is drawn: For fixed x_1 and variable x_2, $P_4(X) = 0$ only if $Y = 0$. This happens for $x_2 = -x_1/2 +$ one of the half-periods $\omega/2 = 0, \omega_1/2, \ldots$, producing four roots of $P_4(X) = 0$, and these must be distinct: In fact, X taken at x_1 and $x_2 = -x_1/2 + x$ is nothing but

$$X(-x_1/2 + x) = \wp'(x_1/2)\,[\wp(x_1/2) - \wp(x)]^{-1}$$

up to an additive constant, and so takes these values doubly.

Exercise 12. Check the display by chasing roots and poles.

Moral 1. $X(-x_1/2 + \omega/2)$ runs through the roots of $P_4(X) = 0$ as ω runs through the periods $0, \omega_1, \omega_1 + \omega_2, \omega_2$.

This is interesting: It represents a **uniformization** of the correspondence between the coefficients of $P_4(X) = X^4 + aX^2 + bX + c \in \mathbb{C}[x]$ and its roots in which (1) $a = -6\wp(x_1)$, (2) $b = \wp'(x_1)$, (3) $c = -3\wp^2(x_1) + g_2$. But how general is P_4? The resolution of the problem of inversion provides the answer: For most values of a, b, c, (1), (2), (3) can be solved (in the opposite order) keeping $\Delta = g_2^3 - 27g_3^2 \neq 0$. (3) determines $g_2 = a^2/12 + c$; (2) determines $g_3 = -a^3/216 - b^2/16 + ac/16$; and for most values of a, b, c (2) and (1) together determine the point $\mathfrak{p} = [\wp(x_1), \wp'(x_1)]$ of the associated cubic. In short, a, b, c produce a marked point on a special cubic. Then the map X produces the roots.

Exercise 13. Check that $P_4(X) = X^4 + aX^2 + bX + c$ has distinct roots if and only if $\Delta = g_2^3 - 27g_3^2 \neq 0$. *Hint*: This can be done by a fatiguing computation of the discriminant of P_4, but don't. Think.

Moral 2. The recipe is valid for any quartic $P_4 \in \mathbb{C}[x]$ with simple roots.

A Final Illustration. The ubiquity of the $\wp$-function is truly remarkable. $\wp(nx) \in \mathbf{K}(\mathbf{X})$ is even, so it can be expressed as a rational function of $\wp(x)$: $\wp(nx) = f_n[\wp(x)]$. Obviously, $f_n \circ f_m = f_{n \bullet m} = f_m \circ f_n$, so these are commuting (rational) maps of the projective line to itself. The extraordinary fact, due to Ritt [1923], is that, up to trivialities, these are all such commuting self-maps of the projective line.[12]

2.15 Abel's Theorem

A function $f \in \mathbf{K}(\mathbf{X})$ has the same number of roots $\mathfrak{q}$ and poles $\mathfrak{p}$. This is its degree d and, in the field $\mathbf{K}(\mathbb{P}^1)$, that is the only restriction: that the roots and poles be equal in number. The field $\mathbf{K}(\mathbf{X})$ is different, as noted in Section 7, item 4: Now you must also match **root sum** $\sum x(\mathfrak{q})$ and **pole sum** $\sum x(\mathfrak{p})$ modulo the lattice, $x(\wp)$ being the covering parameter. This is the cheap half of Abel's theorem [1827]. The other half states that no further restriction is needed: *If the* $\mathfrak{q}$s *and* $\mathfrak{p}$s *are equal in number and if root sum and pole sum match mod* $\mathbb{L}$*, then there is a function from* $\mathbf{K}$ *having these roots and poles, and no others.*

Proof in the style of Jacobi [1838]. Take $\omega_1 = 1$ for simplicity, as can be arranged by a harmless modification of the lattice $\mathbb{L} \mapsto \mathbb{L}/\omega_1$, inducing a conformal map of tori. The period ratio $\omega = \omega_2$ has positive imaginary part, so the sum

$$\vartheta(x) = \sum_{n \in \mathbb{Z}} (-1)^n \exp\left[2\pi\sqrt{-1}nx + \pi\sqrt{-1}n(n+1)\omega\right]$$

converges fast to an entire function of $x \in \mathbb{C}$, of period 1. This is Jacobi's **theta-function**, the subject of Chapter 3.

Exercise 1. ϑ is not constant so it cannot belong to $\mathbf{K}$ (being entire), but it is doing its best: $\vartheta(x+1) = \vartheta(x)$ and $\vartheta(x+\omega) = -\exp\left[-2\pi\sqrt{-1}(x+\omega)\right]\vartheta(x)$. Prove this. *Hint*: The substitution $n \mapsto n-1$ is carried out in the sum.

Exercise 2. Check that $\vartheta(0) = 0$. *Hint*: $\vartheta(-x) = -\exp(-2\pi\sqrt{-1}x)\vartheta(x)$.

Exercise 3. Show that the integral of $(2\pi\sqrt{-1})^{-1} d\log\vartheta(x)$ about the perimeter of the fundamental cell is $+1$. Why? *Hint*: See ex. 1.

[12] A. Veselov pointed out this reference.

The moral of exs. 1–3 is that $\vartheta(x) = 0$ *has a simple root at* $x = 0$ *and no others in the cell.* Now pick the covering parameters of the supposed roots and poles so that

$$\sum x(\mathfrak{q}_i) = \sum x(\mathfrak{p}_i)$$

without any additive periods.

Then the product

$$\frac{\vartheta(x - x(\mathfrak{q}_1))}{\vartheta(x - x(\mathfrak{p}_1))} \times \cdots \times \frac{\vartheta(x - x(\mathfrak{q}_n))}{\vartheta(x - x(\mathfrak{p}_n))}$$

belongs to the function field **K** since it is of period $\omega_1 = 1$ and changes, under the substitution $x \mapsto x + \omega$, by the factor

$$\exp\left(2\pi\sqrt{-1}[\text{root sum} \; - \; \text{pole sum}]\right) = 1.$$

The proof is finished by noting that the function so defined has the right roots and poles.

Exercise 4. Prove that

$$\wp(x) = -\left[\frac{\vartheta'(0)}{\vartheta(x_0)}\right]^2 \frac{\vartheta(x - x_0)\vartheta(x + x_0)}{\vartheta^2(x)},$$

where x_0 is a root of $\wp(x) = 0$ in the fundamental cell.

Amplification. There seems to be no simple way to express the roots $\pm x_0$ of $\wp(x) = 0$. Eichler and Zagier [1982] showed that

$$x_0 = m + \frac{1}{2} + n\omega$$
$$\pm\left[\frac{\log(5 + 2\sqrt{6})}{2\pi\sqrt{-1}} + \frac{512}{9}\pi^{10}\sqrt{-1}\int_{\omega}^{\infty}(\omega' - \omega)\Delta g_3^{-3/2} d\omega'\right].$$

The integral is taken on the vertical line $\text{Re}\,(\omega') = \text{Re}\,(\omega)$. A remarkable formula!

A Mechanical Illustration. Fix $a > b > 0$ and consider the family of confocal ellipses

$$\mathbf{E}_c\colon \frac{x_1^2}{a^2 - c^2} + \frac{x_2^2}{b^2 - c^2} = 1 \quad (0 < c < b)$$

with common foci $(\pm\sqrt{a^2 - b^2}, 0)$. The straight-line motion inside the big ellipse $\mathbf{E}_0$, with perfect reflection at the perimeter, is intimately connected to the confocal family: If a trajectory does not pass between the foci, then each straight piece of it is tangent to one and the same ellipse $\mathbf{E} = \mathbf{E}_c$, as you may verify (with tears). Poncelet [1822] proved that for a fixed $0 < c < b$ either each trajectory grazing $\mathbf{E}$ closes after a fixed number of reflections or else none of them closes at all. The following proof of this pretty mechanical fact is based upon Abel's theorem. The idea goes back to Jacobi [1881–91 (1): 278–93]. Griffiths [1976: 345–8] puts the whole matter in historical perspective; see Bos et al. [1987] for more details.

Proof. Any segment of a trajectory is specified by its initial point $x \in \mathbf{E}_0$ and its point of tangency $y \in \mathbf{E}$, this being subject to the **incidence relation**

$$0 = (y_1 - x_1)\frac{y_1}{a^2 - c^2} + (y_2 - x_2)\frac{y_2}{b^2 - c^2} = 1 - \frac{x_1 y_1}{a^2 - c^2} - \frac{x_2 y_2}{b^2 - c^2},$$

expressing the perpendicularity of tangent to normal at the point of contact; see Fig. 2.22. The proof exploits the fact that the three relations $x \in \mathbf{E}_0$, $y \in \mathbf{E}$,

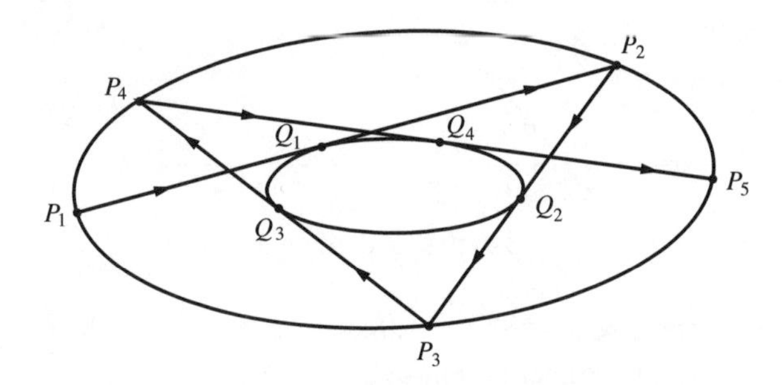

Figure 2.22. Poncelet's trajectories.

and *incidence* together determine an elliptic curve $\mathbf{X}$. To see this, it is simplest to uniformize $\mathbf{E}_0$ and $\mathbf{E}$ in the manner of Section 1.10 so that they appear as projective lines with parameters X and Y:

$$\mathbf{E}_0: \qquad x_1 = a\frac{X^2 - 1}{X^2 + 1}, \qquad x_2 = b\frac{2X}{X^2 + 1},$$

$$\mathbf{E}: \qquad y_1 = \sqrt{a^2 - c^2}\frac{Y^2 - 1}{Y^2 + 1}, \quad y_2 = \sqrt{b^2 - c^2}\frac{2Y}{Y^2 + 1}.$$

Then the incidence relation takes the form

$$P(X, Y) = \frac{a}{\sqrt{a^2 - c^2}}(X^2 - 1)(Y^2 - 1) + \frac{4b}{\sqrt{b^2 - c^2}}XY - (X^2 + 1)(Y^2 + 1) = 0.$$

This defines a curve $\mathbf{X}_0$ in $\mathbb{P}^2$ with coordinates $(X, Y, Z) \in \mathbb{C}^3 - 0$, subject to the usual projective identifications. A simple computation reveals the presence of two singular points $(1, 0, 0)$ and $(0, 1, 0)$ at infinity (and only these) at which $\mathbf{X}_0$ crosses itself in the simple manner of Fig. 2.23, together with four ramifications over the 2:1 projection of $\mathbf{X}_0$ to $\mathbb{P}^1$ via X/Z. $\mathbf{X}_0$ is desingularized as in

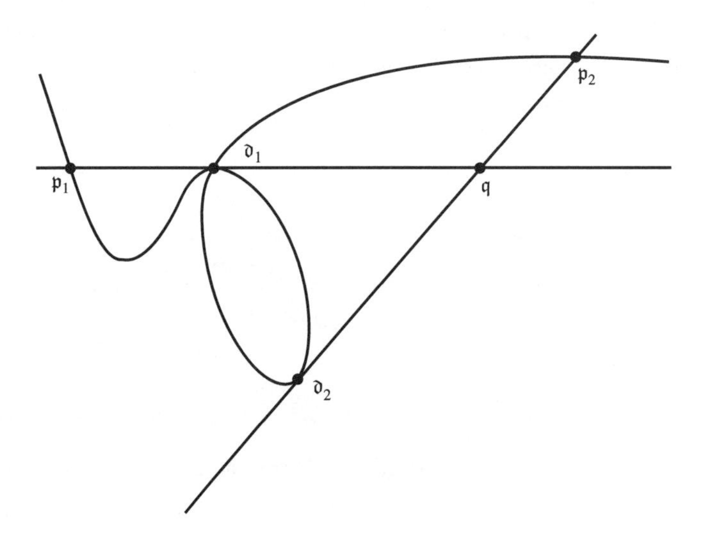

Figure 2.23. Self-crossings of $\mathbf{X}_0$.

Section 1.10. Then the Riemann–Hurwitz count is $4 = 2(2\text{sheets} + \text{genus} - 1)$; in short, the desingularized curve $\mathbf{X}$ is of genus 1, that is, it is an elliptic curve. Now the first segment of the trajectory corresponds to a point $\mathfrak{p}_1 = (X_1, Y_1, 1)$ of $\mathbf{X}_0$ from which you may determine (1) the second root X_2 of $P(\bullet, Y_1) = 0$ and (2) the second root Y_2 of $P(X_2, \bullet) = 0$, P being quadratic both in X and in Y. The point $\mathfrak{p}_2 = (X_2, Y_2, 1)$ corresponds to the second segment of the trajectory, and it is clear that the line joining $\mathfrak{p}_1 = (X_1, Y_1, 1)$ to $\mathfrak{q} = (X_2, Y_1, 1)$ cuts $\mathbf{X}_0$ twice more at the infinite point $\mathfrak{d}_1 = (1, 0, 0)$; likewise, the line joining $\mathfrak{q} = (X_2, Y_1, 1)$ to $\mathfrak{p}_2 = (X_2, Y_2, 1)$ cuts it twice at the other infinite point $\mathfrak{d}_2 = (0, 1, 0)$; see Fig. 2.23 in which points are labeled by their uniformizers. Let $f_1 = Y - Y_1 Z$ and $f_2 = X - X_2 Z$ describe these lines by their vanishing. Then $f = f_1/f_2$ is a function of rational character on $\mathbf{X}_0$ with simple root $\mathfrak{p}_1$, double root $\mathfrak{d}_1$, simple pole $\mathfrak{p}_2$, and double pole $\mathfrak{d}_2$, and Abel's theorem applied to the desingularized curve $\mathbf{X}$ asserts that, up to periods, $x(\mathfrak{p}_1) + 2x(\mathfrak{d}_1) = x(\mathfrak{p}_2) + 2x(\mathfrak{d}_2)$, in which x is the parameter of the universal cover $\mathbb{C}$ of $\mathbf{X}$ and $2x(\mathfrak{d})$ is the sum of the covering coordinates of the two points of $\mathbf{X}$ into which $\mathfrak{d} \in \mathbf{X}_0$ splits. In short,

$$x(\mathfrak{p}_2) - x(\mathfrak{p}_1) = -2x(\mathfrak{d}_2) + 2x(\mathfrak{d}_1) \bmod \mathbb{L}.$$

Now it is clear that *every trajectory closes after the same number of reflections, or that none of them ever closes, according as* $2x(\mathfrak{d}_1) - 2x(\mathfrak{d}_2)$ *is or is not a fractional period of* $\mathbf{X}$. This is Poncelet's theorem.

The result can be spelled out by means of the Jacobian form of $\mathbf{X}$; compare Section 16. The torus is now identified with the (projective) quartic $\mathbf{y}^2 = (1-\mathbf{x}^2)(1-k^2\mathbf{x}^2)$ with

$$\mathbf{x} = \frac{1}{\sqrt{k}}\frac{X}{Z},$$

$$\mathbf{y} = \left[\frac{a}{c}(X^2 - Z^2) - \frac{\sqrt{a^2-c^2}}{c}(X^2+Z^2)\right]\frac{Y}{Z^3} + \frac{2b}{c}\frac{\sqrt{a^2-c^2}}{\sqrt{b^2-c^2}}\frac{X}{Z},$$

and, modulo $0 < k^2 < 1$,

$$k = -d + \sqrt{d^2-1} < 0, \qquad d = (2a^2 - b^2 - c^2)(b^2-c^2)^{-1} > 1.$$

$\mathbf{X}_0$ is desingularized by this substitution; in fact, $\mathfrak{d}_1 = (1,0,0)$ splits into the two distinct poles of $\mathbf{x} = \sin \operatorname{amp}$ while $\mathfrak{d}_2 = (0,1,0)$ splits into the two points

$$(\mathbf{x},\mathbf{y}) = \pm\left(a - \sqrt{a^2-c^2}\right)^{-1} \times \left(\frac{c}{\sqrt{k}}, -2b\sqrt{\frac{a^2-c^2}{b^2-c^2}}\right).$$

The closing of trajectories is now expressed by the condition that the number

$$2x(\mathfrak{d}_1) - 2x(\mathfrak{d}_2) = 2\int_{\sqrt{-1}r_0}^{\sqrt{-1}\infty} [(1-r^2)(1-k^2r^2)]^{-1/2}dr,$$

with $r_0 = (a - \sqrt{a^2-c^2})^{-1}c/\sqrt{k}$, stand in a rational relation to the imaginary period

$$\omega_2 = 2\sqrt{-1}K'(k) = 2\sqrt{-1}\int_1^{1/k} [(r^2-1)(1-k^2r^2)]^{-1/2}dr.$$

It follows that the values of c producing closed trajectories are densely distributed between 0 and b: If not, $2x(\mathfrak{d}_1) - 2x(\mathfrak{d}_2)$ would be a constant multiple m of ω_2 for $0 < c < b$, and this is ruled out by considering the extreme values $c = 0+$ $(m = 0)$ and $c = b-$ $(m = 1/2)$.

Exercise 5. Give all the details necessary to the discussion.

Cayley [1853,1861][13] found an elegant criterion to decide if the trajectory closes or not. Let **A** and **B** be two conics (e.g., ellipses) viewed in the real projective space $\mathbb{RP}^2$,[14] the second inside the first, and let them be described by the vanishing of quadratic forms $x \bullet Ax$ and $x \bullet Bx$ with 3×3 real matrices A and B. The expansion

$$\sqrt{\det(A + B\mathbf{x})} = \sqrt{\det A}\left[1 + c_1\mathbf{x} + c_2\mathbf{x}^2 + \cdots\right]$$

tells the story: Trajectories reflecting at **A** and grazing **B** close in n steps if and only if $\det\left[c_{4i+j}\colon 0 \le i,\, j \le n-1\right] = 0$ for odd n, or $\det[c_{3i+j}\colon 0 \le i,\, j \le n-2] = 0$ for even n; see Griffiths and Harris [1978a] for the geometric explanation of this remarkable fact.

Exercise 6. Check that, in Poncelet's case, $\mathbf{y}^2 = \det(A + B\mathbf{x})$ is nothing but the elliptic curve in disguise.

2.16 Jacobian Functions: Reprise

The whole story of the function field **K(X)** could have been told from the Jacobian viewpoint, starting with the inversion of the incomplete integral $\int_0^x \left[(1-x^2)\,(1-k^2x^2)\right]^{-1/2} dx$, as in Section 15. The Weierstrassian viewpoint adopted in Sections 6–14 is more direct in some respects, but it is important to understand them both.

Equivalence of Quartics and Cubics. This was discussed episodically in Section 2 and ex. 11.5. To put it in a clean form, note first that the general quartic $\mathbf{y}^2 = (\mathbf{x} - e_0)(\mathbf{x} - e_1)(\mathbf{x} - e_2)(\mathbf{x} - e_3)$ with distinct roots reduces to the cubic $\mathbf{y}^2 = (\mathbf{x} - e_1')(\mathbf{x} - e_2')(\mathbf{x} - e_3')$ via the substitutions $\mathbf{x} \mapsto e_0 + (\mathbf{x} - a)^{-1}$ and $\mathbf{y} \mapsto b\mathbf{y}(\mathbf{x} - a)^{-2}$ with $b^2 = (1/4)(e_0 - e_1)(e_0 - e_2)(e_2 - e_3)$ and new roots $e' = a + (e - e_0)^{-1}$; the latter can be made to satisfy $e_1' + e_2' + e_3' = 0$ by choice of $a = -(1/3)\left[(e_1 - e_0)^{-1} + (e_2 - e_0)^{-1} + (e_3 - e_0)^{-1}\right]$. Naturally, the cubic can be converted back into the quartic. The correspondence is conformal. This identifies the nonsingular quartic with a torus, as promised in Section 5.

Exercise 1. Check the substitution and the conformality, including at infinity, that is, in the projective style of Section 11.

[13] A. Veselov kindly contributed this reference.

[14] $\mathbb{RP}^2$ is the space $\mathbb{R}^3 - 0$ subject to the (real) projective identification of points $x = (x_1, x_2, x_3)$.

The reduction of the quartic to Jacobian form $\mathbf{y}^2 = (1-\mathbf{x}^2)(1-k^2\mathbf{x}^2)$ with $k^2 \neq 0, 1$ was explained in step 2 of Section 2; it is effected by a substitution

$$\mathbf{x} \mapsto \frac{a\mathbf{x}+b}{c\mathbf{x}+d}, \quad \mathbf{y} \mapsto \frac{\mathbf{y}}{(c\mathbf{x}+d)^2}$$

plus a minor adjustment: The new roots are $e' = (de-a)(-ce+a)^{-1}$ and can be matched with $1, -1, 1/k, -1/k$ by choice of $a, b, c, d \in \mathbb{C}$ with $ad - bc = 1$ provided k is chosen so that

$$\frac{e_1-e_2}{e_2-e_3} \bullet \frac{e_3-e_4}{e_4-e_1} = \frac{4k}{(1+k)^2}.$$

Function Field. The identity of projective cubics and quartics means that after a (conformal) magnification, any complex torus $\mathbf{X} = \mathbb{C}/\mathbb{L}$ carries the function sn produced by inversion of Jacobi's incomplete elliptic integral, as in Section 5; in particular, *any period ratio can be obtained as the quotient* $\sqrt{-1}K'/2K$ *of complete Jacobian integrals by choice of the complex modulus* $k^2 \in \mathbb{C}-0-1$, *and conversely every such modulus arises in this way*. This resolves the Jacobian form of the **problem of inversion**. The upshot is that the whole story of $\mathbf{K}(\mathbf{X})$ could have been told from the Jacobian viewpoint, especially

$$\mathbf{K}(\mathbf{X}) = \mathbb{C}(\mathbf{x})\left[\sqrt{(1-\mathbf{x}^2)(1-k^2\mathbf{x}^2)}\right]$$

with an indeterminate $\mathbf{x}$ (= sin amp) and a suitable modulus $k^2 \neq 0, 1$; in particular, it must be possible to express the $\wp$-function in terms of sn.

Explicitly,

$$\mathrm{sn}(K-x, k) = \frac{\wp(x')-a}{\wp(x')-b},$$

with $a = \wp(\omega_1/4)$, $b = \wp(\omega_1/4+\omega_2/2)$, $x' = cx/2$, $c = \sqrt{2(1-k^2)/(a-b)}$, $2K = \omega_1/c$, and $\sqrt{-1}K' = \omega_2/c$. Begin with $f = \mathrm{sn}(K-x,k)$ on $\mathbf{X} = \mathbb{C}$ modulo the lattice $\mathbb{L} = 4K\mathbb{Z} \oplus 2\sqrt{-1}K'\mathbb{Z}$. What does $(f')^2 = (1-f^2)(1-k^2f^2)$ say about the (for the moment unknown) function $\wp(x')$ defined by the last display? An easy computation produces

$$(\wp')^2 = 4\left(\wp - \frac{a+b}{2}\right)\left(\wp - \frac{b-ak}{1-k}\right)\left(\wp - \frac{b+ak}{1+k}\right),$$

and the requirement that

$$\left(e_1 = \frac{a+b}{2}\right) + \left(e_2 = \frac{b-ak}{1-k}\right) + \left(e_3 = \frac{b+ak}{1+k}\right) = 0$$

forces $k^2 = (a+5b)/(5a+b)$. It also identifies $\wp$ as the Weierstrassian function of $\mathbf{X}' = \mathbb{C}/\mathbb{L}'$, with $\mathbb{L}' = \omega_1\mathbb{Z} \oplus \omega_2\mathbb{Z}$, in view of the fact that its only pole in the cell is located where $\mathrm{sn}(K - x, k) = 1$, that is, at $x = 0$; the pole must be double since $\mathrm{sn} = 1$ is a double root, its form can only be $dx^{-2} + 0 \cdot x^{-1} + \cdots$, and $d = 1$ is not hard to come by.

Exercise 2. Check the identity from scratch, by chasing roots and poles, and use it to confirm the relation $k^2 = (e_2 - e_3)/(e_1 - e_3)$ employed in ex. 12.3 to define k^2.

The Addition Theorem. The Jacobian form of the addition theorem is

$$\mathbf{x}_3 = \frac{\mathbf{x}_1\mathbf{y}_2 + \mathbf{x}_2\mathbf{y}_1}{1 - k^2\mathbf{x}_1^2\mathbf{x}_2^2} = \frac{\mathbf{x}_1\sqrt{(1-\mathbf{x}_2^2)(1-k^2\mathbf{x}_2^2)} + \mathbf{x}_2\sqrt{(1-\mathbf{x}_1^2)(1-k^2\mathbf{x}_1^2)}}{1 - k^2\mathbf{x}_1^2\mathbf{x}_2^2},$$

in which $\mathbf{x}_1 = \mathrm{sn}(x_1, k)$, $\mathbf{x}_2 = \mathrm{sn}(x_2, k)$, $\mathbf{x}_3 = \mathrm{sn}(x_1 + x_2, k)$. It shows that $\mathbf{x}_3$ is of degree 4 over $\mathbb{C}(\mathbf{x}_1, \mathbf{x}_2)$. The supplementary rule

$$\mathbf{y}_3 = \frac{\mathbf{y}_1\mathbf{y}_2 + \mathbf{x}_2[-(1+k^2)\mathbf{x}_1 + 2k^2\mathbf{x}_1^3]}{1 - k^2\mathbf{x}_1^2\mathbf{x}_2^2} + \mathbf{x}_3\frac{2k^2\mathbf{x}_1\mathbf{y}_1\mathbf{x}_2^2}{1 - k^2\mathbf{x}_1^2\mathbf{x}_2^2}$$

is easily deduced.

Exercise 3. Do it. Confirm, also, that $\mathbf{y}_3$ is symmetric in $\mathfrak{p}_1 = (\mathbf{x}_1, \mathbf{y}_1)$ and $\mathfrak{p}_2 = (\mathbf{x}_2, \mathbf{y}_2)$ despite appearances.

Exercise 4. $d\mathbf{x}_1/\mathbf{y}_1 + d\mathbf{x}_2/\mathbf{y}_2 = d\mathbf{x}_3/\mathbf{y}_3$ may be verified with tears. Deduce the addition theorem from that. This is Euler's proof [1760].[15] Do it first for $k = 0$ so that $\mathbf{x}_3 = \mathbf{x}_1\sqrt{1-\mathbf{x}_2^2} + \mathbf{x}_2\sqrt{1-\mathbf{x}_1^2}$. What is that? Akhiezer [1970] helps if you get stuck.

Amplification. The incomplete integral of $[(1-x^2)(1-k^2x^2)]^{-1/2}$ takes every complex value, so it is obvious that

$$\int_0^a \frac{dx}{\sqrt{(1-x^2)(1-k^2x^2)}} + \int_0^b \frac{dx}{\sqrt{(1-x^2)(1-k^2x^2)}} = \int_0^c \frac{dx}{\sqrt{(1-x^2)(1-k^2x^2)}}$$

[15] Euler [1911–76 (20): 235–55].

for any a, b with the proper choice of c and of the paths of integration. What is *not obvious* is that a, b, c are *algebraically related*:

$$c = \frac{a\sqrt{(1-b^2)(1-k^2b^2)} + b\sqrt{(1-a^2)(1-k^2a^2)}}{1-k^2a^2b^2}.$$

This is the content of the addition theorem from Euler's point of view. Lagrange (1770) studied special cases in connection with gravitational attraction by two fixed bodies;[16] see Jacobi [1866: 221–31] and Arnold [1978: 261–5] for more on that subject.

Doubling the Lemniscate. The earliest instance of the addition theorem is due to the Italian count Fagnano [1750]; it is the case $a = b$ of the lemniscatic addition theorem with $k^2 = -1$:

$$2\int_0^a \frac{dx}{\sqrt{1-x^4}} = \int_0^c \frac{dx}{\sqrt{1-x^4}}$$

in which

$$c = \frac{2a\sqrt{1-a^4}}{1+a^4}.$$

This has an amusing geometric content: It tells how to double the length of the lemniscatic arc; in particular, it shows that the doubled arc can be constructed from the original by the extraction of square roots, that is, *by ruler and compass*. This was Fagnano's discovery. Ayoub [1984] gives a pleasant account of all this; complete details may be found in Siegel [1969].

Exercise 5. The quantity $\mathbf{y} = \sqrt{1-\mathbf{x}^2}$ with $\mathbf{x} = \text{sn}(x,k)$ obeys the rule

$$\left[(1-\mathbf{x}^2)(1-k^2\mathbf{x}^2)\right]^{-1/2} d\mathbf{x} + \left[(1-\mathbf{y}^2)(1-k^2\mathbf{y}^2)\right]^{-1/2} d\mathbf{y} = 0.$$

This is the function **cosinus amplitudinus** $\text{cn}(x,k)$ of Jacobi [1827]; it inverts the incomplete integral $\int_r^1 [(1-r^2)(1-k^2-k^2r^2)]^{-1/2}dr$. The function $\text{dn}(x,k) = \sqrt{1-k^2\mathbf{x}^2}$ inverts $\int_r^1 [(1+r^2)(1-k^2+r^2)]^{-1/2}dr$. Prove that $\mathbf{x} = \text{sn}$, $\mathbf{y} = \text{cn}$, and $\mathbf{z} = \text{dn}$ are characterized by $\mathbf{x}' = \mathbf{yz}$, $\mathbf{y}' = -\mathbf{xz}$, and $\mathbf{z}' = -k^2\mathbf{xy}$, subject to $\mathbf{x} = 0$, $\mathbf{y} = 1$, and $\mathbf{z} = 1$ at $x = 0$. The addition theorem may now be expressed as

$$\text{sn}(x_1+x_2) = \frac{\text{sn}x_1\text{cn}x_2 + \text{sn}x_2\text{cn}x_2}{1-k^2\text{sn}^2x_1\text{sn}^2x_2}.$$

[16] Lagrange [1868: 67–121].

Exercise 6. Jacobi's **imaginary transformation** is expressed as $\mathrm{sn}(\sqrt{-1}x, k) = \sqrt{-1}\mathrm{sn}(x, k')/\mathrm{cn}(x, k')$. Check it. What does it say about K?

Exercise 7. The identity

$$\mathrm{sn}\left[(1+k)x, \frac{2\sqrt{k}}{1+k}\right] = (1+k)\mathrm{sn}(x,k) \times \left[1 + k^2\mathrm{sn}^2(x,k)\right]^{-1}$$

is to be checked. What does *that* say about K?

Exercise 8. This outlines a proof of the addition theorem in the style of Section 14: The (projective) quartic $\mathbf{y}^2 = (1-\mathbf{x}^2)(1-k^2\mathbf{x}^2)$ has four intersections with the quadratic $\mathbf{y} = 1 + a\mathbf{x} + b\mathbf{x}^2$. One is located at $(\mathbf{x}, \mathbf{y}) = (0, 1)$. Let the other three have covering parameters x_1, x_2, x_3. Check that their sum is independent of a and b and identify it, mod $\mathbb{L}$, as either 0 or $2K$. Now compute $\mathbf{x}_1 + \mathbf{x}_2 + \mathbf{x}_3 = 2ab/c$ and $\mathbf{x}_1\mathbf{x}_2\mathbf{x}_3 = 2a/c$ with $c = k^2 - b^2$ and conclude that $\mathbf{x}_1\mathbf{y}_2 + \mathbf{x}_2\mathbf{y}_1 = \pm\mathbf{x}_3(1 - k^2\mathbf{x}_1^2\mathbf{x}_2^2)$. What is the ambiguous sign?

2.17 Covering Tori

It is a common and interesting situation that one torus covers another. This means that there is a (conformal) projection from the cover onto the base, of degree d, say. The case $d = 2$ is of special importance: It leads to a new relation between Jacobian and Weierstrassian functions quite different from that of Section 16; see especially ex. 16.2.

Exercise 1. Check that such a projection of tori $\mathbf{X}_1 = \mathbb{C}/\mathbb{L}_1 \mapsto \mathbf{X}_2 = \mathbb{C}/\mathbb{L}_2$ lifts to an automorphism of $\mathbb{C}$ in its role as (common) universal cover. Deduce that $\mathbb{L}_2 \subset \mathbb{L}_1$. Check that the degree d is (1) the index of $\mathbb{L}_2$ in $\mathbb{L}_1$, (2) the ratio of the areas of the fundamental cells, and (3) the quotient of the period ratios.

Double Covers. The story of the double covers begins with the torus $\mathbf{X} = \mathbb{C}/\mathbb{L}$ and the radical $\wp_1(x) = \sqrt{\wp(x) - e_1}$. This function has a single-valued branch in the vicinity of $x = \omega_1/2$, $\wp(\omega_1/2) = e_1$ being taken on twice, but cannot belong to $\mathbf{K}$ as it has only one pole per cell and $\mathbf{K}$ contains no functions of degree 1. The fact is that its continuation is an elliptic function with primitive periods ω_1 and $2\omega_2$; as such, it belongs to a *double cover* of $\mathbf{X}$.

Proof. The radical is single-valued near $x = 0$ as well, so the continuation is unobstructed in the plane and leads to a single-valued function $\wp_1 = \sqrt{\wp - e_1}$, by the monodromy theorem of Section 1.11. It has simple poles at $\mathbb{L}$ and may be standardized by fixing the residue at the origin as $+1$; also, it is odd and $\wp_1(x+\omega) = \pm\wp_1(x)$ for $\omega \in \mathbb{L}$, so from the fact that $\wp_1(-\omega_2/2) = -\wp_1(\omega_2/2)$ does not vanish ($e_3 \neq e_1$), ω_2 is seen to be a half-period of $\wp_1$, that is, $\wp_1(x+\omega_2) = -\wp(x)$. The final point is that ω_1 is a full period of $\wp_1$; if not, you would have $\wp_1(x+\omega_1) = -\wp_1(x)$, and evaluation at $x = \omega/2$ with $\omega = \omega_1 + \omega_2$ is contradictory:

$$\wp_1(\omega/2) = -\wp_1(\omega/2 - \omega_1) = +\wp_1(\omega/2 - \omega_1 - \omega_2) = \wp_1(-\omega/2),$$

violating the oddness of $\wp_1$ and the fact that $\wp_1(\omega/2)$ does not vanish.

Exercise 2. Check that $\wp_2 = \sqrt{\wp - e_2}$ has periods $2\omega_1$ and $2\omega_2$; similarly, $\wp_3 = \sqrt{\wp - e_3}$ has periods $2\omega_1$ and ω_2.

Relating $\wp$ to sn. This is easy. The ratio $f = \sqrt{e_1 - e_3}/\wp_3(x)$ is viewed as a function of $x' = x\sqrt{e_1 - e_3}$, the radical being chosen so that $f(x') = +1$ at $x = \omega_1/2$. Now

$$\begin{aligned}(f')^2 &= \frac{(\wp - e_1)(\wp - e_2)(\wp - e_3)}{(\wp - e_3)^2} = \left(1 - \frac{e_1 - e_3}{\wp - e_3}\right)\left(1 - \frac{e_2 - e_3}{\wp - e_3}\right) \\ &= (1 - f^2)(1 - k^2 f^2)\end{aligned}$$

with $k^2 = (e_2 - e_3)/(e_1 - e_3)$, and since f takes the values $0, 1, 1/k, \infty$ as in Fig. 2.7 in Section 2.5 for sn, so the identification

$$\operatorname{sn}(x', k^2) = \sqrt{\frac{e_1 - e_3}{\wp(x) - e_3}}$$

follows from the identification of periods:

$$\frac{\omega_1}{2}\sqrt{e_1 - e_3} = \int_0^1 \left[(1 - f^2)(1 - k^2 f^2)\right]^{-1/2} df = K(k),$$

$$\frac{\omega_2}{2}\sqrt{e_1 - e_3} = \int_1^{1/k} \left[(1 - f^2)(1 - k^2 f^2)\right]^{-1/2} df = \sqrt{-1}K'(k).$$

Exercise 3. The substitution $x \mapsto e_3 + (e_1 - e_3)/x^2$ converts the incomplete integral

$$\int [4(x - e_1)(x - e_2)(x - e_3)]^{-1/2}\, dx$$

into

$$-(e_1 - e_3)^{-1/2} \int \left[(1 - x^2)(1 - k^2 x^2)\right]^{-1/2} dx$$

with $k^2 = (e_2 - e_3)/(e_1 - e_3)$ as before. Deduce

$$\wp(x) = e_3 + \sqrt{e_1 - e_3}\,\text{sn}^{-2}(x\sqrt{e_1 - e_3}, k^2).$$

This provides a second proof.

Warning. $k^2 = (e_2 - e_3)/(e_1 - e_3)$ exhibits k^2 as a function of the *Weierstrassian* period ratio $\omega = \sqrt{-1}K'/K$, by inspection of the sums defining $e = e_1, e_2, e_3$. This is not the corresponding *Jacobian* period ratio $\sqrt{-1}K'/2K = \omega/2$. Keep this in mind; it can be confusing and lead to tears.

Exercise 4. Check that k^2 is a function of the period ratio, directly from the complete integrals K and K' for $0 < k < 1$.

Landen's Transformation. This was the subject of Section 3. The present proof is in a very different style, with the help of double covers. Fix $\mathbb{L} = \omega_1\mathbb{Z} \oplus \omega_2\mathbb{Z}$ and let $\mathbb{L}^*$ be the lattice formed with the primitive periods $\omega_1/2$ and ω_2, having double the period ratio of $\mathbb{L}$, so that $\mathbf{X}^* = \mathbb{C}/\mathbb{L}^*$ is a double cover of $\mathbf{X} = \mathbb{C}/\mathbb{L}$. Now the function $\wp^*(x) = \wp(x) + \wp(x + \omega_1/2) - e_1$ is of period $\omega_1^* = \omega_1/2$ and also of period $\omega_2^* = \omega_2$, with one pole per cell of $\mathbb{L}^*$, located at $x = 0$, and having an expansion of the form $x^{-2} + 0 \cdot x^{-1} + 0 \cdot x^0 + \cdots$, as you will check. In short, it is the $\wp$-function of $\mathbf{X}^*$, as its name suggests, and it is easy to determine the roots of the associated cubic:

$$\begin{aligned} e_1^* &= e_1 + 2\sqrt{(e_1 - e_2)(e_1 - e_3)}, \\ e_2^* &= e_1 - 2\sqrt{(e_1 - e_2)(e_1 - e_3)}, \\ e_3^* &= -2e_1. \end{aligned}$$

Exercise 5. Check these values. *Hint*:

$$\wp(\omega_1/4) = e_1 + \sqrt{(e_1 - e_2)(e_1 - e_3)}$$

producing e_1^*; see ex. 14.6.

The change in Jacobi's modulus k^2 is computed from these values. Starting with $k^2 = (e_3 - e_2)(e_3 - e_1)^{-1}$, as in ex. 16.2, you find

$$(k^*)^2 = \frac{e_2^* - e_3^*}{e_1^* - e_3^*} = \frac{e_2^* + (e_1^* + e_2^*)}{e_1^* + (e_1^* + e_2^*)} = \frac{3e_1 - 2\sqrt{(e_1 - e_2)(e_1 - e_3)}}{3e_1 + 2\sqrt{(e_1 - e_2)(e_1 - e_3)}},$$

or, in terms of the complementary modulus $(k')^2 = 1-k^2 = (e_1-e_2)(e_1-e_3)^{-1}$,

$$(k^*)^2 = \frac{e_1 - e_3 + e_1 - e_2 - 2\sqrt{(e_1-e_2)(e_1-e_3)}}{e_1 - e_3 + e_1 - e_2 + 2\sqrt{(e_1-e_2)(e_1-e_3)}} = \left(\frac{1-k'}{1+k'}\right)^2,$$

as you will check. Landen's transformation in the form

$$K(k^*) = \tfrac{1}{2}(1+k')K(k)$$

is immediate: $K(k) = (\omega_1/2)\sqrt{e_1 - e_3}$ and $K(k^*) = (\omega_1/4)\sqrt{e_1^* - e_3^*}$, so

$$\begin{aligned}\frac{1+k'}{2} \bullet \frac{K(k)}{K^*(k)} &= \left(1 + \sqrt{\frac{e_1-e_2}{e_1-e_3}}\right)\sqrt{\frac{e_1-e_2}{e_1^*-e_3^*}} \\ &= \frac{\sqrt{e_1-e_3}+\sqrt{e_1-e_2}}{[e_1-e_3+2\sqrt{(e_1-e_3)(e_1-e_2)}+e_1-e_2]^{1/2}} \\ &= 1.\end{aligned}$$

Amplification. The identity $k^* = (1-k')(1+k')^{-1}$ is an algebraic relation between $k^* = k(2\omega)$ and the original modulus $k(\omega)$ itself:

$$k(2\omega) = \left[1 - \sqrt{1-k^2(\omega)}\right]\left[1 + \sqrt{1-k^2(\omega)}\right]^{-1}.$$

It is the so-called **modular equation of level 2**; modular equations of higher level, relating $k(n\omega)$ to $k(\omega)$ for $n \geq 3$, will be studied in Chapter 4 and applied to Hermite's solution of the quintic [1859] in Chapter 5 and to imaginary quadratic number fields in Chapter 6. This is the general question of **division of periods**, initiated by Gauss (1800),[17] Legendre (1797),[18] and Abel [1827].

2.17.1 Some Particular Moduli

Example 1: Square lattice. The associated cubic $y^2 = x^3 - x$ was noted in example 9.1. Its roots are $1, 0, -1$, so $k^2(\sqrt{-1}) = (0-(-1))/(1-(-1)) = 1/2$. The further value $k(2\sqrt{-1})(\sqrt{2}-1)(\sqrt{2}+1)^{-1} = 3 - 2\sqrt{2}$ comes from the modular equation; it is the modulus of the *Jacobian* square lattice. The value $k^2(\sqrt{-1}) = 1/2$ can also be elicited in this way: $1/\sqrt{2}$ is its own complement, so $\omega = \sqrt{-1}K'/K = \sqrt{-1}$, that is, $k(\sqrt{-1}) = 1/\sqrt{2}$. Now let e_1, e_2, e_3 be the roots of the cubic that produces, not just the period ratio $\sqrt{-1}$, but the particular periods $\omega_1 = 1$ and $\omega_2 = \sqrt{-1}$. The lattice admits complex multiplication by

[17] Gauss [1870–1929 (10): 160–6].

[18] Legendre [1825 (1): 262].

$c = \sqrt{-1}$ and $\wp(cx) = c^{-2}\wp(x) = -\wp(x)$, so $e_3 = -e_1$, $e_2 = -e_1 - e_3 = 0$, $g_2 = 4e_1^2$, and $g_3 = 0$, with the result that $K(1/\sqrt{2}) = (1/2)\sqrt{e_1 - e_3} = \sqrt{e_1/2}$. Besides,

$$\omega_1 = 1 = \int_{e_1}^{\infty} \frac{dx}{\sqrt{x^3 - e_1^2 x}} = \frac{1}{\sqrt{e_1}} \int_1^{\infty} \frac{dx}{\sqrt{x^3 - x}},$$

so

$$K(1/\sqrt{2}) = \frac{1}{\sqrt{2}} \int_1^{\infty} \frac{dx}{\sqrt{x^3 - x}} = 2^{-3/2} \int_0^1 x^{-3/4}(1-x)^{-1/2} dx$$
$$= 2^{-3/2} \frac{\Gamma(1/4)\Gamma(1/2)}{\Gamma(3/4)},$$

in conformity with the values found in Section 4.

Example 2. $k = \sqrt{2} - 1$ is fixed under $k' \mapsto k^*$ so, by period doubling,

$$2\frac{K'}{K}(k) = \frac{K'}{K}(k^*) = \frac{K}{K'}(k),$$

that is,

$$\frac{K'}{K}(k) = \frac{1}{\sqrt{2}},$$

which is to say that the period ratio is $1/\sqrt{2}$; in short,

$$k^2(1/\sqrt{2}) = 1 - (\sqrt{2} - 1)^2 = 2(\sqrt{2} - 1).$$

Example 3: Triangular lattice. The discussion is similar to example 1. The period ratio is $\omega = e^{2\pi\sqrt{-1}/3} = (1/2)(-1 + \sqrt{-3})$, and the associated cubic $y^2 = 4(x^3 - 1)$ of example 9.2 has roots $e = 1, \omega, \omega^2$, so

$$k^2\left(\frac{-1 + \sqrt{-3}}{2}\right) = \frac{\omega^2 - \omega}{1 - \omega} = -\omega = e^{-\sqrt{-1}\pi/3} = \frac{1 - \sqrt{-3}}{2}.$$

Exercise 6. Compute the associated complete integral $K(k)$ in the style of example 1 using the complex multiplication by ω. *Answer*:

$$6K(k) = \sqrt{(1/2)[3 - \sqrt{-3}]} \times \Gamma(1/6)\Gamma(1/2)/\Gamma(2/3).$$

Ramanujan's Example. The special moduli of examples 1–3 are specimens of a wider class. Let $k(\omega/n) = k'(\omega)$ for some integral $n = 2, 3, \ldots$ Then

$$\frac{\omega}{n} = \sqrt{-1}\frac{K'}{K} \quad \text{taken at} \quad k(\omega/n) = \sqrt{-1}\frac{K}{K'} \quad \text{taken at} \quad k(\omega) = -1/\omega,$$

so $\omega = \sqrt{-n}$ is a quadratic irrationality. Abel [1827] proved that, in such a case, the modulus k can be expressed by radicals over the ground field of rational numbers. Hardy [1940] quotes a spectacular example from a letter of Ramanujan: If $n = 210 = 2 \times 3 \times 5 \times 7$, then

$$\sqrt{k} = (1-\sqrt{2})^2(2-\sqrt{3})(3-\sqrt{10})^2(4-\sqrt{15})^2(6-\sqrt{35})$$
$$\times(8-3\sqrt{7})(\sqrt{7}-\sqrt{6})^2(\sqrt{15}-\sqrt{14}).$$

Amplification. Siegel [1949] proved that if ω is not a quadratic irrationality, then $k(\omega)$ is transcendental; see Baker [1975] for such matters.

2.18 Finale: Higher Genus

The present section provides a glimpse of curves $\mathbf{X}$ of higher genus $g \geq 2$. Bliss [1933], Clemens [1980], Kirwan [1992], Shafarevich [1977: 411–23], Siegel [1971], and Weyl [1955] are recommended for additional information.

Dissection. $\mathbf{X}$ is a sphere with g handles. To clarify its topology, let a_i and $b_i (1 \leq i \leq g)$ be closed cycles of $\mathbf{X}$ passing through a common point $\mathfrak{o}$: a_i runs once about the ith hole and b_i runs once about the ith handle, as in Fig. 2.24

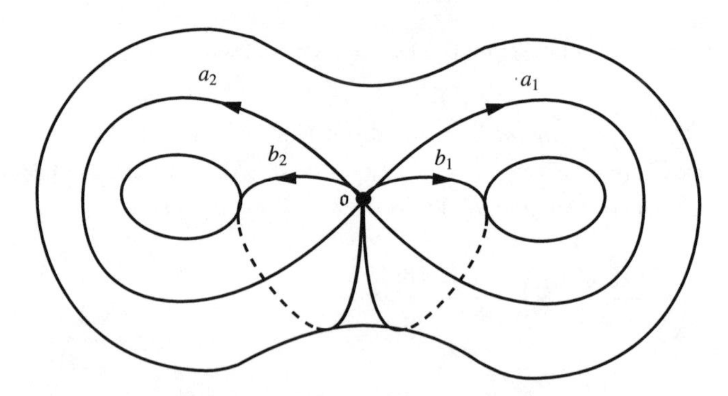

Figure 2.24. The closed cycles.

($g = 2$). Cuts along a_1 and b_1 permit the curve to be opened up as in Fig. 2.25, and further cuts along a_2 and b_2 permit the curve to be laid out flat on the plane as in Fig. 2.26. This figure is the **dissection** of the curve. It follows that the

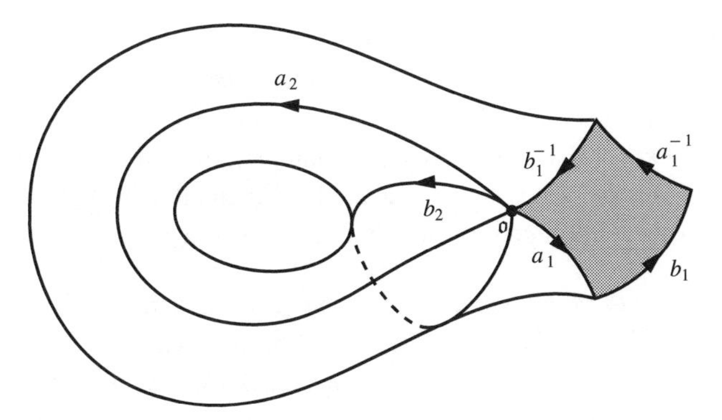

Figure 2.25. The first cut.

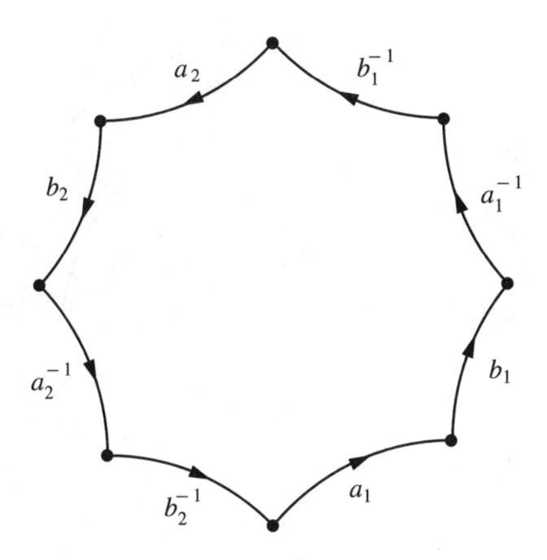

Figure 2.26. The dissection of the curve.

fundamental group Γ of **X** is a free group on $2g$ letters with the single relation

$$a_1 b_1 a_1^{-1} b_1^{-1} \cdots a_g b_g a_g^{-1} b_g^{-1} = 1.$$

Universal Cover. This is the upper half-plane $\mathbb{H}$. The fundamental group Γ is now interpreted as the group of covering maps of Section 1.11. These are conformal automorphisms of the half-plane (= rigid hyperbolic motions), so Γ appears as a subgroup of $PSL(2,\mathbb{R})$; in particular, the quotient $\mathbb{H}/\Gamma$ is a copy of **X**. More concretely, Γ has a fundamental cell $\mathfrak{F}$ in the form of a $2g$-sided hyperbolic polygon as in Fig. 2.26; with suitable conventions at its sides, every orbit of Γ cuts $\mathfrak{F}$ just once, so **X** is just the cell with sides identified. The images of $\mathfrak{F}$ under the action of Γ **tessellate** $\mathbb{H}$, that is, they cover it simply; see Section 4.3 for similar pictures.

Function Field. The function field **K(X)** is the splitting field over $\mathbb{C}(x)$ of the (irreducible) polynomial $P(x, y) \in \mathbb{C}[x, y]$ whose vanishing defines the curve. It can also be viewed as the class of functions of rational character on the universal cover that are invariant (= automorphic) under the action of Γ. The second viewpoint goes back to Klein [1882] and Poincaré (1880).[19] Ford [1972] is the best introduction, though it needs to be supplemented.

Differentials of the First Kind. **X** has g independent **differentials of the first kind** (DFK). Klein explained this in his Cambridge lectures [1893] by means of a hydrodynamical picture; compare Section 1.13. For $g = 1, \mathbf{X} = \mathbb{C}/\mathbb{L}$ admits one conjugate[20] pair of steady, irrotational, incompressible flows, as in Fig. 2.27. They are described by a conjugate pair of real harmonic differentials

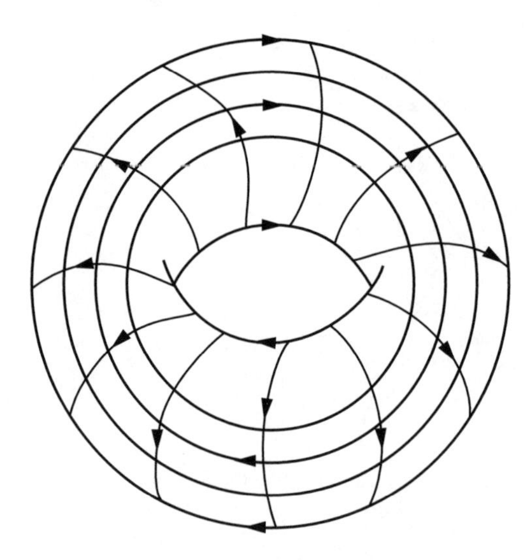

Figure 2.27. The hydrodynamical picture.

da and db, $dx = da + \sqrt{-1}db$ being *the* complex differential of the first kind associated with the parameter $x = x(\mathfrak{p})$ of the universal cover. The same idea works for higher genus: A pair of conjugate flows for genus 2 are depicted in Fig. 2.28; each pair produces one differential of the first kind, and there are as many pairs as handles.[21]

[19] See Poincaré [1985].

[20] The adjective conjugate signifies that the streamlines of one flow are the lines of constant potential of the other.

[21] The black spots on the rim of the second hole are stagnation points. The flow of Fig. 2.27 moves in the same direction underneath, that of Fig. 2.28 in the opposite direction.

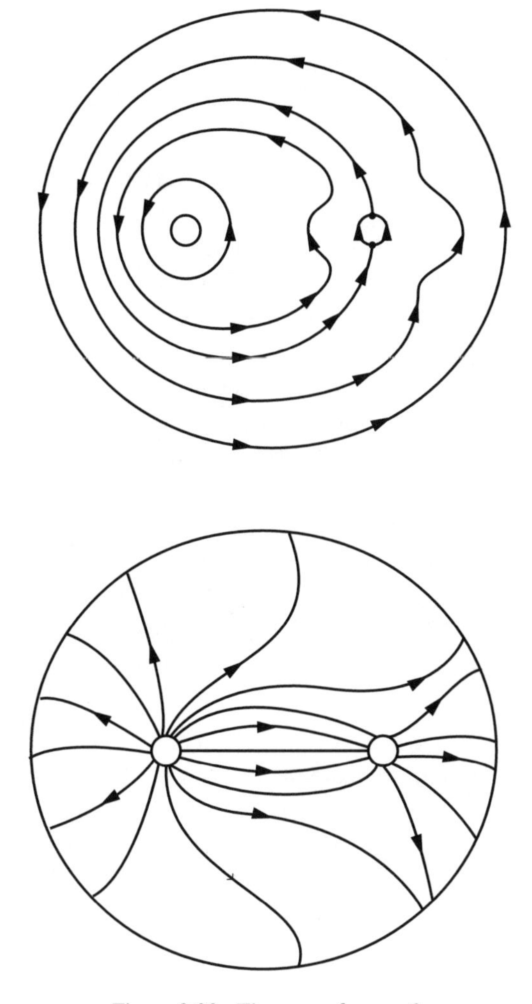

Figure 2.28. The case of genus 2.

Exercise 1. Let **X** be the hyperelliptic curve

$$y^2 = (x - e_0)(x - e_1) \cdots (x - e_{2g})$$

of genus $g \geq 1$. Check that $\omega_k = x^{k-1}y^{-1}dx$ is a differential of the first kind for $1 \leq k \leq g$. Do not forget the point at infinity. Prove by hand that these span DFK.

Riemann–Roch. A function $f \in \mathbf{K}(\mathbf{X})$ has an equal number of roots and poles; this is its **degree**. The theorem of Riemann [1857] and Roch [1865] states (in part) that *if* $\mathfrak{p}_i (1 \le i \le n)$ *are points of* $\mathbf{X}$, *if* $\mathfrak{L}$ *is the class of functions from* $\mathbf{K}$ *with these poles or softer (that is, the multiplicity at the pole is not greater than the number of repetitions of the latter), and if* $\mathfrak{D}$ *is the class of differentials of the first kind with these roots or harder, then*

$$\dim \mathfrak{L} = n + 1 - g - \dim \mathfrak{D}.$$

Exercise 2. What does Riemann–Roch say for $g = 0$? For $g = 1$? Compare ex. 1.3.4. *Hints for* $g = 1$: $\mathbb{C} \subset \mathfrak{L}$ so the dimension of $\mathfrak{L}$ is at least 1. If $n = 3$, then $f \in \mathfrak{L}$ is of degree 0, 2, or 3: No poles means that f is constant; two poles gives two extra dimensions since, for each pair from $\mathfrak{p}_1, \mathfrak{p}_2, \mathfrak{p}_3$, there is a function f with simple poles at those places and residues ± 1, but $f_{12} + f_{23} + f_{31} = 0$ with a self-evident notation; three poles contributes nothing new, so the total dimension is 3, as Riemann–Roch predicts.

Jacobi Variety. Let $\omega_k (1 \le k \le g)$ be a basis of differentials of the first kind, let ω stand for the vector $(\omega_1, \ldots, \omega_g)$, and fix a point $\mathfrak{o}$ of $\mathbf{X}$. The map of the **divisor** $\mathfrak{P} = (\mathfrak{p}_1, \ldots, \mathfrak{p}_g)$ of variable points of $\mathbf{X}$ to the sum

$$\int_{\mathfrak{O}}^{\mathfrak{P}} \omega = \sum_{i=1}^{g} \int_{\mathfrak{o}}^{\mathfrak{p}_i} \omega = \mathbf{x}(\mathfrak{P})$$

covers $\mathbb{C}^g$; naturally, the map is equivocal in that the paths of integration are unspecified. The **periods** of $\mathbf{X}$ are the points $\mathbf{x} \in \mathbb{C}^g$ produced by the choice, in each summand, of some closed path starting and ending at the base point $\mathfrak{o}$. These form a lattice $\mathbb{L} \subset \mathbb{C}^g$. The map of divisors is rendered unambiguous by viewing the image of $\mathbf{x}(\mathfrak{P})$ as a point of the quotient $\mathbb{C}^g/\mathbb{L}$. The latter is the **Jacobi variety J** of $\mathbf{X}$; it is a complex torus of dimension g, and more: It is an (irreducible) algebraic variety, as the name suggests. Shafarevich [1977: 155] explains such matters. For $g = 1$, it is a copy of the curve. For $g \ge 2$, it is a more or less faithful copy of the g-fold symmetric product of the curve with itself, but only in the top dimension; the correspondence goes bad in codimension 1, as it must on topological grounds.

Abel's theorem. *This states that the divisors* $\mathfrak{P} = (\mathfrak{p}_1, \ldots, \mathfrak{p}_n)$ *and* $\mathfrak{Q} = (\mathfrak{q}_1, \ldots, \mathfrak{q}_n)$ *of* $\mathbf{X}$ *are the poles and roots of a function in* $\mathbf{K}$ *if and only if*

$$\int_{\mathfrak{Q}}^{\mathfrak{P}} \omega = \sum_{i=1}^{n} \int_{\mathfrak{q}_i}^{\mathfrak{p}_i} \omega$$

vanishes modulo $\mathbb{L}$ *for every differential* ω *of the first kind.* For $g = 1$, ω is the differential dx of the covering coordinate, and the statement reduces to that of Section 15. The cheap half of the full statement is proved by integration of $\int_{\mathfrak{q}}^{\mathfrak{p}} \omega \times d \log f(\mathfrak{p})$ about the boundary of a dissection of the curve and comparison with the sum of residues inside.

Riemann's Theta Function. Let the loops $a_i, b_i (1 \le i \le g)$ be placed as in Fig. 2.24, and let the basis $\omega_j (1 \le j \le g)$ of differentials of the first kind be taken so that $a_i(\omega_j) = 1$ or 0 according as $i = j$ or not. Then the **period matrix** $B = \left[b_i(\omega_j)\colon 1 \le i, j \le g\right]$ is (real) symmetric and has positive imaginary part; it is the counterpart of the period ratio. Riemann [1857] introduced the **theta function** $\vartheta(x)$ formed by summing $\exp[2\pi\sqrt{-1}n \cdot x + \pi\sqrt{-1}B(n)]$ over the points n of the g-dimensional integral lattice $\mathbb{Z}^g$. Here, $B(n)$ is the quadratic form $n \cdot Bn$ and x is a point of $\mathbb{C}^g$. Riemann [1857] located the roots of $\vartheta(x) = 0$ (the **vanishing theorem**) and used this for the proof of the hard half of Abel's theorem. Lefschetz [1921] used ϑ to embed the Jacobian variety $\mathbf{J}$ in a high-dimensional projective space. The embedding displays the fact that $\mathbf{J}$ really is a variety; compare Section 3.4 for $g = 1$. See Narasimhan [1992] for a modern description of these matters.

The Addition Theorem. As before let $\mathfrak{P}, \mathfrak{P}'$, and $\mathfrak{P}''$ denote divisors of $\mathbf{X}$ comprising g points apiece. The sums $\int_{\mathfrak{O}}^{\mathfrak{P}} \omega$ cover $\mathbb{C}^g$ so the sum of any two is a third:

$$\int_{\mathfrak{O}}^{\mathfrak{P}'} \omega + \int_{\mathfrak{O}}^{\mathfrak{P}''} \omega = \int_{\mathfrak{O}}^{\mathfrak{P}} \omega.$$

The content of the **addition theorem** is that the upper limit $\mathfrak{P}$ of the third integral is rationally expressible in terms of the upper limits $\mathfrak{P}'$ and $\mathfrak{P}''$; compare Section 16. Jacobi [1866:231–7] gave a remarkable mechanical proof of part of the statement in the special case of hyperelliptic curves: Straight-line motion at constant speed in $\mathbb{R}^{g+1}$ is expressed in ellipsoidal coordinates and then integrated via the Hamilton–Jacobi equation of classical mechanics. Compare the *mechanical illustration* of Section 10; see Arnold [1978] for details of this and Siegel [1971] for the general case.

Conformal Structures. The existence of many different complex structures for higher handlebodies can be illustrated, much as in Sections 1.12 and 1.14, by the fact that the tubes extending from the several cut copies of $\mathbb{P}^1$ seen in Fig. 1.20 can be twisted before being pasted to their associates. Actually, the matter is more subtle than that; indeed, Riemann [1857] found that the conformal structure of a topological handlebody of genus $g \ge 2$ depends upon $3g - 3$ complex parameters ($=$ **moduli**), as opposed to a single such modulus ($=$ the period ratio) for $g = 1$ and a single complex structure for $g = 0$ (see Section 1.14).

Imayoshi and Taniguchi [1992] provide a nice introduction to these ideas. It is a by-product that the higher handlebodies have many geometries of curvature -1. This may be contrasted to the extraordinary fact that in $\mathbb{R}^3$ *a compact 3-dimensional manifold of constant curvature* -1 *is completely specified by its fundamental group*! This is Mostow's celebrated **rigidity theorem** [1966].

3
Theta Functions

Theta functions appear in Bernoulli's *Ars Conjectandi* [1713] and in the number-theoretic investigations of Euler [1773: 71–84] and Gauss [1801], but come into full flower only in Jacobi's *Fundamenta Nova* [1829]. They have marvelous properties. The whole story of elliptic curves can be based upon them, but that road is not taken here; rather, they will be developed only to such a point that their connection to the Weierstrassian and Jacobian functions of Chapter 2 can be made and a few of their arithmetic and geometric applications are accessible. The whole chapter is an interlude, really. Mumford [1983], Krazer [1903], and Lawden [1989] are recommended for fuller information.

3.1 Jacobi's Theta Functions

Jacobi [1829] introduced four functions of the variables $p = e^{\pi\sqrt{-1}x}$ and $q = e^{\pi\sqrt{-1}\omega}$, x being the usual covering coordinate of the curve $\mathbf{X} = \mathbb{C}/\mathbb{L}$ and ω its period ratio, with the familiar standardization that the imaginary part of ω is positive. $\mathbb{L}$ is taken to be $\mathbb{Z} \oplus \omega\mathbb{Z}$ for simplicity. Jacobi's **theta functions** are as follows:

$$\vartheta_1(x) = \vartheta_1(x\,|\,\omega) = \sqrt{-1}\sum(-1)^n p^{2n-1} q^{(n-1/2)^2},$$

$$\vartheta_2(x) = \vartheta_2(x\,|\,\omega) = \sum p^{2n-1} q^{(n-1/2)^2},$$

$$\vartheta_3(x) = \vartheta_3(x\,|\,\omega) = \sum p^{2n} q^{n^2},$$

$$\vartheta_4(x) = \vartheta_4(x\,|\,\omega) = \sum(-1)^n p^{2n} q^{n^2}.$$

The sums are taken over $n \in \mathbb{Z}$; their rapid convergence is assured by $|q| = \exp(-\pi \times \text{Im}\,\omega) < 1$. Obviously, they are entire functions of $x \in \mathbb{C}$, so they cannot have two independent periods, being pole-free and not constant, but

Table 3.1.1. *Addition of half-periods*

	$x+\frac{1}{2}$	$x+\frac{\omega}{2}$	$x+\frac{1}{2}+\frac{\omega}{2}$	$x+1$	$x+\omega$	$x+1+\omega$
ϑ_1	ϑ_2	$\sqrt{-1}a\vartheta_4$	$a\vartheta_3$	$-\vartheta_1$	$-b\vartheta_1$	$b\vartheta_1$
ϑ_2	$-\vartheta_1$	$a\vartheta_3$	$-\sqrt{-1}a\vartheta_4$	$-\vartheta_2$	$b\vartheta_2$	$-b\vartheta_2$
ϑ_3	ϑ_4	$a\vartheta_2$	$\sqrt{-1}a\vartheta_1$	ϑ_3	$b\vartheta_3$	$b\vartheta_3$
ϑ_4	ϑ_3	$\sqrt{-1}a\vartheta_1$	$a\vartheta_2$	ϑ_4	$-b\vartheta_4$	$-b\vartheta_4$

they are doing their best in that they transform simply ϑ_1 is the basic function; the other ones are produced by addition of half-periods. $\vartheta = -\sqrt{-1}a\vartheta_1$ is the function of Section 2.15. The facts are recorded in Table 3.1.1 with the abbreviations $a = p^{-1}q^{-1/4}$ and $b = p^{-2}q^{-1}$.

Sample proof.

$$\begin{aligned}\vartheta_3(x+\omega/2) &= \sum [e^{\pi\sqrt{-1}(x+\omega/2)}]^{2n} \cdot [e^{\pi\sqrt{-1}\omega}]^{n^2} \\ &= e^{-\pi\sqrt{-1}\omega/4} \sum e^{2\pi\sqrt{-1}nx} \cdot e^{\pi\sqrt{-1}\omega(n+1/2)^2} \\ &= e^{-\pi\sqrt{-1}(x+\omega/4)} \sum p^{2n-1} q^{(n-1/2)^2} \\ &= a\vartheta_2(x),\end{aligned}$$

the shift $n \mapsto n-1$ being used in line 3.

Exercise 1. Check the rest of Table 3.1.1.

Exercise 2. $\vartheta_1, \vartheta_2, \vartheta_3, \vartheta_4$ all obey $\vartheta(x+2) = \vartheta(x)$ and $\vartheta(x+2\omega) = (pq)^{-4}$ $\vartheta(x)$, as you will check. Discuss the general solution of the two identities (a) $f(x+2) = f(x)$ and (b) $f(x+2\omega) = e^{ax+b}f(x)$. *Hint*: Expand f in a Fourier series $\sum \hat{f}(n)p^n$ and use (b) to pin down $\hat{f}(n)$. You will see thetas coming out.

The (simple) roots of the theta functions are easily located; for example, ϑ_1 is odd about $x = 0$, as in ex. 2.15.2, and the integral of $(2\pi\sqrt{-1})^{-1}d\log\vartheta_1$ about the perimeter of the fundamental cell is $+1$, as in ex. 2.15.3, so this function

Table 3.1.2. *Location of the zeros*

$\vartheta_1(x) = 0$	$\vartheta_2(x) = 0$	$\vartheta_3(x) = 0$	$\vartheta_4(x) = 0$
$x \in \mathbb{L}$	$x \in \mathbb{L} + 1/2$	$x \in \mathbb{L} + 1/2 + \omega/2$	$x \in \mathbb{L} + \omega/2$

has a *simple* root at $x = 0$ and no others in the cell. Now look at Table 3.1.1 to confirm the root patterns displayed in Table 3.1.2.

Exercise 3. ϑ_1 is odd about $x = 0$, as noted. Check that $\vartheta_2, \vartheta_3, \vartheta_4$ are even about $x = 0$.

Exercise 4. Prove that $c_1\vartheta_1(x) + c_2\vartheta_2(x) + c_3\vartheta_3(x) + c_4\vartheta_4(x)$ vanishes identically only if $c_1 = c_2 = c_3 = c_4 = 0$ and has, otherwise, just four roots in the fundamental cell of the *doubled* lattice $2\mathbb{L}$. *Hints*: Use Table 3.1.1 for the first part, and ex. 2.15.3 as a model for the second part.

3.2 Some Identities

Theta functions obey a bewildering number and variety of identities. Mumford [1983: 20–3] and Krazer [1903] are recommended for the full treatment. What will be needed here is simpler; it is all due to Jacobi [1829].

Identity 1. $\vartheta_2^2(0)\vartheta_2^2(x) + \vartheta_4^2(0)\vartheta_4^2(x) = \vartheta_3^2(0)\vartheta_3^2(x)$.

Proof. $[\vartheta_4(0)\vartheta_4(x)]^{-2} \times \left[\vartheta_3^2(0)\vartheta_3^2(x) - \vartheta_2^2(0)\vartheta_2^2(x)\right]$ is periodic relative to $\mathbb{L}$ (see Table 3.1.1) and has just one (simple) pole per cell, of degree ≤ 2, located at the root $x = \omega/2$ of $\vartheta_4(x) = 0$ (see Table 3.1.2). But $\vartheta_3(\omega/2) = a\vartheta_2(0)$ and similarly $\vartheta_2(\omega/2) = a\vartheta_3(0)$, so the numerator vanishes, the degree of the pole is ≤ 1, and the function is constant in view of its periodicity. Now take $x = 1/2$ and use Table 3.1.1 once more to elicit its value ($= 1$).

Exercise 1.

$$\vartheta_2^2(0)\vartheta_1^2(x) + \vartheta_4^2(0)\vartheta_3^2(x) = \vartheta_3^2(0)\vartheta_4^2(x),$$

$$\vartheta_4^2(0)\vartheta_1^2(x) + \vartheta_3^2(0)\vartheta_2^2(x) = \vartheta_2^2(0)\vartheta_3^2(x),$$

$$\vartheta_3^2(0)\vartheta_1^2(x) + \vartheta_4^2(0)\vartheta_2^2(x) = \vartheta_2^2(0)\vartheta_4^2(x)$$

can be proved in the same style, but use Table 3.1.1 instead.

Aside 1. Identity 1 specialized to $x = 0$ states that $\vartheta_2^4(0) + \vartheta_4^4(0) = \vartheta_3^4(0)$, which is to say

$$\left[\sum_{\mathbb{Z}} q^{(n-1/2)^2}\right]^4 + \left[\sum_{\mathbb{Z}} (-1)^n q^{n^2}\right]^4 = \left[\sum_{\mathbb{Z}} q^{n^2}\right]^4,$$

or, what is the same,

$$\sum_{\mathbb{Z}^4} q^{(\mathbf{n}-\mathbf{1/2})^2} + \sum_{\mathbb{Z}^4} (-1)^{\mathbf{n}\cdot\mathbf{1}} q^{\mathbf{n}^2} = \sum_{\mathbb{Z}^4} q^{\mathbf{n}^2},$$

in which $\mathbf{n} = (n_1, n_2, n_3, n_4) \in \mathbb{Z}^4$, $\mathbf{n}^2 = n_1^2 + n_2^2 + n_3^2 + n_4^2$, $\mathbf{1} = (1, 1, 1, 1)$, and similarly for $\mathbf{1/2}$. Move the second sum to the right. Now the identity reads

$$\sum_{\mathbb{Z}^4} q^{(\mathbf{n}-\mathbf{1/2})^2} = 2 \sum_{\substack{\mathbf{n}\in\mathbb{Z}^4 \\ \mathbf{n}\cdot\mathbf{1}\ \text{odd}}} q^{\mathbf{n}^2},$$

and in this form it can be checked by hand. The orthogonal matrix

$$A = \frac{1}{2}\begin{pmatrix} 1 & 1 & 1 & 1 \\ 1 & -1 & 1 & -1 \\ 1 & 1 & -1 & -1 \\ 1 & -1 & -1 & 1 \end{pmatrix}$$

maps the sublattice $\{\mathbf{n}\cdot\mathbf{1}\ \text{even}\} \subset \mathbb{Z}^4$ 1:1 onto itself and carries $\mathbf{1/2}$ to $(1, 0, 0, 0)$, so with the notation $\mathbf{n} \mapsto \mathbf{n}^*$ for its action, you find

$$\begin{aligned}
\sum_{\mathbf{n}\cdot\mathbf{1}\ \text{even}} q^{(\mathbf{n}-\mathbf{1/2})^2} &= \sum_{\mathbf{n}\cdot\mathbf{1}\ \text{even}} q^{(\mathbf{n}^*-\mathbf{1/2}^*)^2} \\
&= \sum_{\mathbf{n}\cdot\mathbf{1}\ \text{even}} q^{(n_1-1)^2+n_2^2+n_3^2+n_4^2} \\
&= \sum_{\mathbf{n}\cdot\mathbf{1}\ \text{odd}} q^{\mathbf{n}^2}.
\end{aligned}$$

But also

$$(\mathbf{n} - \mathbf{1/2})^2 = (\mathbf{n}\cdot\mathbf{1})^2 - \mathbf{n}\cdot\mathbf{1} + 1 + (\text{the coprojection of } \mathbf{n} \text{ and } \mathbf{1})^2$$

so

$$\sum_{\mathbf{n}\cdot\mathbf{1}\ \text{odd}} q^{(\mathbf{n}-\mathbf{1/2})^2} = \sum_{\mathbf{n}\cdot\mathbf{1}\ \text{even}} q^{(\mathbf{n}-1/2)^2} = \sum_{\mathbf{n}\cdot\mathbf{1}\ \text{even}} q^{\mathbf{n}^2} = \sum_{\mathbf{n}\cdot\mathbf{1}\ \text{odd}} q^{\mathbf{n}^2}.$$

Now add the two displays.

Identity 2. The so-called **null values** $\vartheta_1'(0)$, $\vartheta_2(0)$, $\vartheta_3(0)$, $\vartheta_4(0)$ play a considerable role in what follows. Jacobi [1829] discovered the remarkable identity

$$\vartheta_1'(0) = \pi\vartheta_2(0)\vartheta_3(0)\vartheta_4(0).$$

Proof. [1] $[\vartheta_1(2x)\vartheta_2(0)\vartheta_3(0)\vartheta_4(0)]^{-1} \times 2\vartheta_1(x)\vartheta_2(x)\vartheta_3(x)\vartheta_4(x)$ is periodic relative to $\mathbb{L}$ (see Table 3.1.1) and pole-free (see Table 3.1.2), so it is constant ($= 1$), as you can see by taking $x = 0$. This proves the **duplication formula**

$$\vartheta_1(2x) = 2\frac{\vartheta_1(x)\vartheta_2(x)\vartheta_3(x)\vartheta_4(x)}{\vartheta_2(0)\vartheta_3(0)\vartheta_4(0)}.$$

Next, take logarithms, differentiate twice by x, and put $x = 0$ to obtain

$$\frac{\vartheta_1'''(0)}{\vartheta_1'(0)} = \frac{\vartheta_2''(0)}{\vartheta_2(0)} + \frac{\vartheta_3''(0)}{\vartheta_3(0)} + \frac{\vartheta_4''(0)}{\vartheta_4(0)}.$$

The computation is elementary but tiresome; it is left to you as ex. 2 to follow. Now each ϑ solves the heat equation $4\pi\sqrt{-1}\partial\vartheta/\partial\omega = \partial^2\vartheta/\partial x^2$, as you can see from the sums, so the preceding display states that $\vartheta_2(0)\vartheta_3(0)\vartheta_4(0)/\vartheta_1'(0)$ is independent of ω. The value π is now obtained by looking at $\omega = \sqrt{-1}\infty$, alias $q = 0$: To leading order,

$$\vartheta_1'(0) = 2\pi q^{1/4}, \ \vartheta_2(0) = 2q^{1/4}, \ \vartheta_3(0) = 1, \quad \text{and} \quad \vartheta_4(0) = 1,$$

by inspection of sums. The proof is finished.

Exercise 2. Check the computation of $\vartheta_1'''(0)/\vartheta_1'(0)$. *Hint*: $\vartheta_1''(0)$, $\vartheta_2'(0)$, $\vartheta_3'(0)$, and $\vartheta_4'(0)$ vanish. Why?

Aside 2. The present identity seems incapable of direct proof in the style of 1; it states that

$$\sum_{\mathbb{Z}^1}(-1)^n(2n-1)q^{n^2-n} = -\sum_{\mathbb{Z}^3}(-1)^{n_3}q^{n_1^2-n_1+n_2^2+n_3^2},$$

which does not even *look* plausible.

Addition-Like Identities. Theta functions are not rational in x, or in a single exponential, nor do they belong to any elliptic function field and so cannot have addition theorems, properly speaking, in view of ex. 2.14.10, but they *do* obey

[1] Whittaker and Watson [1963: 490].

similar identities; for example,

$$\vartheta_1(x_1 - x_2)\vartheta_2(x_1 + x_2) + \vartheta_1(x_1 + x_2)\vartheta_2(x_1 - x_2)$$
$$= \frac{\vartheta_1(x_1)\vartheta_1(2x_2)\vartheta_2(x_1)\vartheta_2(0)}{\vartheta_1(x_2)\vartheta_2(x_2)}.$$

Exercise 3. Check this in the style used for identity 1.

Watson's Identities. Watson [1929] found a pretty variant with the doubled period ratio in it:

$$\vartheta_1(x_1|\omega)\vartheta_1(x_2|\omega) = \vartheta_3(x_1 + x_2|2\omega)\vartheta_2(x_1 - x_2|2\omega)$$
$$- \vartheta_2(x_1 + x_2|2\omega)\vartheta_3(x_1 - x_2|2\omega).$$

Proof. Write out the left side as a sum over $\mathbf{n} \in \mathbb{Z}^4$ (line 1); make the substitution $n_1 + n_2 = n_1', n_1 - n_2 = n_2'$, recognizing that n_1' and n_2' are now of like parity (lines 2 and 3); and split the sum so produced according to that parity (line 4):

$$\sum_{\mathbb{Z}^2}(-1)^{n_1+n_2} p^{(2n_1-1)x_1+(2n_2-1)x_2} q^{(n_1-1/2)^2+(n_2-1/2)^2}$$

$$= - \sum_{n_1' \equiv n_2' (\text{mod } 2)} (-1)^{n_1'} p^{(n_1'+n_2'-1)x_1+(n_1'-n_2'-1)x_2} q^{(1/4)[(n_1'+n_2'-1)^2+(n_1'-n_2'-1)^2]}$$

$$= - \sum_{n_1 \equiv n_2 (\text{mod } 2)} (-1)^{n_1} p^{(n_1-1)(x_1+x_2)} q^{\frac{1}{2}(n_1-1)^2} \times p^{n_2(x_1-x_2)} q^{\frac{1}{2}n_2^2}$$

$$= + \sum_{\mathbb{Z}} p^{2n(x_1+x_2)} q^{2n^2} \times \sum_{\mathbb{Z}} p^{(2n-1)(x_1-x_2)} q^{2(n-1/2)^2}$$

$$- \sum_{\mathbb{Z}} p^{(2n-1)(x_1+x_2)} q^{2(n-1/2)^2} \times \sum_{\mathbb{Z}} p^{2n(x_1-x_2)} q^{2n^2}.$$

Now read it off.

Exercise 4.

(a) $$\vartheta_1(x_1|\omega)\vartheta_2(x_2|\omega) = \vartheta_4(x_1 + x_2|2\omega)\vartheta_1(x_1 - x_2|\omega)$$
$$+ \vartheta_1(x_1 + x_2|2\omega)\vartheta_4(x_1 - x_2|\omega),$$

(b) $$\vartheta_2(x_1|\omega)\vartheta_2(x_2|\omega) = \vartheta_3(x_1 + x_2|2\omega)\vartheta_2(x_1 - x_2|\omega) + \vartheta_2(x_1 + x_2|2\omega)\vartheta_3(x_1 - x_2|\omega),$$

(c) $$\vartheta_3(x_1|\omega)\vartheta_4(x_2|\omega) = \vartheta_4(x_1 + x_2|2\omega)\vartheta_4(x_1 - x_2|\omega) + \vartheta_1(x_1 + x_2|2\omega)\vartheta_1(x_1 - x_2|\omega)$$

can be proved in the same way, but use Table 3.1.1 instead.

Exercise 5. Differentiate (a) by x_1 and put $x_1 = x_2 = 0$, evaluate (b) and (c) at the same place, and combine to confirm that $\vartheta_2(0)\vartheta_3(0)\vartheta_4(0)/\vartheta_1'(0)$ is unchanged by doubling the period ratio. Now iterate. This is Watson's [1929] pretty proof of identity 2.

3.3 The Jacobi and Weierstrass Connections

The purpose of this section is to express all the important Jacobian and Weierstrassian quantities by means of theta functions.

The Jacobi Connection. The function $f(x) = \vartheta_1(x)/\vartheta_4(x)$ is periodic relative to the lattice $\mathbb{L}^* = 2\mathbb{Z} \oplus \omega\mathbb{Z}$ (see Table 3.1.1) with simple roots at $x = 0$ and $x = 1$, simple poles at $x = \omega/2$ and $x = \omega/2 + 1$, and no others in the fundamental cell (see Table 3.1.2). It is a candidate for sin amp. Now $f'(x) = \pi\vartheta_4^2(0)\vartheta_2(x)\vartheta_3(x)\vartheta_4^{-2}(x)$, as you will check in ex. 1, to follow, and it follows from ex. 2.1 that

$$(f')^2 = \left[\vartheta_2^2(0) - \vartheta_3^2(0)f^2\right]\left[\vartheta_3^2(0) - \vartheta_2^2(0)f^2\right],$$

which confirms the idea; in fact, you have the identities:

$$\mathrm{sn}(x,k) = \frac{\vartheta_3(0)}{\vartheta_2(0)}\frac{\vartheta_1}{\vartheta_4}(x'|\omega) \quad \text{with} \quad x' = x/\pi\vartheta_3^2(0),$$

$$k^2 = \frac{\vartheta_2^4(0)}{\vartheta_3^4(0)},$$

$$(k')^2 = \frac{\vartheta_4^4(0)}{\vartheta_3^4(0)},$$

$$K(k) = \frac{\pi}{2}\vartheta_3^2(0),$$

$$\sqrt{-1}K'(k) = \frac{\pi}{2}\vartheta_3^2(0) \times \omega.$$

Exercise 1. Check the details. *Hints*: Exs. 1.3 and 2.1 are helpful. For example, to verify $f'(x) = \pi\vartheta_4^2(0)\vartheta_2(x)\vartheta_3(x)\vartheta_4^{-2}(x)$, you will want to know that $\vartheta_1'\vartheta_4 - \vartheta_1\vartheta_4'$ vanishes at the roots of $\vartheta_2\vartheta_3 = 0$; this follows from ex. 1.3 and Table 3.1.2.

Exercise 2. The arithmetic–geometric mean $M(a, b)$ of Section 2.3 is to be expressed by null values. *Answer*: $M\left(\vartheta_4^2(0)/\vartheta_3^2(0), 1\right) = \vartheta_3^{-2}(0)$. *Hint*: See identity 2.1.

This finishes the Jacobian part of the story.

The Weierstrass Connection. ϑ_1 vanishes simply on the lattice $\mathbb{L}$, so $-(\log\vartheta_1)''$ has at $x = 0$ a pole with principal part $x^{-2} + 0\cdot x^{-1}$; it is also periodic with respect to $\mathbb{L}$ since $(\log b)'' = 0$ (see Table 1.1). In short, it is the $\wp$-function of $\mathbf{X} = \mathbb{C}/\mathbb{L}$, up to an additive constant.

Exercise 3. Confirm $\wp(x) = -[\log\vartheta_1(x)]'' + e_1 + [\log\vartheta_1]''(1/2)$.

This is one type of connection. The following connections are of more interest:

$$\begin{aligned}\wp(x) &= e_1 + \left[\frac{\vartheta_1'(0)}{\vartheta_1(x)}\cdot\frac{\vartheta_2(x)}{\vartheta_2(0)}\right]^2\\ &= e_2 + \left[\frac{\vartheta_1'(0)}{\vartheta_1(x)}\cdot\frac{\vartheta_3(x)}{\vartheta_3(0)}\right]^2\\ &= e_3 + \left[\frac{\vartheta_1'(0)}{\vartheta_1(x)}\cdot\frac{\vartheta_4(x)}{\vartheta_4(0)}\right]^2.\end{aligned}$$

Sample proof. $\vartheta_2^2(x)/\vartheta_1^2(x)$ is periodic relative to $\mathbb{L}$ (see Table 3.1.1), it has a pole of degree 2 at $x = 0$ and no others in the fundamental cell (see Table 3.1.2), and it is even. The rest will be plain from the fact that $\vartheta_2(1/2) = 0$.

Null Values. The null values $\vartheta_1'(0), \vartheta_2(0), \vartheta_3(0), \vartheta_4(0)$ are related to the roots e_1, e_2, e_3, to the discriminant $\Delta = 16(e_1 - e_2)^2(e_2 - e_3)^2(e_3 - e_1)^2$, and to the absolute invariant $j = g_2^3/\Delta$ by a series of pretty identities:

$$\begin{aligned}\sqrt{e_1 - e_2} &= \frac{\vartheta_1'(0)\vartheta_4(0)}{\vartheta_2(0)\vartheta_3(0)} = \pi\vartheta_4^2(0),\\ \sqrt{e_1 - e_3} &= \frac{\vartheta_1'(0)\vartheta_3(0)}{\vartheta_2(0)\vartheta_4(0)} = \pi\vartheta_3^2(0),\\ \sqrt{e_2 - e_3} &= \frac{\vartheta_1'(0)\vartheta_2(0)}{\vartheta_3(0)\vartheta_4(0)} = \pi\vartheta_2^2(0),\end{aligned}$$

$$e_1 = \frac{1}{3}\pi^2[\vartheta_3^4(0) + \vartheta_4^4(0)],$$

$$e_2 = \frac{1}{3}\pi^2[\vartheta_2^4(0) - \vartheta_4^4(0)],$$

$$e_3 = -\frac{1}{3}\pi^2[\vartheta_2^4(0) + \vartheta_3^4(0)],$$

$$\Delta = 16\pi^{12}[\vartheta_2(0)\vartheta_3(0)\vartheta_4(0)]^8 = 16\pi^4[\vartheta_1'(0)]^8,$$

$$j = \frac{g_2^3}{\Delta} = \frac{1}{54} \times \frac{[\vartheta_2^8(0) + \vartheta_3^8(0) + \vartheta_4^8(0)]^3}{\vartheta_2^8(0)\vartheta_3^8(0)\vartheta_4^8(0)}.$$

Sample proof. The second presentation of the $\wp$-function is used at $x = 1/2$ to obtain

$$\sqrt{e_1 - e_2} = \frac{\vartheta_1'(0)}{\vartheta_1(1/2)}\frac{\vartheta_3(1/2)}{\vartheta_3(0)} = \frac{\vartheta_1'(0)\vartheta_4(0)}{\vartheta_2(0)\vartheta_3(0)}$$

with the help of Table 3.1.1. The alternative form $\pi\vartheta_4^2(0)$ comes from identity 2.2. The next set of formulas follows from identity 2.1 taken at $x = 0$ to produce $\vartheta_2^4(0) + \vartheta_4^4(0) = \vartheta_3^4(0)$. The expression for the discriminant $\Delta = g_2^3 - 27g_3^2 = 16(e_1 - e_2)^2(e_2 - e_3)^2(e_3 - e_1)^2$ will be obvious.

Exercise 4. Check that $g_2 = 2(e_1^2+e_2^2+e_3^2) = (2/3)\pi^4\left[\vartheta_2^8(0) + \vartheta_3^8(0)+\vartheta_4^8(0)\right]$.

The expression for the absolute invariant is immediate from that. Now the Weierstrassian part of the story is finished, too.

3.4 Projective Embedding of Tori

Theta functions provide an alternative proof of the fact that *a complex torus is an algebraic curve*. This was the content of Section 2.11. The new proof is outlined in a series of easy steps.

Step 1. Take $\mathbf{X} = \mathbb{C}/\mathbb{L}$ with $\mathbb{L} = \mathbb{Z} \oplus \omega\mathbb{Z}$. The map

$$J\colon x \in \mathbb{C}/\mathbb{L} \to [\vartheta_1(2x), \vartheta_2(2x), \vartheta_3(2x), \vartheta_4(2x)]$$

is of period 1; it is also of period ω up to the (nonvanishing) multiplicative factor b^2, by Table 3.1.1; finally, the four theta functions involved have no common roots, by Table 3.1.2, so J can be construed as a map of $\mathbf{X} = \mathbb{C}/\mathbb{L}$ into the 3-dimensional projective space $\mathbb{P}^3$.

Step 2. The map is 1:1. If not, you could find distinct points x_1 and x_1' in the fundamental cell of $\mathbb{L}$ with $J(x_1) = J(x_1')$ and, by Table 3.1.1, you could produce a second such pair x_2, x_2' from the first by addition of some half-period, keeping the four points x_1, x_1', x_2, x_2' distinct in $\mathbb{C}/\mathbb{L}$. Clearly, $J(x_2) = J(x_2')$ by columns 4 and 5 of Table 3.1.1. Now fix a fifth point x_3 distinct from x_1, x_1', x_2, x_2' mod $\mathbb{L}$ and choose constants c_1, c_2, c_3, c_4, not all vanishing, so as to make $\vartheta = c_1\vartheta_1 + c_2\vartheta_2 + c_3\vartheta_3 + c_4\vartheta_4$ vanish at $x = 2x_1, 2x_2$, and $2x_3$. This can be done because there are three equations in four unknowns, but is contradictory. ϑ vanishes at the *five* distinct points $2x_1, 2x_1', 2x_2, 2x_2'$, and $2x_3$ of the fundamental cell of the doubled lattice $2\mathbb{L}$ and that is too many, as only four are permitted by ex. 1.4.

Step 3. Let $\mathbf{x}_n = \vartheta_n(2x_n)$, for $n = 1, 2, 3, 4$, and recall identity 1 from Section 2 and its companions from ex. 2.1. They state that the image of $\mathbf{X} = \mathbb{C}/\mathbb{L}$ in $\mathbb{P}^3$ lies in the intersection V of the two quadrics

$$\mathbf{x}_3^2\vartheta_3^2(0) - \mathbf{x}_2^2\vartheta_2^2(0) = \mathbf{x}_4^2\vartheta_4^2(0),$$

$$\mathbf{x}_3^2\vartheta_2^2(0) - \mathbf{x}_2^2\vartheta_3^2(0) = \mathbf{x}_1^2\vartheta_4^2(0).$$

Step 4. Confirm that the image is precisely V, that is, $J(\mathbf{X}) = V \subset \mathbb{P}^3$.

Proof. The equation $c_1\vartheta_1(2x) + c_2\vartheta_2(2x) + c_3\vartheta_3(2x) + c_4\vartheta_4(2x) = 0$ has precisely four roots in the fundamental cell of $2\mathbb{L}$, by ex. 1.4. This means that the projective plane

$$c_1\mathbf{x}_1 + c_2\mathbf{x}_2 + c_3\mathbf{x}_3 + c_4\mathbf{x}_4 = 0$$

intersects the image of $\mathbf{X}$ in precisely four points. The proof is now finished by invoking an elementary fact of geometry: *A plane meets a pair of quadrics in* $\mathbb{P}^3$ *in at most* four *points.* This forces the image to fill up V. Why?

Exercise 1. The rule of intersections in $\mathbb{P}^3$ is an instance of Bezout's theorem. Shafarevich [1977: 198] or Kirwan [1992] can be consulted for the general statement. Give a proof for the present purpose along the following lines: First change coordinates so that the plane is $\mathbf{x}_4 = 0$. This reduces the problem to checking that two quadrics in $\mathbb{P}^2$ have at most four intersections. Next reduce one quartic to the projective circle $\mathbf{x}_1^2 + \mathbf{x}_2^2 + \mathbf{x}_3^2 = 0$ and uniformize it by a projective line $\mathbb{P}^1$ via the substitution $\mathbf{x}_1 = (1/2)(x + x^{-1})$, $\mathbf{x}_2 = (\sqrt{-1}/2)(x - x^{-1})$, and $\mathbf{x}_3 = \sqrt{-1}$ of Section 1.10. The second quadric is now expressed by the vanishing of a rational function of degree at most 4 on $\mathbb{P}^1$. The rest will be plain.

Exercise 2. Check the conformality of the correspondence between the torus $\mathbf{X}$ and the intersection V of the two quadrics.

Punch Line. $\mathbf{X} = \mathbb{C}/\mathbb{L}$ is presented as the common roots of two polynomials, that is, it is an **algebraic variety**.

3.5 Products

The next item of business is Jacobi's [1829] expression of the ϑ-functions as infinite products; these will be applied to number-theoretic questions presently. The formulas are reminiscent of Euler's [1743] product

$$\sin \pi x = \pi x \prod_{n=1}^{\infty} (1 - x^2/n^2)$$

Exercise 1. Prove that in the following style: $(1-x^2/n^2) = (1-x/n)(1+x/n)$, so the product $P(x)$ satisfies $P(x+1) = -P(x)$. $Q(x) = \sin \pi x / P(x)$ is now seen to be a pole-free function of $z = \exp(2\pi\sqrt{-1}x)$. $Q(x) = o(z)$ at $z = \infty$. What about $z = 0$? What then?

Jacobi's products are now displayed with $p = e^{\pi\sqrt{-1}x}$ and $q = e^{\pi\sqrt{-1}\omega}$, as before, and $C = \prod_{n=1}^{\infty} (1 - q^{2n})$:

$$\sqrt{-1}\vartheta_1(x) = Cq^{1/4}(p - p^{-1}) \prod_{n=1}^{\infty} (1 - q^{2n}p^2)(1 - q^{2n}p^{-2}),$$

$$\vartheta_2(x) = Cq^{1/4}(p + p^{-1}) \prod_{n=1}^{\infty} (1 + q^{2n}p^2)(1 + q^{2n}p^{-2}),$$

$$\vartheta_3(x) = C \prod_{n=1}^{\infty} (1 + q^{2n-1}p^2)(1 + q^{2n-1}p^{-2}),$$

$$\vartheta_4(x) = C \prod_{n=1}^{\infty} (1 - q^{2n-1}p^2)(1 - q^{2n-1}p^{-2});$$

evaluating at $x = 0$ supplies products for the null values:

$$\pi^{-1}\vartheta_1'(0) = 2Cq^{1/4} \prod_{n=1}^{\infty} (1 - q^{2n})^2,$$

$$\vartheta_2(0) = 2Cq^{1/4}\prod_{n=1}^{\infty}(1+q^{2n})^2,$$

$$\vartheta_3(0) = C\prod_{n=1}^{\infty}(1+q^{2n-1})^2,$$

$$\vartheta_4(0) = C\prod_{n=1}^{\infty}(1-q^{2n-1})^2.$$

Proof. $\vartheta_3(x) = 0$ has simple roots at the shifted lattice $\mathbb{L} + 1/2 + \omega/2$ (see Table 3.1.2), that is, when $p^2 = -q^{2n+1}$ or $p^2 = -1/q^{2n+1}$. Now $|q| < 1$, so the product $f(p) = \prod(1+q^{2n+1}p^2)(1+q^{2n+1}p^{-2})$ converges fine, it has the same roots as ϑ_3, it is of period 1 since $p^2 = e^{2\pi\sqrt{-1}x}$ is such, and it reacts to the shift $x \mapsto x+\omega$ $(p \mapsto pq)$ as

$$\begin{aligned} f(p) \mapsto f(pq) &= \prod(1+q^{2n+3}p^2)(1+q^{2n-1}p^{-2}) \\ &= \frac{1+q^{-1}p^{-2}}{1+qp^2} f(p) = q^{-1}p^{-2}f(p). \end{aligned}$$

ϑ_3 transforms in the same way (see Table 3.1.1) so the ratio ϑ_3/f is a pole-free elliptic function, which is to say it is a constant C. The other three products, with the *same* constant C, are deduced from columns 1 and 2 of Table 3.1.1. The value of C is easily determined from identity 2.2, the products for the null values, and the evaluation

$$\prod_{n=1}^{\infty}(1+q^{2n})(1-q^{2(2n-1)}) \equiv P(q) = 1.$$

Exercise 2. Prove that $P(q) = \prod(1+q^n)(1-q^{2n-1})$. Deduce $P(q) = P(q^2) = P(q^4)$, and so on, and so confirm that $P(q) = 1$. *Hint*: $(1+q^n) = (1-q^{2n})/(1-q^n)$.

Now, by identity 2.2 and the products for null values,

$$\begin{aligned} 1 = \frac{\vartheta_2(0)\vartheta_3(0)\vartheta_4(0)}{\pi^{-1}\vartheta_1'(0)} &= C^2\prod_{n=1}^{\infty}(1+q^{2n})^2(1+q^{2n-1})^2(1-q^{2n-1})^2(1-q^{2n})^{-2} \\ &= C^2\prod_{n=1}^{\infty}(1+q^{2n})^2(1-q^{2(2n-1)})^2(1-q^{2n})^{-2} \\ &= C^2\prod_{n=1}^{\infty}(1-q^{2n})^{-2}, \end{aligned}$$

so $C = \pm\prod(1-q^{2n})$, and the correct sign is found to be $+1$ upon evaluation of $\vartheta_3(0)$ at $\omega = \sqrt{-1}\infty$ ($q = 0$). The proof is finished.

The sum for ϑ_3 is written out in all its glory as Jacobi's **triple product**:

$$\vartheta_3(x) = \sum_{\mathbb{Z}} p^{2n}q^{n} = \prod_{n=1}^{\infty}(1-q^{2n})(1+q^{2n-1}p^2)(1+q^{2n-1}p^{-2}).$$

Exercise 3. $\vartheta_1(x)[\sin(\pi x)]^{-1}$ is an entire function $F(x)$. Check that $F''(0)/F(0) = 8\pi^2\sum q^{2n}(1-q^{2n})^{-2}$.

Exercise 4. Check the identity

$$\sum_{n=0}^{\infty}(-1)^n(2n+1)q^{n(n+1)/2} = \prod_{n=1}^{\infty}(1-q^n)^3.$$

Hint: Jacobi's triple product is multiplied by p and differentiated with regard to that variable at $p = \sqrt{-q}$.

Exercise 5. The **Fermat curve** $x_0^3 + x_1^3 + x_2^3 = 0$ is of genus 1. Prove that; compare ex. 2.11.5. Now uniformize it by means of $x_0 = P(q), x_1 = \rho^{1/3}P(\rho q)$, $x_2 = \rho^{2/3}P(\rho^2 q)$ with $\rho = e^{2\pi\sqrt{-1}/3}$ and $P(q) = \prod(1-q^n)$, this being the cube root of the product of ex. 4. *Hint*: $n(n+1)/2 \equiv 0$ or 1 (but not 2) mod 3, so the sum of ex. 4 satisfies $S(q) + \rho S(\rho q) + \rho^2 S(\rho^2 q) = 0$. This clever proof is due to Garvan [1995]. The fact appears for the first time in Farkas and Kra [1993]. It seems remarkable that it was not known before.

A whole series of pretty expressions for Weierstrassian and Jacobian quantities can be obtained from the ϑ-products via the identities of Section 3. Jacobi [1829] produced a vast number of these; a few samples are presented here.

Sample 1.

$$k^2 = \frac{\vartheta_2^4(0)}{\vartheta_3^4(0)} = 16q\prod\left(\frac{1+q^{2n}}{1+q^{2n-1}}\right)^8$$

and

$$(k')^2 = \frac{\vartheta_4^4(0)}{\vartheta_3^4(0)} = \prod\left(\frac{1-q^{2n-1}}{1+q^{2n-1}}\right)^8,$$

so, with the temporary notation $k^* = k/4\sqrt{q}$,

$$\frac{k'}{\sqrt{k^*}} = \prod \left(\frac{1+q^{2n-1}}{1+q^{2n}}\right)^2 \left(\frac{1-q^{2n-1}}{1+q^{2n-1}}\right)^4 = \prod (1-q^{2n-1})^4(1+q^n)^{-2}$$
$$= \prod (1-q^{2n})^6,$$

by ex. 2.

Exercise 6. Check the allied identities:

$$\prod (1+q^{2n-1})^6 = 1/\sqrt{k^* k'},$$
$$\prod (1+q^{2n})^6 = k^*/\sqrt{k'},$$
$$\prod (1-q^{2n})^6 = C^6 = 8\pi^{-3} k^* k' K^3(k).$$

Sample 2.

$$\operatorname{sn}(2K(k)x, k) = \frac{\vartheta_3(0)}{\vartheta_2(0)} \frac{\vartheta_1(x)}{\vartheta_4(x)}$$
$$= 2q^{1/4} \cdot \frac{\sin \pi x}{\sqrt{k}} \prod \frac{1 - 2q^{2n} \cos 2\pi x + q^{4n}}{1 - 2q^{2n-1} \cos 2\pi x + q^{2(2n-1)}}.$$

Sample 3.

$$\frac{\vartheta_3(x)}{\vartheta_4(x)} = \prod \left[(1+q^{2n})/(1-q^{2n})\right]^2 \times \sum_{\mathbb{Z}} \frac{e^{2\pi\sqrt{-1}nx}}{\cosh(n\pi\omega)}.$$

Proof.[2] For simplicity, take ω purely imaginary, so that $0 < q < 1$, and view

$$\frac{\vartheta_3(x)}{\vartheta_4(x)} = \prod \frac{(1+q^{2n-1}p^2)(1+q^{2n-1}p^{-2})}{(1-q^{2n-1}p^2)(1-q^{2n-1}p^{-2})}$$

as a function $F(z)$ of $z = p^2$. It has a simple pole at $z = q$, with residue r, and no others in the annulus $q^3 < |z| < q^{-1}$. Now expand F in $q < |z| < q^{-1}$ as $\sum_{\mathbb{Z}} F_n z^{-n}$ with $F_n = (2\pi\sqrt{-1})^{-1} \oint F z^{n-1} dz$, the integral being taken about $|z| = 1$. $F(q^2 z) = -F(z)$, by Table 3.1.1, so deformation of the contour

[2] Schoenberg [1981].

produces

$$\begin{aligned}
F_n - r \times q^{n-1} &= \frac{1}{2\pi\sqrt{-1}} \int_{|z|=q^2} F(z) z^{n-1} dz \\
&= \frac{1}{2\pi\sqrt{-1}} \int_{|z|=1} F(q^2 z) z^{n-1} dz \times q^{2n} \\
&= \frac{-q^{2n}}{2\pi\sqrt{-1}} \int_{|z|=1} F(z) z^{n-1} dz \\
&= -q^{2n} F_n,
\end{aligned}$$

which is to say $F_n = r(1+q^{2n})^{-1} q^{n-1}$. The evaluation of the residue as $2q \prod (1+q^{2n})^2 (1-q^{2n})^{-2}$ completes the proof:

$$\frac{\vartheta_3(x)}{\vartheta_4(x)} = \prod \left(\frac{1+q^{2n}}{1-q^{2n}} \right)^2 \times \left[\sum_{\mathbb{Z}} \frac{2p^{-2n}}{q^n + q^{-n}} = \sum_{\mathbb{Z}} \frac{e^{2\pi\sqrt{-1}nx}}{\cosh(\pi n \omega)} \right].$$

Sample 4.

$$\Delta = (2\pi)^4 \left[\vartheta_1'(0) \right]^8 = (2\pi)^{12} C^8 q^2 \prod_{n=1}^{\infty} (1-q^{2n})^{16} = (2\pi)^{12} q^2 \prod_{n=1}^{\infty} (1-q^{2n})^{24}.$$

The final product may be expanded as $(2\pi)^{12} \sum_1^\infty \tau(n) q^{2n}$ with rational integral coefficients $\tau(n)$. These are the **Ramanujan numbers** and have marvelous arithmetic properties. Ramanujan [1916] conjectured and Mordell [1917] proved that $\tau(n)$ is a *multiplicative* function, that is, $\tau(mn) = \tau(m)\tau(n)$ if m, n are coprime; see Section 4.8 for more information.

Exercise 7. Compute $\tau(1) = 1,\ \tau(2) = -24,\ \tau(3) = 252$, and $\tau(4) = -1472$.

Exercise 8. Use the product for the discriminant and ex. 2 to check that

$$\Delta\left(\frac{\omega}{2}\right) \Delta\left(\frac{\omega+1}{2}\right) \Delta(2\omega) = -\Delta^3(\omega),$$

$$\Delta\left(\frac{\omega}{3}\right) \Delta\left(\frac{\omega+1}{3}\right) \Delta\left(\frac{\omega+2}{3}\right) \Delta(3\omega) = \Delta^4(\omega).$$

Exercise 9. Prove that

$$\Delta(\omega)^{1/6} \Delta(2\omega)^{1/12} \Delta(4\omega)^{1/6} = \prod_{n=1}^{\infty} (1-q^n)^{2+2m},$$

in which $m = 4,\ 2$, or 1 according as 4 divides n, or only 2, or n is odd. Zagier [1992] presents more formulas of this type.

Sample 5. $g_2 = (2/3)\pi^4 \left[\vartheta_2^8(0) + \vartheta_3^8(0) + \vartheta_4^8(0)\right]$ by ex. 3.4, so

$$j = \frac{g_2^3}{\Delta} = \frac{(2/3)^3\pi^4 \left[\vartheta_2^8(0) + \vartheta_3^8(0) + \vartheta_4^8(0)\right]^3}{2^{12}q^2 \prod(1-q^{2k})^{24}}$$

$$= 24^{-3}q^{-2}\left[2^8q^2\prod(1+q^{2k})^{16} + \prod(1+q^{2k-1})^{16} + \prod(1-q^{2k-1})^{16}\right]^3.$$

This may be expanded in the form $1728^{-1}q^{-2}(1 + c(1)q^2 + c(2)q^4 + \cdots.)$ with rational integral coefficients, as you will easily check. Zuckerman [1939] computed the first 24 coefficients and van Wijngaarden [1953] went on to

$$c_{100} = 83798831110707476912751950384757452703801918339072000,$$

but *who's counting*? Peterson [1932] obtained the asymptotic law $c(n) \simeq 2^{-1/2}n^{-3/4}\exp(4\pi\sqrt{n})$. Like Ramanujan numbers, these too have intricate arithmetic properties; see, for example, Van der Pol [1951] and Atkin and O'Brien [1967]. What is most startling is their appearance in connection with the celebrated **monster group**: They are the dimensions of its irreducible representations! Conway and Norton [1979] and Thompson [1979] explain this; see also Koike [1994].

3.6 Sums of Two Squares

The problem of expressing a whole number as sum of perfect squares goes back to antiquity. Neugebauer [1957] traced it back to Babylonian times. Pythagoras (520 B.C.) understood the special case $a^2 + b^2 = c^2$. Fermat (1640) resolved the question of which primes are sums of two squares, but it was Jacobi [1829] who produced the most remarkable results via his theta functions, and it is his method that is followed here. The case of four squares is postponed to the next section. Grosswald [1984a] gives the full story, including the more difficult case of sums of three squares. The present story begins with the identity of Section 3

$$\sqrt{\wp(x) - e_3} = \frac{\vartheta_1'(0)\vartheta_4(x)}{\vartheta_1(x)\vartheta_4(0)}$$

and the expansion of this quotient into partial fractions

$$\frac{1}{2\pi\sqrt{-1}} \times \left[\sum_{n=0}^{\infty} \frac{q^n p^{-1}}{1 - q^{2n}p^{-2}} - \sum_{n=1}^{\infty} \frac{q^n p}{1 - q^{2n}p^2}\right].$$

This looks more complicated than it is. The quotient is periodic relative to the lattice $2\mathbb{Z} \oplus \omega\mathbb{Z}$, by Table 3.1.1, as is the sum, by inspection, and they have common simple roots, at $x = 0$ and $x = 1$, with the same residues $+1$ and -1, respectively. It follows that they differ only by a constant, which is seen to vanish by evaluation at the root $x = \omega/2$ of $\vartheta_4(x) = 0$. To proceed, take $x = 1/2$ and use

$$\frac{q^n}{1+q^{2n}} = q^n \frac{1-q^{2n}}{1-q^{4n}} = (q^n - q^{3n}) \sum_{l=0}^{\infty} q^{4nl}$$

to produce

$$\begin{aligned} \pi^{-1}\sqrt{e_1 - e_3} &= 1 + 4\sum_{n=1}^{\infty} q^n (1+q^{2n})^{-1} \\ &= 1 + 4\sum_{n=1}^{\infty}\sum_{l=0}^{\infty} q^{n(4l+1)} - 4\sum_{n=1}^{\infty}\sum_{l=0}^{\infty} q^{n(4l+3)}. \end{aligned}$$

Now comes the punch line. By Section 3, you recognize the left-hand side as

$$\vartheta_3^2(0) = \left[\sum_{\mathbb{Z}} q^{n^2}\right]^2 = \sum_{\mathbb{Z}^2} q^{n_1^2+n_2^2} = \sum_{m=0}^{\infty} r_2(m) q^m,$$

in which $r_2(m)$ is the number of representations of the whole number m as a sum of two squares, whereupon an arithmetic theorem of Jacobi [1829] emerges by comparison of like powers: $r_2(m) = 4d_1(m) - 4d_3(m)$, which is to say that the number of representations of a whole number as a sum of two squares is four times the excess (if any) of its positive divisors $d \equiv 1 \bmod 4$ over its positive divisors $d \equiv 3 \bmod 4$. Fermat's theorem (1640), that odd primes $\equiv 1 \bmod 4$ are sums of two squares but primes $\equiv 3 \bmod 4$ are not, is a special case.

Exercise 1. Check the details of Fermat's statement; in particular, show that $r_2(p) = 8$ for $p \equiv 1 \bmod 4$, so the representation $p = n_1^2 + n_2^2$ is unique, up to trivialities.

Exercise 2. Confirm the more general statement that the whole number m (prime or not) is a sum of two squares if and only if all its prime factors $\equiv 3 \bmod 4$ appear in even powers.

Hardy and Wright [1979] and Nagell [1964] provide elementary proofs of these and many related results.

3.7 Sums of Four Squares

The investigation is based upon the null value:

$$\vartheta_3^4(0) = \sum_{\mathbb{Z}^4} q^{n_1^2+n_2^2+n_3^2+n_4^2} = \sum_{m=0}^{\infty} r_4(m) q^m,$$

in which $r_4(m)$ is now the number of representations of the whole number m as a sum of four squares.

Step 1. The identity $\sqrt{\wp(x) - e_3} = \vartheta_1'(0)\vartheta_4(x)/\vartheta_1(x)\vartheta_4(0)$ is used to verify that e_3 is the null value $-\vartheta_4''(0)/\vartheta_4(0)$. This is done by expanding both sides in the form $x^{-1} + 0 \cdot 1 + c \cdot x$, plus terms that it is needless to record, keeping in mind that $\wp(x) = x^{-2} + 0 \cdot x^{-1} + 0 \cdot 1 + \cdots$, that ϑ_1 is odd, and that ϑ_4 is even: Up to terms in x,

$$\sqrt{\wp(x) - e_3} = \frac{1}{x} - \frac{1}{2} e_3 x$$

and

$$\begin{aligned}\frac{\vartheta_1'(0)\vartheta_4(x)}{\vartheta_1(x)\vartheta_4(0)} &= \frac{\vartheta_1'(0)}{\vartheta_4(0)} \frac{\vartheta_4(0) + (x^2/2)\vartheta_4''(0)}{x\vartheta_1'(0)} \\ &= \frac{1}{x} + \frac{1}{2}\frac{\vartheta_4''(0)}{\vartheta_4(0)} x.\end{aligned}$$

Now compare the coefficients of x on both sides.

Exercise 1. Check that $e_1 = -\vartheta_2''(0)/\vartheta_2(0)$ in the same style.

Step 2. The identity $\vartheta_3^4(0) = \pi^{-2}(e_1 - e_3)$ and the heat equation[3] $\vartheta'' = 4\pi\sqrt{-1}\dot{\vartheta}$ figuring in the proof of identity 2.2 now lead to

$$\vartheta_3^4(0) = \pi^{-2}\left[\frac{\vartheta_4''(0)}{\vartheta_4(0)} - \frac{\vartheta_2''(0)}{\vartheta_2(0)}\right] = \frac{4\sqrt{-1}}{\pi}\left[\frac{\dot{\vartheta}_4(0)}{\vartheta_4(0)} - \frac{\dot{\vartheta}_2(0)}{\vartheta_2(0)}\right],$$

which can also be expressed as $4q\times$ the logarithmic derivative of $\vartheta_2(0)/\vartheta_4(0)$ with regard to q.

Step 3. Express $\vartheta_2(0)/\vartheta_4(0)$ by the products of Section 5 and perform the stated differentiation:

$$\frac{\vartheta_2(0)}{\vartheta_4(0)} = 2q^{1/4}\prod_{n=1}^{\infty}(1+q^{2n})^2(1-q^{2n-1})^{-2} = 2q^{1/4}\prod_{n=1}^{\infty}(1-q^{4n})^2(1-q^n)^{-2},$$

[3] $\dot{}$ means $\partial/\partial\omega$.

so

$$\begin{aligned}\vartheta_3^4(0) &= 4q \times \frac{\partial}{\partial q} \log\, 2q^{1/4} \prod_{n=1}^{\infty} (1-q^{4n})^2 (1-q^n)^{-2} \\ &= 1 - \sum_{n=1}^{\infty} \frac{32nq^{4n}}{1-q^{4n}} + \sum_{l=1}^{\infty} \frac{8lq^l}{1-q^l} \\ &= 1 - 32 \sum_{n=1}^{\infty} n \sum_{l=1}^{\infty} q^{4ln} + 8 \sum_{n=1}^{\infty} n \sum_{l=1}^{\infty} q^{ln}.\end{aligned}$$

Step 4. Compare the start to the finish, with the result that $r_4(n) = 8 \times$ *the sum of divisors d of n indivisible by* 4. This is due to Jacobi [1829]. For example, $3 = 0^2 + (\pm 1)^2 + (\pm 1)^2 + (\pm 1)^2$ in various orders, for the total count of $r_4(3) = 4 \times 2^3 = 32 = 8 \times (1 + 3)$. Note that $d = 1$ always divides n so $r_4(n)$ cannot vanish, that is, *every whole number is a sum of four squares*. Bachet (1621) and Lagrange (1770) proved this by elementary means; see Nagell [1964: 191–5]. Jacobi [1829] considered also sums of six and eight squares; $r_6(n)$ is a bit complicated, but $r_8(n)$ is $16\times$ the sum of cubes of divisors d of n if n is odd, and $16\times$ the excess of the even cubes over the odd cubes if n is even. Eisenstein [1847] found elementary proofs of these remarkable results. Hirschhorn [1985, 1987] used Jacobi's product for ϑ_3 to give simpler proofs for sums of two and four squares. Hardy and Wright [1979: 132–60] has a fascinating essay on the whole subject; see also Roberts [1977: 167–84] for a simpler account.

3.8 Euler's Identities: *Partitio Numerorum*

Euler delighted in the manipulation of series and proved a host of remarkable identities with deep number-theoretic implications. Here we present a sampling of these using theta functions for most of the proofs.

Pentagonal Numbers. The product

$$\vartheta_3 = \prod_{n=1}^{\infty} (1-q^{2n})(1+q^{2n-1}p^2)(1+q^{2n-1}p^{-2})$$

is now employed with $\sqrt{-1}q^{1/4}$ in place of p and $q^{3/2}$ in place of q to produce Euler's **pentagonal number identity** [1748]:

$$\sum_{n\in\mathbb{Z}} (-1)^n q^{n(3n+1)/2} = \prod_{n=1}^{\infty} (1-q^n).$$

The name will be explained in a moment, but first expand the right-hand product in powers of q:

$$\begin{aligned}\sum_{n\in\mathbb{Z}}(-1)^n q^{n(3n+1)/2} &= \prod_{n=1}^{\infty}(1-q^n)\\ &= \sum_{d=0}^{\infty}(-1)^d \sum_{n_1,\ldots,n_d\geq 1} q^{n_1+\cdots+n_d}\\ &= 1+\sum_{m=1}^{\infty} q^m \sum_{n_1+\cdots+n_d=m}(-1)^d\\ &= 1+\sum_{m=1}^{\infty} q^m\left[p_{\rm e}(m)-p_{\rm o}(m)\right],\end{aligned}$$

in which $p_{\rm e}(m)$, respectively $p_{\rm o}(m)$, is the number of **partitions** of the whole number m into an even, or into an odd, number of parts $n \geq 1$. This produces the beautiful result of Euler that $p_{\rm e}(m) - p_{\rm o}(m) = (-1)^m$ *or* 0 *according as* m *is a pentagonal number* $m = (1/2)n(3n \pm 1)$ *or not*; compare Hardy and Wright [1979: 83–100] for a series of enlightening proofs in different styles and also Shanks [1951] for a beautiful elementary proof. The adjective *pentagonal* is justified by the diagram and the next exercise.

Exercise 1. Check that $(1/2)n(3n-1)$ is the number of vertices in Fig. 3.1. *Hint*: The increment $(1/2)n(3n-1) - (1/2)(n-1)[3(n-1)-1] = 3n-2$ corresponds to the adjunction of three lines, with two vertices apiece, two of which are shared, whence the -2.

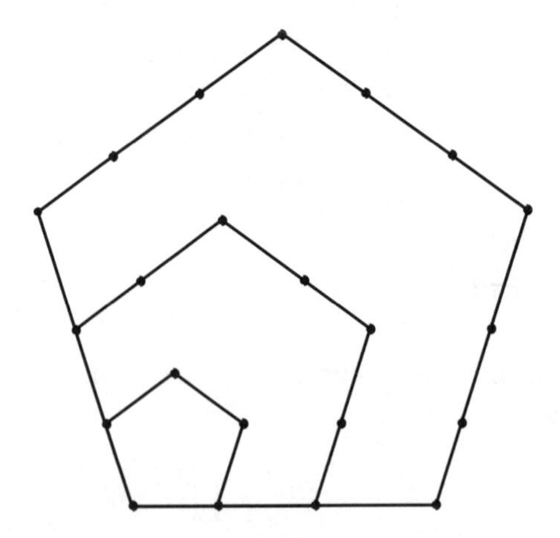

Figure 3.1. Pentagonal numbers.

Amplification. In the identity

$$\prod_{n=1}^{\infty}(1-q^n)^{-1} = 1 + \sum_{d=1}^{\infty}\sum_{n_1,\dots,n_d\geq 1} q^{n_1+\cdots+n_d} = 1 + \sum_{n=1}^{\infty} p(n)q^n,$$

the coefficient $p(n)$ is the number of partitions of the whole number $n \geq 1$ into positive parts without regard to order; for example, $4 = 4 = 1+3 = 2+2 = 1+1+2 = 1+1+1+1$, for a total count $p(4) = 5$. It is of rapid growth: $p(n) \simeq \exp(\pi\sqrt{2n/3})/4n\sqrt{3}$; see Knopp [1970] for this and Andrews [1976] for more information. It also has remarkable arithmetic properties such as $p(5n+4) \equiv 0 \bmod 5$, $p(7n+5) \equiv 0 \bmod 7$, $p(11n+6) \equiv 0 \bmod 11$, for which see Watson [1938] and, for elementary proofs, Hardy and Wright [1979: 273–97] and Winquist [1969].

Related Identities. Euler proved other beautiful identities between similar products and sums; for example,

$$\prod_{n=1}^{\infty}(1+q^{2n-1}) = 1 + \sum_{n=1}^{\infty} q^{n^2}[(1-q^2)(1-q^4)\cdots(1-q^{2n})]^{-1}$$

and

$$\prod_{n=1}^{\infty}(1+q^{2n}) = 1 + \sum_{n=1}^{\infty} q^{n(n+1)}[(1-q^2)(1-q^4)\cdots(1-q^{2m})]^{-1}.$$

Proof. $F(x) = \prod_{n=1}^{\infty}(1+q^n x)$ satisfies $(1+qx)F(qx) = F(x)$, which serves to determine the series $F(x) = \sum F_n x^n$ by the rule $F_n = q^n(1-q^n)^{-1}F_{n-1}$ $(n \geq 1)$ with $F_0 = 1$. In short,

$$\prod_{n=1}^{\infty}(1+q^n x) = \sum_{n=0}^{\infty} q^{n(n+1)/2}[(1-q)\cdots(1-q^n)]^{-1}x^n.$$

Now replace q by q^2 and take $x = q^{-1}$ or 1.

Exercise 2. Check that

$$\prod_{n=1}^{\infty}(1+q^n x)^{-1} = \sum_{n=0}^{\infty}(-1)^n q^n[(1-q)\cdots(1-q^n)]^{-1}x^n.$$

Exercise 3. Use Jacobi's product for ϑ_3 with $q^{5/2}$ in place of q and $p^2 = -q^{3/2}$ or $-q^{1/2}$ to obtain

$$\prod_{n=0}^{\infty}(1-q^{5n+1})(1-q^{5n+4})(1-q^{5n+5}) = \sum_{n=-\infty}^{\infty}(-1)^n q^{(5n+3)n/2},$$

$$\prod_{n=0}^{\infty}(1-q^{5n+2})(1-q^{5n+3})(1-q^{5n+5}) = \sum_{n=-\infty}^{\infty}(-1)^n q^{(5n+1)n/2}.$$

The deeper identities of Rogers [1894] and Ramanujan [1927: no. 6] state that

$$\prod_{n=0}^{\infty}\frac{1}{(1-q^{5n+1})(1-q^{5n+4})} = \sum_{n=0}^{\infty} q^{n^2}\left[(1-q)\cdots(1-q^n)\right]^{-1},$$

$$\prod_{n=0}^{\infty}\frac{1}{(1-q^{5n+2})(1-q^{5n+3})} = \sum_{n=0}^{\infty} q^{n(n+1)}\left[(1-q)\cdots(1-q^n)\right]^{-1}.$$

Andrews [1989] and Berndt [1991: 77] present the history of these identities and compare the several known proofs; none of them is simple and they are not presented here. Hardy [1940: 83–100] discusses the whole circle of ideas and explains its significance for partitions; see also Rademacher [1973].

Exercise 4. The first Rogers–Ramanujan identity states that the number of partitions of a whole number m into equal or unequal parts $\equiv 1 \bmod 5$ is the same as the number of its partitions into m equal parts differing by 2 or more. Prove it. *Hints*: The first part is self-evident from the product. The second employs the fact that there is a natural pairing between the partitions of m into parts not exceeding n and the partitions of $m + n^2$ into parts differing by at least 2. This will appear from Fig. 3.2 in which $m = 16$ is partitioned into $5 + 5 + 3 + 2 + 1$, as depicted in the right-hand dotted columns, and $41 = 16 + 5^2 = 16 + 1 + 3 + 5 + 7 + 9$ is partitioned into $3 + 5 + 8 + 11 + 14$, as depicted in the rows.

The number and variety of such identities is endless. Watson's **quintuple product identity** [1929]

$$\prod_{n=1}^{\infty}(1-q^n)(1-q^n p)(1-q^{n-1}p^{-1})(1-q^{2n-1}p^2)(1-q^{2n-1}p^{-2})$$

$$= \sum_{\mathbb{Z}}\left(p^{3n} - p^{-3n-1}\right) q^{n(3n+2)/2}$$

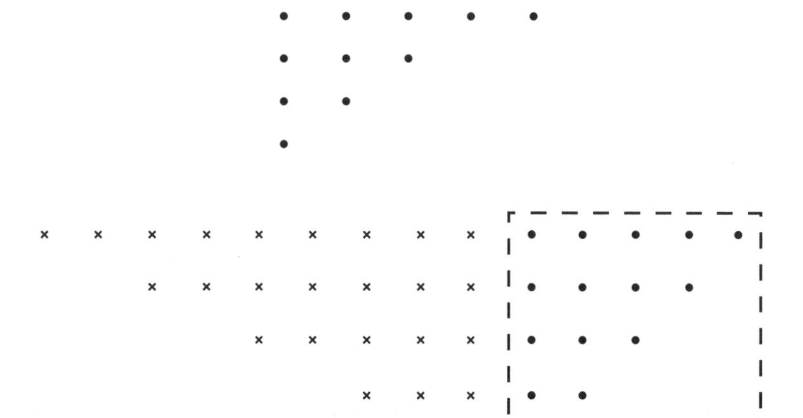

Figure 3.2. Partitions.

will serve as a finale. Watson discovered it in the course of proving certain statements of Ramanujan; see Carlitz and Subbarao [1972] for an easy proof and Berndt [1991: 83] for its history.

3.9 Jacobi's and Higher Substitutions

The dependence of the ϑ-functions upon the period ratio ω has been largely neglected to date. The rule $\vartheta_3(x|\omega+1) = \vartheta_4(x|\omega)$ provides the simplest example.

Exercise 1. Check it.

Jacobi's substitution. This describes the effect upon $\vartheta_3(x|\omega) = \sum p^{2n}q^{n^2}$ of the inversion $\omega \mapsto -1/\omega$:

$$\vartheta_3\left(\frac{x}{\omega}\middle| -\frac{1}{\omega}\right) = \sqrt{\omega}e^{-\pi\omega/4}e^{\pi\sqrt{-1}x^2/\omega}\vartheta_3(x|\omega).$$

This is interesting in itself; it is also preparatory to further arithmetic applications, namely, the law of quadratic reciprocity of Section 10 to follow. The other ϑ-functions obey similar rules, deducible from Table 3.1.1.

Table 3.9.1. *Transformations of null values*

	$\omega+1$	$-1/\omega$
ϑ_1'	$e^{\pi\sqrt{-1}/4}\vartheta_1'$	$\omega^{3/2}e^{-3\pi\sqrt{-1}/4}\vartheta_1'$
ϑ_2	$e^{\pi\sqrt{-1}/4}\vartheta_2$	$\omega^{1/2}e^{-\pi\sqrt{-1}/4}\vartheta_4$
ϑ_3	ϑ_4	$\omega^{1/2}e^{-\pi\sqrt{-1}/4}\vartheta_3$
ϑ_4	ϑ_3	$\omega^{1/2}e^{-\pi\sqrt{-1}/4}\vartheta_2$

Proof. Let ϑ_3^* be the left-hand function in Jacobi's rule, with its variable reversed: $\vartheta_3^*(x) = \vartheta_3(-x/\omega| - 1/\omega)$. It vanishes simply on

$$\omega \times [\mathbb{Z} \oplus (1/\omega)\mathbb{Z} + 1/2 + 1/2\omega] = \mathbb{Z} \oplus \omega\mathbb{Z} + 1/2 + \omega/2,$$

as does $\vartheta_3(x|\omega)$ itself (see Table 3.1.2), so $e^{-\pi\sqrt{-1}x^2/\omega}\vartheta_3^*(x)/\vartheta_3(x)$ is a root-free function of $x \in \mathbb{C}$. It is also periodic relative to $\mathbb{Z} \oplus \omega\mathbb{Z}$, as you will check from Table 3.1.1, and so must be constant ($= c$). Further use of Table 3.1.1 produces similar results for ϑ_1, ϑ_2, and ϑ_4:

$$c\vartheta_1 = -\sqrt{-1}f\vartheta_1^*, c\vartheta_2 = f\vartheta_4^*, c\vartheta_3 = f\vartheta_3^*, c\vartheta_4 = f\vartheta_2^*$$

with *the same* constant c and the common factor $f = \exp(-\pi\sqrt{-1}x^2/\omega)$; in particular, $c\vartheta_1'(0|\omega) = (\sqrt{-1}/\omega)\vartheta_1'(0| - 1/\omega)$. Identity 2.2 is now applied to the starred and unstarred null values:

$$\begin{aligned} c^3\vartheta_1'(0|\omega) &= c^3\pi\vartheta_2(0)\vartheta_3(0)\vartheta_4(0) \\ &= \pi\vartheta_2^*(0)\vartheta_3^*(0)\vartheta_4^*(0) \\ &= \vartheta_1'(0| - 1/\omega) = -\sqrt{-1}\omega c\vartheta_1'(0|\omega); \end{aligned}$$

that is, $c^2 = -\sqrt{-1}\omega$, which is to say, $c = \pm e^{-\pi\sqrt{-1}/4}\sqrt{\omega}$, with the principal branch of the root. The correct sign ($+1$) is elicited by evaluation at the fixed point $\omega = \sqrt{-1}$ of the involution $\omega \mapsto -1/\omega$.

Modular Substitutions in General. Table 3.9.1 shows how the null values transform under the substitutions $\omega \mapsto \omega + 1$ and $\omega \mapsto -1/\omega$. These two substitutions generate the **modular group** $PSL(2, \mathbb{Z})$ discussed in Chapter 4.

Exercise 2. Check that.

Exercise 3. For the general modular substitution $[ab/cd] \in PSL(2, \mathbb{Z})$,

$$\vartheta_1\left(\frac{x}{c\omega+d}\Big|\frac{a\omega+b}{c\omega+d}\right) = (c\omega+d)^{1/2} \exp\left(\frac{\pi\sqrt{-1}cx^2}{c\omega+d}\right)\vartheta_1(x|\omega)$$

up to an eighth root of unity. The eighth root is a character of the modular group. Why? Fricke [1928: 483–91] gives its value; it is complicated and will not be repeated here.

Exercise 4. Jacobi's substitution applied to the null value $\vartheta_3(0)$ with positive imaginary $\omega = \sqrt{-1}t$ yields the pretty fact, known earlier to Gauss, that

$$\sum_{\mathbb{Z}} e^{-\pi n^2 t} = \frac{1}{\sqrt{t}}\sum_{\mathbb{Z}} e^{-\pi n^2/t}.$$

Check that for $t = 0.01$, the evaluation of the left-hand side to one significant figure requires 21 summands whereas one summand of the right-hand side yields about 130 significant digits! This opens the way to the very effective computation of theta functions.

Exercise 5. Prove the identity of Landsberg [1893] and Schaar [1890]: For positive integral p and q,

$$\frac{1}{\sqrt{p}}\sum_{n=0}^{p-1} e^{2\pi\sqrt{-1}n^2 q/p} = \frac{e^{\pi\sqrt{-1}/4}}{\sqrt{2q}}\sum_{n=0}^{2q-1} e^{-\pi\sqrt{-1}n^2 p/2q}.$$

Hint: Use the null value formula of ex. 4 with $t - 2\sqrt{-1}q/p$ in place of t. Then make t approach 0 via positive values, using the fact that $\exp(2\pi\sqrt{-1}n^2q/p)$ is of period p as a function of $n \in \mathbb{Z}$. This pretty proof is due to Pólya [1927].

Poisson's Summation Formula. A very different proof of Jacobi's substitution may be obtained from Poisson's summation formula; see Dym and McKean [1972] for a brief discussion, and Rademacher [1973] for the full treatment. The idea is simple. Let $f(x)$ vanish rapidly at infinity and likewise its transform $\hat{f}(k) = \int_{\mathbb{R}} e^{-2\pi\sqrt{-1}kx} f(x)dx$. Then $f_1(x) = \sum_{\mathbb{Z}} f(x+m)$ is of period 1 and its expansion into a Fourier series $\sum_{\mathbb{Z}} c_n e^{2\pi\sqrt{-1}nx}$ produces Poisson's formula $\sum_{\mathbb{Z}} f(m) = f_1(0) = \sum_{\mathbb{Z}} \hat{f}(n)$ upon making the identification

$$c_n = \int_0^1 e^{-2\pi\sqrt{-1}nx} f_1(x)dx = \sum_{\mathbb{Z}} \int_m^{m+1} e^{-2\pi\sqrt{-1}nx} f(x)dx = \hat{f}(n).$$

The application is to $f(x) = e^{\pi\sqrt{-1}\omega x^2}$ with $\hat{f}(k) = [-\sqrt{-1}\omega]^{-1/2} e^{-\pi\sqrt{-1}k^2/\omega}$. Jacobi's substitution drops out.

Exercise 6. Check it.

Landen's Transformation: Reprise. The rule $K(k_1) = (1/2)(1 + k')K(k)$ has to do with doubling the period ratio: If $k = k(\omega)$, then $k(2\omega) = (1 - k')(1 + k')^{-1} \equiv k_1$; see Sections 2.3 and 2.17. It may be expressed in terms of null values: $k' = \vartheta_4^2(0)/\vartheta_3^2(0)$ and $K(k) = (\pi/2)\vartheta_3^2(0)$, so $\vartheta_3^2(0|2\omega) = (1/2)\left[\vartheta_3^2(0|\omega) + \vartheta_4^2(0|\omega)\right]$ is what you want. The substitution $\omega \mapsto 2\omega$ is the simplest of the so called integral substitutions $[ab/cd]$ of higher level having determinant 2 or more; see Section 4.11 and also Fricke [1916 (2): 483–91] for further information about that. Landen's transformation in its new guise is a by-product of the identity

$$\vartheta_1(2x|2\omega) = \frac{\vartheta_1\vartheta_2(x|\omega)}{\vartheta_3\vartheta_4(0|\omega)}.$$

Proof. The quotient of $\vartheta_1(2x|2\omega)$ by $\vartheta_1\vartheta_2(x|\omega)$ is periodic relative to the lattice $\mathbb{Z} \oplus \omega\mathbb{Z}$ and pole-free, by Tables 3.1.1 and 3.1.2, so it can only be constant. The rest follows from evaluation at $x = 0$ and the products for null values.

Exercise 7. Give details.

Exercise 8. Deduce these formulas for null values:

$$\vartheta_1'(0|2\omega) = \vartheta_1'(0|\omega)\vartheta_2(0|\omega)/2\vartheta_3(0|\omega)\vartheta_4(0|\omega),$$

$$\vartheta_2^2(0|2\omega) = \frac{1}{2}\left[\vartheta_3^2(0|\omega) - \vartheta_4^2(0|\omega)\right],$$

$$\vartheta_3^2(0|2\omega) = \frac{1}{2}\left[\vartheta_3^2(0|\omega) + \vartheta_4^2(0|\omega)\right].$$

The last is Landen's transformation in its new form.

Exercise 9. $\vartheta_4^2(0|2\omega) = \vartheta_3\vartheta_4(0|\omega)$ is to be checked directly from the relevant products. Now compare this and Landen's transformation in its new form to Gauss's arithmetic–geometric mean in Section 2.3.

Exercise 10. Check that $\vartheta_1(4x|4\omega)$ is a constant multiple of $\vartheta_1(x|\omega)\vartheta_1(1/4 - x|\omega)\vartheta_2(x|\omega)\vartheta_1(1/4 + x|\omega)$. What is the constant?

3.10 Quadratic Reciprocity

Let p be an odd prime. The integer $0 \le k < p$ is a **quadratic residue** ($k \in Q$) if the congruence $x^2 \equiv k \bmod p$ is solvable; otherwise it is a **quadratic nonresidue** ($k \in Q'$). Define the **residue symbol** $\left(\frac{k}{p}\right)$ to be $+1$ or -1 according as $k \in Q$ or not. Gauss's **law of quadratic reciprocity** states that if p and q are odd primes, then

$$\left(\frac{p}{q}\right) \cdot \left(\frac{q}{p}\right) = (-1)^{\frac{1}{2}(p-1) \cdot \frac{1}{2}(q-1)}.$$

Gauss himself gave eight different proofs. The present one is based upon the Landsberg–Schaar identity of ex. 9.5; see Roberts [1977: 192–210] and Nagell [1964: 132–45] for more conventional accounts. Ireland and Rosen [1990] describe "higher" reciprocity laws obtained by Eisenstein [1847] by means of elliptic functions. Wyman [1972] presents a simple account of the whole subject.

The proof of the law of quadratic reciprocity now begins; it is broken into six easy steps.

Step 1. $\{1^2, 2^2, \dots, (p-1)^2\}$ is a two-fold list of Q since $j^2 \equiv (p-j)^2 \bmod p$ and $j^2 \equiv i^2$ if and only if $j \equiv i$ or $j \equiv -i \equiv p - i \bmod p$. In particular, there are $(1/2)(p-1)$ residues and the same number of nonresidues.

Step 2. $\left(\frac{k}{p}\right)$ is a character of the multiplicative group $\mathbb{Z}_p^\times$ of integers modulo p.

Proof. This is the same as saying that (a) $Q \cdot Q \subset Q$, (b) $Q \cdot Q' \subset Q'$, (c) $Q' \cdot Q' \subset Q$. Item (a) is self-evident. Item (b) follows from the count in step 1 and from the fact that multiplication by $1 \le j < p$ is an automorphism of $\mathbb{Z}_p^\times$, which is to say that, as n runs over a full set of residue classes modulo p, so does jn. Item (c) is proved the same way.

Step 3. Introduce the **Gaussian sum**

$$G_q(e^{2\pi\sqrt{-1}n/q}) = \sum_{j=0}^{q-1} e(j^2 n/q)$$

with $q \ge 1$, prime or not, $1 \le n < q$, and $e(x) = \exp(2\pi\sqrt{-1}x)$, and note the connection with the residue symbol: For odd prime p and $1 \le n < p$,

$$\begin{aligned}
G_p(e^{2\pi\sqrt{-1}n/p}) &= \sum_{j=0}^{p-1} e(j^2 n/p) \\
&= 1 + 2\sum_{k\in Q} e(kn/p) - \sum_{k=0}^{p-1} e(kn/p) \\
&= \sum_{k\in Q} e(kn/p) - \sum_{k\in Q'} e(kn/p) \\
&= \sum_{k=1}^{p-1} \left(\frac{k}{p}\right) e(kn/p) \\
&= \left(\frac{n}{p}\right) \sum_{k=1}^{p-1} \left(\frac{kn}{p}\right) e(kn/p) \\
&= \left(\frac{n}{p}\right) \sum_{k=1}^{p-1} \left(\frac{k}{p}\right) e(k/p) = \left(\frac{n}{p}\right) G_p(e^{2\pi\sqrt{-1}/p}),
\end{aligned}$$

in which line 2 employs the fact that the second sum vanishes (why?) and line 6 the fact that multiplication by n is an automorphism of $\mathbb{Z}_p^\times$. In short,

$$G_p(e^{2\pi\sqrt{-1}n/p}) = \left(\frac{n}{p}\right) G_p(e^{2\pi\sqrt{-1}/p})$$

for any odd prime p.

Step 4. $\left(\frac{p}{q}\right)\left(\frac{q}{p}\right) = G_{pq}(e^{2\pi\sqrt{-1}/pq})/G_p(e^{2\pi\sqrt{-1}/p})G_q(e^{2\pi\sqrt{-1}/q})$ for odd primes p and q.

Proof. $\left(\frac{p}{q}\right)$ is the ratio of $G_p(e^{2\pi\sqrt{-1}q/p})$ to $G_p(e^{2\pi\sqrt{-1}/p})$, by step 3, so you have only to check that

$$G_p(e^{2\pi\sqrt{-1}/p})G_q(e^{2\pi\sqrt{-1}/q}) = G_{pq}(e^{2\pi\sqrt{-1}/pq}).$$

This is easy: As i runs from 0 to $p-1$ and j runs from 0 to $q-1$, $k = ip + jq$ runs *once* over $0 \le k < pq$, so

$$\begin{aligned}
G_{pq}(e^{2\pi\sqrt{-1}/pq}) &= \sum_{k=0}^{pq-1} e(k^2/pq) \\
&= \sum_{i=0}^{p-1}\sum_{j=0}^{q-1} e(i^2 p/q + 2ij + j^2 q/p).
\end{aligned}$$

Step 5. Next is the crucial evaluation of the Gaussian sum:

$$\frac{1}{\sqrt{p}} G_p(e^{2\pi\sqrt{-1}/p}) = \frac{1}{\sqrt{2}} e^{\pi\sqrt{-1}/4} \left(1 + e^{-\pi\sqrt{-1}p/2}\right)$$

for p prime or not.

Proof. Put $q = 1$ in the Landsberg–Schaar identity of ex. 9.5:

$$\frac{1}{\sqrt{p}} \sum_{n=0}^{p-1} e(n^2 q/p) = \frac{1}{\sqrt{2q}} e^{\pi\sqrt{-1}/4} \sum_{n=0}^{2q-1} e(-n^2 p/4q).$$

Step 6. Put this evaluation back into step 4, with the result that

$$\left(\frac{p}{q}\right)\left(\frac{q}{p}\right) = \sqrt{2} e^{-\pi\sqrt{-1}/4} \frac{1 + e^{-\pi\sqrt{-1}pq/2}}{(1 + e^{-\pi\sqrt{-1}p/2})(1 + e^{-\pi\sqrt{-1}q/2})}.$$

Exercise 1. $e^{-\pi\sqrt{-1}n/2} = -\sqrt{-1}$ or $+\sqrt{-1}$ according as $n \equiv 1 \bmod 4$ or $n \equiv 3 \bmod 4$. Use this to reduce the display to $+1$ if either $p \equiv 1 \bmod 4$ or $q \equiv 1 \bmod 4$, and to -1 in the opposite case, and check that this is the same as $(-1)^{(1/2)(p-1)\times(1/2)(q-1)}$.

The proof is finished.

Exercise 2. Prove the **supplementary laws**

$$\left(\frac{-1}{p}\right) = (-1)^{(p-1)/2} \text{ and } \left(\frac{2}{p}\right) = (-1)^{(1/8)(p^2-1)}.$$

Hint for the first: $\left(\frac{-1}{p}\right) G_p$ is the complex conjugate of G_p. *Hint for the second*: Use ex. 9.5 with $p = 2$.

Example. 666 is a quadratic residue of the prime $p = 2137$. $666 = 2 \cdot 3^2 \cdot 37$, so

$$\left(\frac{666}{2137}\right) = \left(\frac{2}{2137}\right)\left(\frac{3}{2137}\right)^2\left(\frac{37}{2137}\right) = \left(\frac{37}{2137}\right),$$

by the second supplementary law, since $2137 = 4 \cdot 534 + 1 \equiv 1 \bmod 4$ makes $(1/8)(p^2 - 1)$ even. Now 37 is prime and $(1/2)(2137 - 1) \times (1/2)(37 - 1)$ is even, so from $2137 = 57 \cdot 37 + 28$, the law of quadratic reciprocity, and so forth, you find

$$\left(\frac{37}{2137}\right)=\left(\frac{2137}{37}\right)=\left(\frac{28}{37}\right)=\left(\frac{2}{37}\right)^2\left(\frac{7}{37}\right)=\left(\frac{37}{7}\right)=\left(\frac{2}{7}\right)$$
$$=(-1)^{(1/8)(49-1)}=+1,$$

as promised.

Exercise 3. Find a solution of $k^2\equiv 666 \bmod 2137$.

3.11 Ramanujan's Continued Fractions

In a letter of 1913 to Hardy, Ramanujan stated the continued fractions

$$\cfrac{1}{1+\cfrac{e^{-2\pi}}{1+\cfrac{e^{-4\pi}}{1+\cdots}}}=\left(\sqrt{\frac{5+\sqrt{5}}{2}}-\frac{\sqrt{5}+1}{2}\right)e^{2\pi/5}$$

and

$$\cfrac{1}{1+\cfrac{e^{-\pi}}{1+\cfrac{e^{-2\pi}}{1+\cdots}}}=\left(\sqrt{\frac{5-\sqrt{5}}{2}}-\frac{\sqrt{5}-1}{2}\right)e^{\pi/5}.$$

Hardy [1937] wrote:

> [These formulas] defeated me completely. I had never seen anything in the least like this before. A single look at them is enough to show that they could only be written down by a mathematician of the highest class. They must be true because no one would have the imagination to invent them.

The actual evaluations are mere curiosities, but the beautiful discovery of Ramanujan upon which they rest is too surprising to omit. It states that the continued fraction

$$x=q^{1/5}\times\left(\frac{1}{1+}\,\frac{q}{1+}\,\frac{q^2}{1+\ \ldots}\right)$$

is a root of

$$\frac{1}{x}-x=1+q^{-1/5}\prod_{n=1}^{\infty}\frac{1-q^{n/5}}{1-q^{5n}}.$$

Watson's [1929] reconstruction of Ramanujan's line of thought will now be described; compare Berndt [1991].

Step 1. The sum $S(p) = \sum_{n=0}^{\infty} p^n q^{n^2} [(1-q)(1-q^2)\cdots(1-q^n)]^{-1}$ converges for fixed $|q| < 1$, by inspection. It satisfies $S(p) = S(pq) + pqS(pq^2)$, so $S_n = S(q^{n+1})/S(q^n)$ satisfies $S_n = [1 + q/S_{n+1}]^{-1}$, and you have

$$\frac{1}{Q(1)} \equiv \frac{S(q)}{S(1)} = \frac{1}{1+}\,\frac{q}{1+}\,\frac{q^2}{1+\ \ldots}$$

in view of $S_n = 1 + o(1)$ for $n \to \infty$. This produces continued fractions of the desired type.

Step 2. This step employs the Rogers–Ramanujan identities of Section 8 to express the continued fraction $1/Q$ as an infinite product:

$$S(q) = 1 + \sum_{n=1}^{\infty} q^{n(n+1)} \left[(1-q)\cdots(1-q^n)\right]^{-1} = \prod_{n=0}^{\infty} \frac{1}{(1-q^{5n+2})(1-q^{5n+3})},$$

$$S(1) = 1 + \sum_{n=1}^{\infty} q^{n^2} \left[(1-q)\cdots(1-q^n)\right]^{-1} = \prod_{n=0}^{\infty} \frac{1}{(1-q^{5n+1})(1-q^{5n+4})},$$

so

$$\frac{1}{Q} = \frac{S(q)}{S(1)} = \prod_{n=0}^{\infty} \frac{(1-q^{5n+1})(1-q^{5n+4})}{(1-q^{5n+2})(1-q^{5n+3})}.$$

Step 3. The scene now shifts to Euler's pentagonal identity of Section 8:

$$\prod_{n=1}^{\infty}(1-q^n) = \sum_{n=-\infty}^{\infty} (-1)^n q^{n(3n+1)/2}.$$

This is applied with $q^{1/5}$ and q^5 in place of q to obtain

$$\prod_{n=1}^{\infty} \frac{(1-q^{n/5})}{(1-q^{5n})} = \frac{\sum (-1)^n q^{(n/10)(3n+1)}}{\sum (-1)^n q^{(5n/2)(3n+1)}}.$$

The power $n(3n+1)/2 \in \mathbb{Z}$ is congruent to 0, 2, 2, 0, 1 mod 5 according as $n \equiv 0, 1, 2, 3, 4 \bmod 5$, so the quotient is expressible in the form $A + q^{1/5}B + q^{2/5}C$, in which A, B, C are sums of whole powers of q. Now $A = 1$ at $q = 0$, and the value $B = -1$ follows from the fact that $(n/2)(3n+1) \equiv 1 \bmod 5$ if and only if $n = -1 + 5m$; in which case $(n/10)(3n+1) = 1/5 + (5m/2)(3m-1)$. This was the goal of step 3. The values of A and C are harder to come by.

Step 4. Check that $C = -1/A$. Start with the identity of ex. 5.4. This is used with $q^{1/5}$ and q^5 in place of q to produce

$$\left(A - q^{1/5} + q^{2/5}C\right)^3 = \frac{\sum(-1)^n(2n+1)q^{(n/10)(n+1)}}{\sum(-1)^n(2n+1)q^{(5n/2)(n+1)}}.$$

Now argue as in step 2: The power $n(n+1)/2 \in \mathbb{Z}$ is congruent to 0, 1, 3, 1, 0 mod 5 according as $n \equiv 0, 1, 2, 3, 4 \bmod 5$, so the absence of powers $\equiv 2/5 \bmod 5$ to the right requires the vanishing of the coefficient $3A(AC+1)$ of $q^{2/5}$ to the left. In short, $C = -1/A$ in view of $A(0) = 1$.

Step 5. Check that $A = Q$. The starting point is the identity

$$\left(A - q^{1/5} - q^{2/5}A^{-1}\right)\sum(-1)^n q^{(5n/2)(3n+1)} = \sum(-1)^n q^{(n/10)(3n+1)},$$

from which the *whole* powers of q are to be extracted. On the left, you find

$$A \times \sum(-1)^n q^{(5n/2)(3n+1)} = A \times \prod_{n=1}^{\infty}(1 - q^{5n})$$

upon reversing the application of Euler's identity; on the right, whole powers appear when $n(3n+1)/2 \equiv 0 \bmod 5$, which is to say $n \equiv 0$ or $3 \bmod 5$, the upshot being

$$A \times \prod_{n=1}^{\infty}(1 - q^{5n}) = \sum_{n \equiv 0 \text{ or } 3(5)} (-1)^n q^{(n/10)(3n+1)}.$$

Now introduce the sum

$$D(p) = \sum_{\mathbb{Z}}(-1)^n q^{(n/2)(15n+1)} p^{3n} + \sum_{\mathbb{Z}}(-1)^n q^{(5n/2-1)(3n-1)} p^{1-3n}$$

and note that it reduces to the preceding sum at $p = 1$. The summands of D cancel in pairs if $p = -1/q$ and, more generally, if $p = -q^{5n-1}$ or $p = \pm q^{5n+3/2}$ for any $n \in \mathbb{Z}$. This suggests comparing D to a product with just these roots:

$$P(p) = \prod_{n=1}^{\infty}(1 + q^{5n-1}p^{-1})(1 + q^{5n-4}p)(1 - q^{10n-7}p^{-2})(1 - q^{10n-3}p^2).$$

The latter is of period 2 in x, as is D, and both are multiplied by $-q^{-8}p^{-3}$ if x is increased by $5\omega (p \to pq^5)$, so the quotient D/P is pole-free and periodic relative to $2\mathbb{Z} \oplus 5\omega\mathbb{Z}$; as such, it can only be constant. Its value is obtained by inspection at $p = q^{-1}$:

$$P(q^{-1}) = \prod_{n=1}^{\infty}(1+q^{5n})(1+q^{5n-5})(1-q^{10n-5})(1-q^{10n-5})$$
$$= 2\prod_{n=1}^{\infty}(1+q^{5n})^2(1-q^{10n-5})^2$$
$$= 2,$$

by ex. 5.2, and

$$D(q^{-1}) = 2\sum_{\mathbb{Z}}(-1)^n q^{(5n/2)(3n+1)} = 2\prod_{n=1}^{\infty}(1-q^{5n}),$$

so $D(p) = P(p) \times \prod_{n=1}^{\infty}(1-q^{5n})$. Now take $p = 1$ to obtain $A = P(1)$ and complete the identification $A = Q$ as follows:

$$P(1) = \prod_{n=1}^{\infty}(1+q^{5n-1})(1+q^{5n-4})(1-q^{10n-7})(1-q^{10n-3})$$
$$= \prod_{n=1}^{\infty}\frac{(1-q^{10n-2})}{(1-q^{5n-1})}\frac{(1-q^{10n-8})}{(1-q^{5n-4})} \times (1-q^{10n-7})\,(1-q^{10n-3})$$
$$= \prod_{n=1}^{\infty}\frac{(1-q^{5n-2})(1-q^{5n-3})}{(1-q^{5n-1})(1-q^{5n-4})}$$
$$= Q(1).$$

Step 6. Now invoke the result of step 4 in the form

$$1 + q^{-1/5}\prod_{1}^{\infty}\frac{1-q^{n/5}}{1-q^{5n}} = q^{-1/5}A - q^{1/5}A^{-1} = \frac{1}{x} - x$$

with

$$x = q^{1/5}A^{-1} = q^{1/5}Q^{-1} = q^{1/5} \times \left(\frac{1}{1+}\,\frac{q}{1+}\,\frac{q^2}{1+}\,\frac{q^3}{1+\cdots}\right).$$

This is Ramanujan's result.

The continued fractions stated at the outset can now be evaluated. The third identity of ex. 5.6,

$$q^{1/2}\prod(1-q^{2n})^6 = 2\pi^{-3}kk'K^3(k),$$

is taken with $q^{1/10}$ and $q^{5/2}$ in place of q to produce

$$q^{-1/5}\prod\frac{(1-q^{n/5})}{(1-q^{5n})}=\left[\frac{kk'K^3 \text{ evaluated at } k(\omega/10)}{kk'K^3 \text{ evaluated at } k(5\omega/2)}\right]^{1/6}.$$

Now split into cases as follows.

Case 1. $q=e^{-2\pi}$. Then $\omega=2\sqrt{-1}$, and $\omega/10=\sqrt{-1}/5$ is related to $5\omega/2=5\sqrt{-1}$ by the substitution $\omega\mapsto -1/\omega$, so $k(\omega/10)=k'(5\omega/2)$ in view of $\omega=\sqrt{-1}K'(k)/K(k)$ and the fact that k^2 is a function of the period ratio only; compare ex. 2.17.4 and the warning of Section 2.17. It follows that the sixth root on the right-hand side of the last display is nothing but

$$\sqrt{\frac{K \text{ at modulus } k'(5\omega/2)}{K \text{ at modulus } k(5\omega/2)}}=\sqrt{\frac{(5/2)\times 2\sqrt{-1}}{\sqrt{-1}}}=\sqrt{5},$$

permitting you to solve $x^{-1}-x=1+\sqrt{5}$ for its (positive) root

$$x=\sqrt{\frac{5+\sqrt{5}}{2}}-\frac{\sqrt{5}+1}{2}.$$

Case 2. $q=e^{-\pi}$. The computation is a little modified. The identity of ex. 5.6

$$q^{1/5}\prod(1+q^{2k-1})^6(1-q^{2k})^6=4\pi^{-3}\sqrt{kk'}K^3(k)$$

is employed in the same style, with $q^{1/5}$ and q^5 in place of q and $\omega=\sqrt{-1}$, to obtain $x^{-1}-x=1-\sqrt{5}$ and *its* positive root

$$x=\sqrt{\frac{5-\sqrt{5}}{2}}-\frac{\sqrt{5}-1}{2}.$$

Exercise 1. Do it.

Now everything is *verified* but how Ramanujan ever discovered it is still obscure.

4
Modular Groups and Modular Functions

The **modular group of first level** $\Gamma_1 = PSL(2, \mathbb{Z})$ made its debut in Section 2.6 in connection with the conformal equivalence of tori: Two such tori are equivalent if and only if their period ratios are related by a substitution of this kind. The quotient of the upper half-plane $\mathbb{H}$ by the action of Γ_1, and by the further modular groups to be introduced later in this section, produces a whole series of interrelated curves whose function fields (modular functions) have deep arithmetic and geometric applications going back to Abel [1827] and Jacobi [1829]. The two most striking of these are Hermite's solution of the general polynomial equation of degree 5 [1859], explained in Chapter 5, and Weber's realization [1891] of Kronecker's youthful dream (1860) of describing the absolute class field of the imaginary quadratic field $\mathbb{Q}(\sqrt{-D})$ for a positive square-free integer D; see Chapter 6. The present section prepares the way. Fricke and Klein [1926] and Shimura [1971] are recommended for information at a more advanced level.

4.1 The Modular Group of First Level

A **modular function** of first level is a function f of rational character in the open upper half-plane $\mathbb{H}$ that is *invariant* under the action of $\Gamma_1 = PSL(2, \mathbb{Z})$, that is, $f((a\omega + b)/(c\omega + d)) = f(\omega)$ for every $\omega \in \mathbb{H}$ and $[ab/cd] \in \Gamma_1$; such a function is of period 1 because $[11/01]\colon \omega \mapsto \omega + 1$ is a modular substitution, so it is possible to expand it in powers of $q = e^{2\pi\sqrt{-1}\omega}$ in the vicinity of $\omega = \sqrt{-1}\infty$ $(q = 0)$, as in $f(\omega) = \sum_{\mathbb{Z}} \hat{f}(n)q^n$. The definition of a modular function is now completed by requiring that it be of rational character in *this* parameter, too, which is to say, that the sum breaks off at some power $n = m > -\infty$.

Exercise 1. Clarify the statement. *Hint*:

$$f(x+\sqrt{-1}y)=\sum_{\mathbb{Z}} e^{2\pi\sqrt{-1}nx}\int_0^1 e^{-2\pi\sqrt{-1}nx'} f(x'+\sqrt{-1}y)dx'.$$

Now check that $e^{2\pi ny}\times$ the integral is independent of $y>0$.

Warning. The present usage $q=e^{2\pi\sqrt{-1}\omega}$ conflicts with the $q=e^{\pi\sqrt{-1}\omega}$ of Chapter 3. The warning will be repeated when confusion could arise.

Example. The absolute invariant $j=g_2/\Delta$ of Section 2.12 is such a modular function: It is a function of the period ratio only, as noted in ex. 2.12.1; it is of rational character on $\mathbb{H}$; its invariance under Γ_1 is the further content of ex. 2.12.1; and the expansion $j(q)=(1728)^{-1}[1+c(1)q+\cdots]$ elicited in sample 3.5.5 confirms its rational character at $\sqrt{-1}\infty$.

Exercise 2. Check that, under the modular substitution $\omega\mapsto\omega'$ effected by $[ab/cd]$, $g_2\mapsto(c\omega+d)^4g_2$, $g_3\mapsto(c\omega+d)^6g_3$, and $\Delta\mapsto(c\omega+d)^{12}\Delta$, substantiating the invariance of $j(\omega)$.

The absolute invariant has also this crucial property: that it separates orbits of Γ_1 in $\mathbb{H}$, that is, it takes the value $j(\omega)$ only on the orbit $\Gamma_1\omega$. This could be deduced from Section 2.12; it will be proved in a different style in Section 9. The map $j:\mathbb{H}\to\mathbb{C}$, completed to a map from $\mathbb{H}\cup\sqrt{-1}\infty$ to $\mathbb{P}^1$ by taking $j(\sqrt{-1}\infty)=\infty$, identifies the quotient $(\mathbb{H}+\sqrt{-1}\infty)/\Gamma_1$ with the Riemann sphere $\mathbb{P}^1$ and permits the conclusion that the field $\mathbf{K}(\mathbb{P}^1)$ of modular functions is nothing but $\mathbb{C}(j)$; that is, every modular function is a rational function of j. This explains the name **absolute invariant**; see Section 9 for a more leisurely tour of these ideas; compare also Section 1.7 on the absolute invariants of the Platonic solids. Dedekind [1877] was the first to introduce the absolute invariant and to emphasize its importance. This late date is surprising as both Abel [1827] and Jacobi [1829] were perfectly clear about the significance of the modulus k^2 as absolute invariant for the modular group of second level; that is the subject of the next section.

4.2 The Modular Group of Second Level

The considerations of Section 1 apply to certain (arithmetic) subgroups of Γ_1. The **modular group** $\Gamma_n\subset\Gamma_1$, **of level** n, is defined by the requirement that $[ab/cd]$ be congruent to the identity $[10/01]\bmod n$; obviously, it is an invariant subgroup of Γ_1. Other, more recondite subgroups will come in later. The present section is devoted to the group Γ_2 of second level and to its invariant functions; the general discussion occupies Section 16.

The key to the story of Γ_2 is the action of Γ_1 on the Jacobi modulus $k^2 = (e_2-e_3)(e_1-e_3)^{-1}$, viewed as a function of the period ratio; see Section 2.17. It turns out that Γ_2 is the *stability group* of $k^2(\omega)$ in Γ_1 and that the quotient group Γ_1/Γ_2 is simply the symmetric group S_3 of permutations of the three letters e_1, e_2, e_3, alias the group of anharmonic ratios of Section 1.5; in particular, Γ_2 is of index 6 in Γ_1. These matters are discussed here.

It is not much of a jump to the notion that k^2 is the *absolute invariant* of Γ_2 or to the certainty that j, as a modular function of Γ_2, must be rational in k^2, of degree 6; compare ex. 2.12.3 and Section 6.

To understand the (fractional linear) action of $\Gamma_1 = PSL(2,\mathbb{Z})$ on k^2, it is necessary to look at the underlying (linear) action of $G = SL(2,\mathbb{Z})$ on the letters e_1, e_2, e_3. G fixes the lattice $\mathbb{L} = \mathbb{Z} \oplus \omega\mathbb{Z}$ and so also the $\wp$-function. What it *does* do is to change the labeling of the half-periods so that the numbers $e_1 = \wp(1/2)$, $e_2 = \wp(1/2 + \omega/2)$, $e_3 = \wp(\omega/2)$ are permuted. For example, the linear action of $[11/01]$ sends 1 to ω and ω to $1+\omega$, so

$$e_1 \mapsto \wp(\omega/2) = e_3,\ \ e_2 \mapsto \wp(1/2) = e_1,\ \ e_3 \mapsto \wp(1/2+\omega/2) = e_2,$$

that is, it effects the permutation $(123) \mapsto (312)$; at the same time, the period ratio changes to $1 + 1/\omega$, the outcome being

$$k^2(1+1/\omega) = \frac{e_1 - e_2}{e_3 - e_2} = 1 - \frac{e_3 - e_1}{e_3 - e_2} = 1 - 1/k^2(\omega).$$

The general computation is just as easy. Any substitution $[ab/cd]$ of level 2 has only a trivial effect: It sends $1/2$ to $(c/2)\omega + d/2 \equiv 1/2 \bmod \mathbb{L}$ and $\omega/2$ to $(a/2)\omega + (b/2) \equiv \omega/2 \bmod \mathbb{L}$, so e_1, e_2, e_3 are fixed and vice versa, which is to say that Γ_2 *is the stability group of* k^2. Now observe that Table 4.2.1 displays all six possible permutations of e_1, e_2, e_3 with six distinct outcomes for k^2. It follows that Γ_1/Γ_2 is a copy of S_3, and more: In the present format, its action on k^2 is nothing but a copy of the group of anharmonic ratios, as is seen from the table; in particular, the index $[\Gamma_1 : \Gamma_2]$ is 6.

Exercise 1. Check the table.

Exercise 2. Think over the identification of Γ_1/Γ_2 carefully.

Exercise 3. Column 3 of Table 4.2.1 can also be verified by means of the formula $k^2 = \vartheta_2^4(0)/\vartheta_3^4(0)$ and the rules of Table 3.1.1 for the transformation of null-values. Do a sample or two.

Table 4.2.1. *Action of* Γ_1 *on* k^2

Coset of Γ_2 in Γ_1	Permutation of e_1, e_2, e_3	Action on k^2
$\begin{pmatrix} 1 & 0 \\ 0 & 1 \end{pmatrix}$	(123)	k^2
$\begin{pmatrix} 0 & -1 \\ 1 & 0 \end{pmatrix}$	(321)	$1 - k^2$
$\begin{pmatrix} 1 & 0 \\ -1 & 1 \end{pmatrix}$	(213)	$1/k^2$
$\begin{pmatrix} 1 & -1 \\ 0 & 1 \end{pmatrix}$	(132)	$k^2/(k^2 - 1)$
$\begin{pmatrix} 0 & 1 \\ -1 & 1 \end{pmatrix}$	(231)	$1/(1 - k^2)$
$\begin{pmatrix} 1 & -1 \\ 1 & 0 \end{pmatrix}$	(312)	$1 - 1/k^2$

4.3 Fundamental Cells

A **fundamental cell** for a subgroup Γ of the modular group $\Gamma_1 = PSL(2, \mathbb{Z})$ is a figure $\mathfrak{F}$ in the open upper half-plane that meets each orbit of Γ in a single point. The word *cell* is used to convey the idea that $\mathfrak{F}$ should be geometrically nice; for instance, it might be a hyperbolic polygon with a finite number of sides; see Ford [1972], Kra [1972], and Lehner [1964] for general information about such matters and numerous examples. The present discussion deals with Γ_1 and Γ_2 only.

First Level. The section $\mathfrak{F}_1$ of Fig. 4.1 is a fundamental cell for Γ_1, as will be proved shortly. The heavy part of the boundary is included in the cell; the rest is not. $\mathfrak{F}_1$ can be viewed as a complete list of the inequivalent complex structures on the topological torus since conformal equivalence of tori is determined by modular equivalence of their period ratios.

Exercise 1. The points on the orbit $\Gamma_1\infty \subset (\mathbb{H}+\sqrt{-1}\infty)+\mathbb{R}$ are called **cusps**. Leaving ∞ aside, check that these are nothing but the rational numbers $\mathbb{Q} \subset \mathbb{R}$.

Exercise 2. What is the hyperbolic area of $\mathfrak{F}_1$? *Answer*: $\pi/3$; compare ex. 1.9.6.

Second Level. Γ_2 is of index 6 in Γ_1, so a fundamental cell for Γ_2 can be formed from the six copies of any fundamental cell $\mathfrak{F}_1$ produced by the action of the six

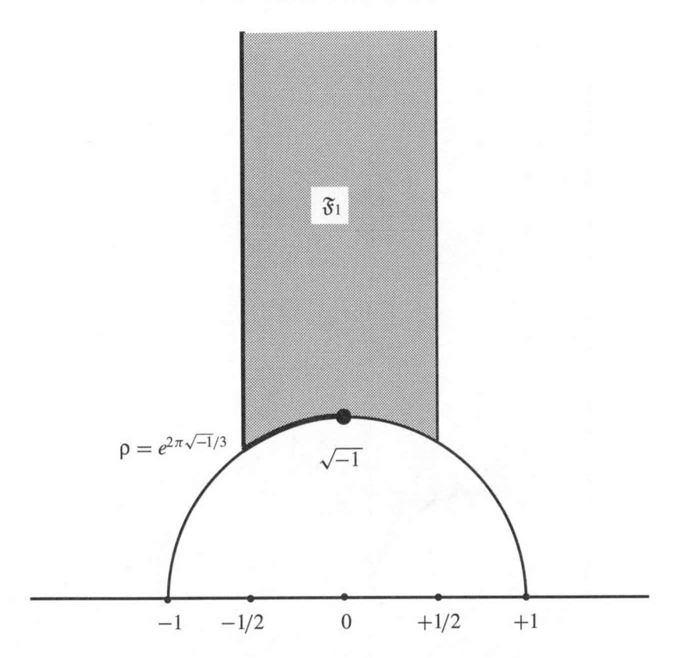

Figure 4.1. Fundamental cell for Γ_1.

substitutions in column 1 of Table 4.2.1. The result of such a construction is depicted in Fig. 4.2. $\mathfrak{F}_1$ is first replaced by its right half, plus inversion of its left half by the substitution $\omega \mapsto -1/\omega$. This figure is labeled 1; it is a perfectly acceptable fundamental cell for Γ_1. The other labels refer to the nontrivial cosets of Γ_2 in their order of appearance in Table 4.2.1. The full figure $\mathfrak{F}_2$ so produced is the part of the half-plane above the two circles of radius $1/2$ centered at $\pm 1/2$. The heavy part of the boundary is retained, the rest is not; in particular, the cusps -1, 0, $+1$, and $\sqrt{-1}\infty$ are excluded.

Exercise 3. Check the figure.

The picture suggests that Γ_1 maps $\mathfrak{F}_1$ to the tiles of a *tessellation* of the half-plane, covering it without holes or overlaps, as in Fig. 4.3. The labels A, B refer to the two substitutions $A\colon \omega \mapsto \omega + 1$ and $B\colon \omega \mapsto -1/\omega$; these generate the whole group. This is just the geometric version of the fact that $\mathfrak{F}_1$ is a fundamental cell. A similar tessellation is obtained by application of Γ_2 to $\mathfrak{F}_2$; for variety this is depicted in Fig. 4.4 after a further map to the unit disk $\mathbb{D}$.

The proof that $\mathfrak{F}_1$ is a fundamental cell of Γ_1 is divided into two parts.

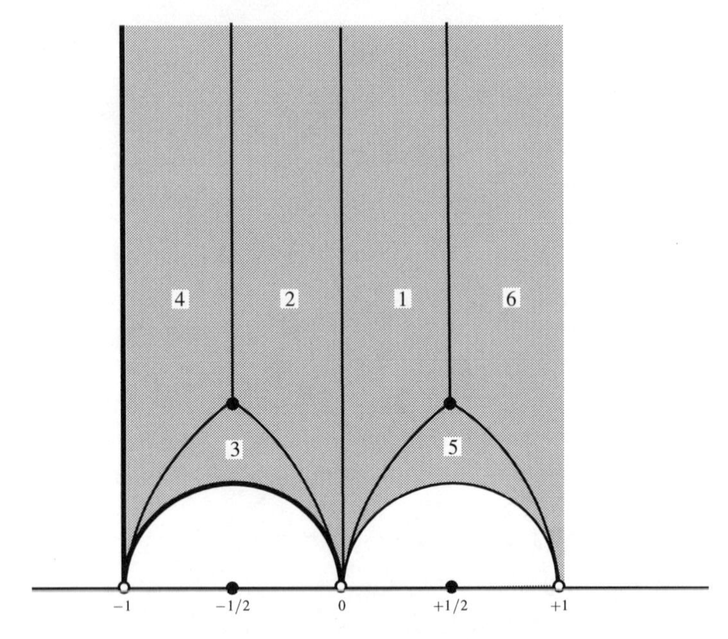

Figure 4.2. Fundamental cell for Γ_2.

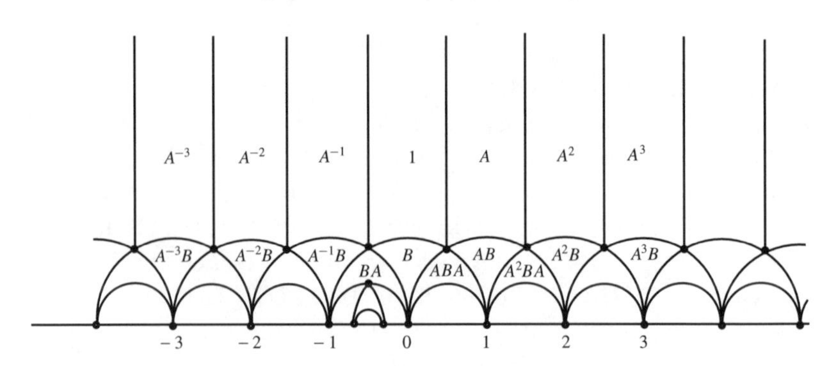

Figure 4.3. Tessellation of $\mathbb{H}$ by Γ_1.

Proof that $\mathfrak{F}_1$ is not too small. This means that every orbit of Γ_1 meets $\mathfrak{F}_1$. The key to the proof is that every orbit contains highest points, the height of ω being its imaginary part. Note first that, for the general modular substitution $\omega' = [ab/cd]\omega$, the height obeys the rule $h(\omega') = |c\omega + d|^{-2} \times h(\omega)$. It follows that $h(\omega') \le c^{-2}/h(\omega) \le 1/h(\omega)$ unless $c = 0$, in which case $d^2 = 1$ and $h(\omega') \le h(\omega)$. In short, the orbit of ω is of limited height. It is also plain that only limited values of the integers c and d enter into the competition for highest point(s). The existence of such points is now clear. Let ω, itself, be such a point

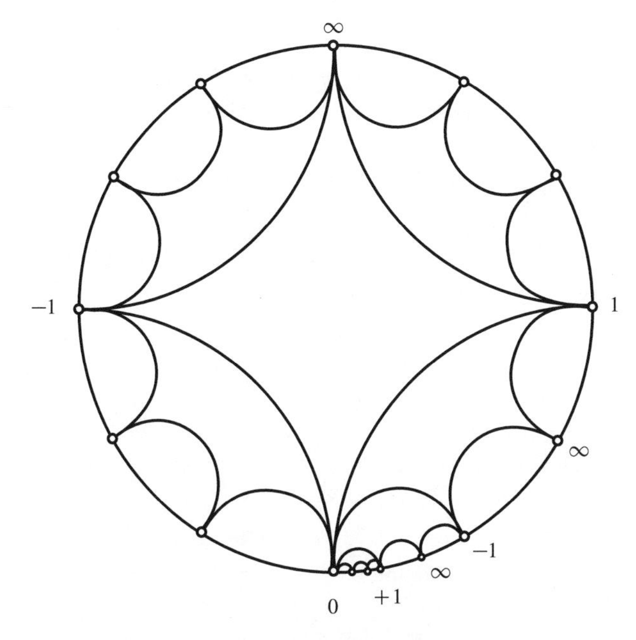

Figure 4.4. Tessellation of the unit disk by Γ_2.

of height h with real part between $-1/2$ (included) and $+1/2$ (excluded), as can be arranged by a (horizontal) translation $[1n/01]$. Then $\omega' = -1/\omega$ is of height $h' = |\omega|^{-2}h \leq h$, so $|\omega| \geq 1$, and either ω belongs to $\mathfrak{F}_1$ or else it falls on the excluded arc $\pi/3 \leq \theta < \pi/2$ of the unit circle and can be exchanged for its involute $-1/\omega$ on the included arc $2\pi/3 > \theta > \pi/2$.

Proof that $\mathfrak{F}_1$ is not too big. This means that no two points of $\mathfrak{F}_1$ are equivalent under Γ_1. Let $\omega \in \mathfrak{F}_1$ be of largest possible height h on its orbit and let $\omega' = x_1 + \sqrt{-1}x_2$ be another point of $\mathfrak{F}_1$ on the same orbit, of height h'. Then ω arises from ω' by some substitution $[ab/cd]$ and $h' \leq h = h'|c\omega' + d|^{-2}$, so

$$\begin{aligned}
1 \geq |c\omega' + d|^2 &= (cx_1 + d)^2 + c^2x_2^2 \\
&= c^2(x_1^2 + x_2^2) + 2cdx_1 + d^2 \\
&\geq c^2 - |cd| + d^2 \\
&\geq (|c| - |d|)^2
\end{aligned}$$

in view of $x_1^2 + x_2^2 \geq 1$ and $-1/2 \leq x_1 < 1/2$.

Case 1. $cd \neq 0$. The final inequality is strict, so $|c| = |d|$, and line 3 states that $c^2 \leq 1$. Then $c^2 = d^2 = 1$, $cd = \pm 1$, and line 2 reads $1 \geq x_1^2 + x_2^2 \pm 2x_1 + 1 \geq 1$, with the result that $x_1^2 + x_2^2 = 1$ and $x_1 = \mp 1/2$. The only possibility for ω' is the corner $\theta = 2\pi/3$; moreover, $|c\omega + d| = 1$, so ω is at height $h = h' = 1/2$, and there is no such point in the cell besides the corner.

Case 2. $cd = 0$. Now $c^2 + d^2 \leq 1$, so either $c = d = 0$, violating $ad - bc = 1$, or $c = \pm 1$ and $d = 0$, or else $c = 0$ and $d = \pm 1$.

Case 2 −. $c = \pm 1$, $d = 0$. The equality $x_1^2 + x_2^2 = 1$ forces ω to lie on the unit circle; also $ad - bc = 1$ implies $b = \pm 1$, so $\omega = \pm a - 1/\omega'$. The only way to keep both points in the cell is to take $a = 0$ and even that does not help: Only one of ω and $-1/\omega$ belongs to $\mathfrak{F}_1$ unless $\omega' = \omega = \sqrt{-1}$.

Case 2 +. $c = 0, d = \pm 1$. The reasoning is simpler. Now $a = \pm 1, \omega = \omega' \pm b$, and the only possibility to keep both points in the cell is to have $b = 0$.

This completes the proof that $\mathfrak{F}_1$ is a fundamental cell for Γ_1.

4.4 Generating the Groups

The method of highest points provides easy access to the structure of Γ_1 and Γ_2. Figure 4.3 suggests that Γ_1 is generated by the substitutions $A\colon \omega \mapsto \omega + 1$ and $B\colon \omega \mapsto -1/\omega$; in fact, the argument of Section 3 shows that $\mathfrak{F}_1$ is a fundamental cell for the group generated by A and B. That does the trick.

Exercise 1. Why so?

Exercise 2. Γ_2 is generated by $X = A^2 = [12/01]$ and $Y = BA^{-2}B = [10/21]$. The method of highest points may be used as at level 1, but prove it without geometric aids, for variety.

Γ_2 is, in fact, the **free group** generated by X and Y; that is, a nontrivial word in these letters is never the identity.

Exercise 3. Prove it. *Hints*: (1) The whole orbit $\Gamma_2\sqrt{-1}$ is of height ≤ 1. (2) If m is arbitrary and $n \neq 0$, then $X^m Y^n$ effects a strict reduction of heights in $(|\text{Re}\,\omega| \geq 1) \times (0 < h \leq 1)$ and reduces all heights to $\leq 1/3$ in $(|\text{Re}\,\omega| < 1) \times (0 < h \leq 1)$. Now think about the shortest word in X and Y that fixes $\sqrt{-1}$.

Γ_2 is a copy of the fundamental group of the thrice-punctured sphere, providing a topological proof of its free character; see Section 9, to follow.

Γ_1 is also generated by $A^{-1}B = [11/-10]$ and $B = [01/-10]$. The first substitution fixes the corner $\theta = 3\pi/2$ of the fundamental cell $\mathfrak{F}_1$ and is of period 3; the second fixes $\sqrt{-1}$ and is of period 2.

Exercise 4. Deduce that Γ_1 can be presented as the free group on two letters X and Y with two relations $X^3 = 1$ and $Y^2 = 1$. This means that the substitutions $X \mapsto A^{-1}B$ and $Y \mapsto B$ map the free group homomorphically onto Γ_1, the kernel being the subgroup generated by X^3 and Y^2. *Hint*: The image of any product of factors XY or X^2Y cannot fix $\sqrt{-1}$: It maps the second quadrant to itself and diminishes heights.

4.5 Gauss on Quadratic Forms

Gauss [1801] used the fundamental cell $\mathfrak{F}_1$ for the arithmetic reduction of quadratic forms. Let $\mathfrak{Q}(x) = Ax_1^2 + Bx_1x_2 + Cx_2^2$ be a positive quadratic form in $x = (x_1, x_2) \in \mathbb{R}^2$ with coprime whole numbers $A > 0$, B, $C > 0$ and fixed **discriminant** $D = B^2 - 4AC < 0$. $\mathfrak{Q}$ is **reduced** if $-A \le B < A \le C$, with the proviso that $B \le 0$ if $A = C$. Note that, for such fixed D, the number of reduced forms is finite. The form $\mathfrak{Q}$ is said to represent the whole number $n = 1, 2, 3, \ldots$ with multiplicity m if the equation $\mathfrak{Q}(x) = n$ has m solutions $x \in \mathbb{Z}^2$; for example, the number of representations of a positive number by $\mathfrak{Q}(x) = x_1^2 + x_2^2$ is the excess (if any) of its positive divisors $d \equiv 1 \bmod 4$ over its positive divisors $d \equiv 3 \bmod 4$. This is Fermat's theorem of Section 3.6. Dirichlet [1894] is the classical account of this circle of ideas.

Two forms $\mathfrak{Q}$ and $\mathfrak{Q}'$ with the same discriminant D are **arithmetically equivalent** if they represent the same numbers with the same multiplicities; they are **geometrically equivalent** if $\mathfrak{Q}'(x) = \mathfrak{Q}(x')$, in which x' is related to x by a modular substitution from $SL(2, \mathbb{Z})$. Gauss [1801] proved that *the two notions of equivalence are the same and that every class of equivalent forms contains just one reduced form.*

Proof. $\mathfrak{Q}[x] = x \cdot Qx$ with positive, symmetric 2×2 matrix $Q = [ab/cd]$ with $a = A, b = c = B/2, d = D$. Let $\sqrt{Q}$ be the positive, symmetric square root of Q so that $\mathfrak{Q}(x) = |\sqrt{Q}x|^2$. The whole numbers n represented by $\mathfrak{Q}$ are the squares of the lengths of the elements of the lattice $\mathbb{L} = \sqrt{Q}\mathbb{Z}^2$. Let $\mathfrak{Q}$ and $\mathfrak{Q}'$ be geometrically equivalent forms. Then $\mathfrak{Q}'(x) = \mathfrak{Q}(gx)$ with a modular substitution $g \in SL(2, \mathbb{Z})$, and $\mathfrak{Q}'(\mathbb{Z}^2) = \mathfrak{Q}(g\mathbb{Z}^2) = \mathfrak{Q}(\mathbb{Z}^2)$, multiplicities and all, in view of the fact that g maps $\mathbb{Z}^2$ 1:1 onto itself. Besides,

$$-(1/4)D(\mathfrak{Q}') = \det Q' = \det g^* Qg = (\det g)^2 \det Q = -(1/4)D(\mathfrak{Q}).$$

In short, $\mathfrak{Q}$ and $\mathfrak{Q}'$ are arithmetically equivalent. Now start from the arithmetically equivalent forms $\mathfrak{Q}$ and $\mathfrak{Q}'$. Then the associated lattices $\mathbb{L} = \sqrt{Q}\mathbb{Z}^2$ and $\mathbb{L}' = \sqrt{Q'}\mathbb{Z}^2$ produce the same lengths with the same multiplicities, and the areas of the corresponding fundamental cells are the same in view of $(\text{area})^2 = \det Q = AC - B^2/4 = -D/4$. It follows that $\mathbb{L}$ and $\mathbb{L}'$ differ by a (possibly improper) plane rotation: First, a proper rotation superimposes the primitive periods closest to the origin; second, the knowledge of multiplicities permits you to specify a circle on which the second primitive period must lie; third, the matching of cell areas means that only an extra reflection may be required to make the two lattices coincide; see Fig. 4.5. The upshot is that $\sqrt{Q'}$ is the same as $O\sqrt{Q}$ with a 2×2 orthogonal matrix O, up to a possible

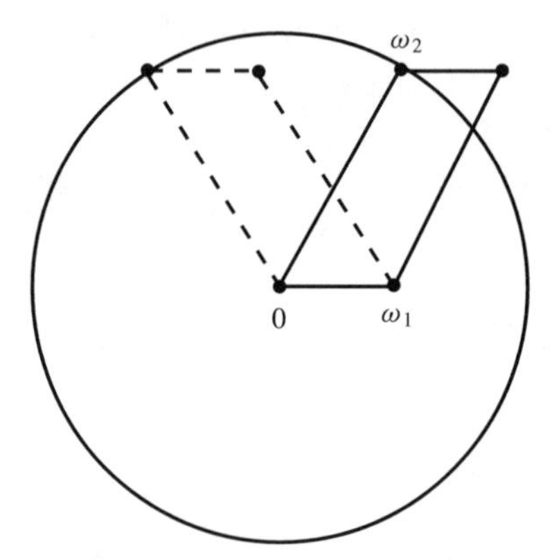

Figure 4.5. Lattices and quadratic forms.

modular substitution $g \in SL(2, \mathbb{Z})$ to the right. The upshot is that $\mathfrak{Q}$ and $\mathfrak{Q}'$ are geometrically equivalent: With $x' = gx$,

$$\mathfrak{Q}'(x) = |\sqrt{Q'}x|^2 = |O\sqrt{Q}gx|^2 = |\sqrt{Q}x'|^2 = \mathfrak{Q}(x')$$

Exercise 1. Think it over.

Exercise 2. Check that the period ratio of $\mathbb{L} = \sqrt{Q}\mathbb{Z}^2$ is $\omega = (2A)^{-1}(B + \sqrt{D})$.

The proof that each class of equivalent forms contains just one reduced form rests upon this fact. Note first that ω determines $\mathfrak{Q}$, D being known, and that $\mathfrak{Q}$ is reduced precisely when ω belongs to the fundamental cell $\mathfrak{F}_1$. Now the equivalence of Q and Q' via the linear action of $SL(2, \mathbb{Z})$ is the same as the equivalence of the corresponding ratios under the fractional action of $PSL(2, \mathbb{Z})$.

Exercise 3. Check that.

The rest is plain.

4.6 The Group of Anharmonic Ratios

Γ_1/Γ_2 is a copy of the symmetric group S_3 of permutations of the three letters e_1, e_2, e_3. It is also a copy of the group $\mathfrak{H}$ of anharmonic ratios:

$$x \mapsto x,\ 1-x,\ \frac{1}{x},\ \frac{x}{x-1},\ \frac{1}{1-x},\ \frac{x-1}{x},$$

as seen in Table 4.2.1, which records the action of Γ_1 on the modulus k^2. A fundamental cell for $\mathfrak{H}$ may be constructed as follows. Draw unit circles centered at 0 and 1 and divide the complex plane into six regions as in Fig. 4.6. Region 1 is mapped to the other five regions by the nonidentical anharmonic substitutions and so may serve, with suitable protocols at its edges, as a

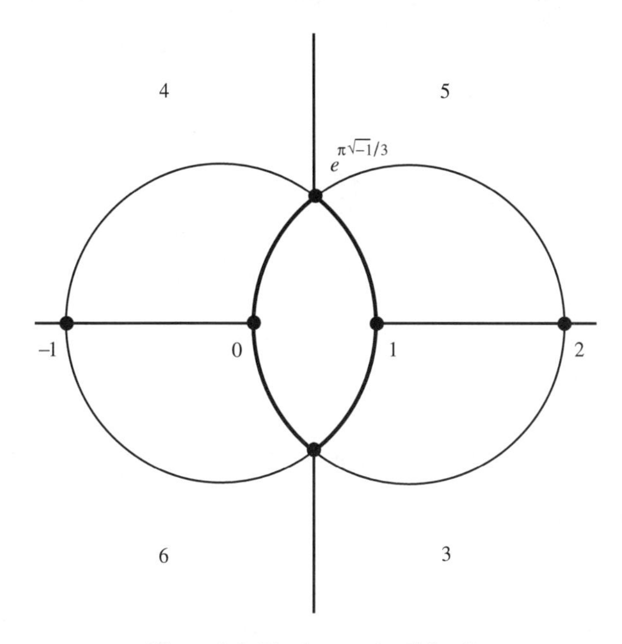

Figure 4.6. Fundamental cell for $\mathfrak{H}$.

fundamental cell for $\mathfrak{H}$: The left-hand arc is included, as is the left half of $0 \le x \le \frac{1}{2}$ of the base; the rest of the boundary is excluded.

Exercise 1. Check the mapping of region 1 onto regions 2, 3, 4, 5, 6.

The associated **absolute invariant** h is to be preserved by the action of $\mathfrak{H}$ and must separate orbits, just as for j and Γ_1. It was derived in ex. 1.7.4. A new method is introduced here, for variety and as a model for subsequent constructions. Divide the fundamental cell in two, as in Fig. 4.7, map the right half 1:1 onto the upper half-plane by the Riemann map with the indicated

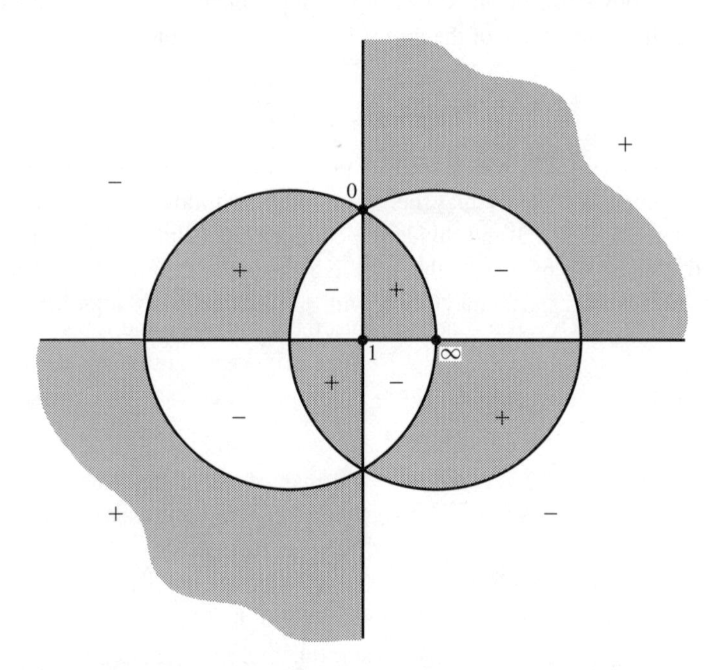

Figure 4.7. Construction of an absolute invariant for $\mathfrak{H}$.

values $0,\ 1,\ \infty$ at the corners, and extend this function to the whole plane by reflection. Do this first for the right half-plane and finish the job by a final reflection in the imaginary axis so as to finesse any question of multiple values. The extended function h is of rational character, with special features at -1, $0, 1/2, 1, 2, \omega = e^{\pi\sqrt{-1}/3}$, and $\omega^{-1} = e^{-\pi\sqrt{-1}/3}$: In detail, h is a rational function of degree $d = 6$ with triple roots at ω and ω^{-1}, double poles at 0 and 1, and another double pole at ∞, for a total count of six apiece, and takes the value $h = 1$ doubly at $-1, 1/2$, and 2, for a further count of six. The value

$h(2) = 1$ pins down the explicit expression

$$h(x) = \frac{4}{27}\frac{(x-\omega)^3(x-\omega^{-1})^3}{x^2(1-x)^2} = \frac{4}{27}\frac{(x^2-x+1)^3}{x^2(1-x)^2}.$$

There are three possible ways to verify that h is invariant under the action of $\mathfrak{H}$.

Method 1. A double reflection reproduces the value of h and represents an anharmonic substitution.

Exercise 2. Check one instance.

Method 2. The substitutions $x \mapsto 1/x$ and $x \mapsto 1-x$ generate $\mathfrak{H}$ and leave h invariant; this was the method of ex. 1.7.4.

Method 3. Let $h_1(x) = x, \ldots, h_6(x) = (x-1)/x$ be the six anharmonic substitutions. Then $h(x, y) = [h_1(x) - y][h_2 - y] \cdots [h_6(x) - y]$ is an invariant function of $\mathfrak{H}$ with poles of degree 2 at 0, 1, and ∞ and roots at the six anharmonic ratios of y. The function h has the same poles and roots at the six anharmonic ratios of $\omega = e^{\pi\sqrt{-1}/3}$, so it is a constant multiple of $h(x, \omega)$ and shares the invariance of that function.

The function h may be regarded as a 1:1 map of the fundamental cell to the projective line $\mathbb{P}^1$; its cubic character at $\omega = e^{\pi\sqrt{-1}/3}$ and its quadratic character at 0 and 1 are necessary to open up the vicinity of such a corner into a little disk. The plane, completed to a projective line at ∞, now appears as a sixfold cover of $\mathbb{P}^1$, ramified triply over ω and ω^{-1} and doubly over the anharmonic ratios $\infty, 0, 1$ of ∞. Any other invariant function of $\mathfrak{H}$ of *rational character* conforms to the rational character of h: It must be of degree divisible by 3 at ω and ω^{-1} and of degree divisible by 2 at the orbit of ∞, as you see from Fig. 4.7. The conclusion is that any such invariant function is of rational character in h or, what is the same, the field $\mathbf{K}(\mathfrak{H})$ of invariant functions of $\mathfrak{H}$ is one and the same as $\mathbb{C}(h)$. In short, h is the *absolute invariant* of $\mathfrak{H}$. This was already proved in ex. 1.7.4 by means of Luroth's theorem.

The isomorphism of $\mathfrak{H}$ and Γ_1/Γ_2 suggests a relation between the absolute invariants h, j, and k^2: Γ_1 subjects k^2 to the six anharmonic substitutions of $\mathfrak{H}$, so $h(k^2)$ is an invariant function of Γ_1. What else could it be but j, up to trivialities. This was already proved in ex. 2.12.3; compare Section 9.

4.7 Modular Forms

A **modular form** of **weight** $2n \in \mathbb{N}$ is an analytic function f on the open half-plane $\mathbb{H}$ that transforms under the modular group Γ_1 according to the rule[1]

$$f\left(\frac{a\omega + b}{c\omega + d}\right) = (c\omega + d)^{2n} f(\omega);$$

such a form is of period 1 and so can be expanded in whole powers of the parameter $q = e^{2\pi\sqrt{-1}\omega}$ in a neighborhood of $\omega = \sqrt{-1}\infty (q = 0)$, as in ex. 1.1, and it is a further requirement that this expansion contains no negative powers of q. In short, f is to be pole-free at ∞, too.

Three examples. $g_2 = 60 \sum'(n + m\omega)^{-4}$, $g_3 = 140 \sum'(n + m\omega)^{-6}$, and $\Delta = g_2^3 - 27g_3^2$ are modular forms of weights 4, 6, 12, respectively, by ex. 1.2.

Note that the absolute invariant $j = g_2^3/\Delta$ is the ratio of modular forms of the *same* weight 12. This exemplifies a general principle exploited by Poincaré [1882] that invariant functions may be constructed as ratios of functions that transform in the same way; compare the construction of elliptic functions from theta functions in Section 2.15.

The letter $\mathfrak{M}_n$ denotes the class of modular forms of weight $2n$; it is a vector space over $\mathbb{C}$. The key to its structure is contained in the

Fundamental Lemma. Let f be a nonvanishing form of weight $2n$ and let $m_\rho, m_{\sqrt{-1}}$, and m_∞ be the multiplicities with which it takes on its values at $\rho = e^{2\pi\sqrt{-1}\omega/3}$, $\sqrt{-1}$, and ∞. Then

$$\frac{1}{3}m_\rho + \frac{1}{2}m_{\sqrt{-1}} + m_\infty + m = \frac{1}{6}n,$$

m being the number of roots of $f(\omega) = 0$ in the fundamental cell, over and above such roots as may occur at the special points ρ, $\sqrt{-1}$, and ∞.

Proof. The identity is established by integration of $(2\pi\sqrt{-1})^{-1} d \log f$ about the boundary of the fundamental cell, modified as in Fig. 4.8 to avoid ρ, $\sqrt{-1}$, and ∞; further indentations may be required to avoid roots of $f = 0$ on the perimeter as for the fundamental lemma of Section 2.7; see Fig. 2.15. The number of roots inside is

$$m = \frac{1}{2\pi\sqrt{-1}} \int d \log f.$$

[1] Some people say f is of weight $-2n$, while others say n, so watch out if you read other accounts.

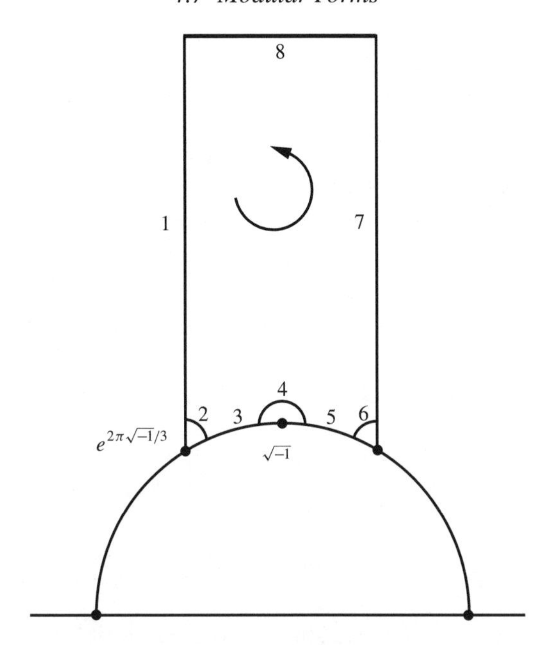

Figure 4.8. The modified contour.

Now f is of period 1, so the integrals along 1 and 7 cancel and it is easy to see that the contributions of $2 + 6, 4$, and 8 tend to $-m_\rho/3$, $-m_{\sqrt{-1}}/2$, and $-m_\infty$ as the indentations at ρ and $\sqrt{-1}$ go to zero and the height to infinity. A final contribution comes from $3 + 5$; this produces $n/6$ in view of $f(-1/\omega) = \omega^{2n} f(\omega)$. Collect the pieces.

The fundamental lemma is now used to determine the structure of $\mathfrak{M}_n$.

Step 1. $\mathfrak{M}_n = 0$ if $n < 0$ since $n/6$ must be ≥ 0.

Step 2. $\mathfrak{M}_0 = \mathbb{C}$: In fact, a nonvanishing form of weight 0 is invariant under Γ_1 and the fundamental lemma shows it to be root-free. Then $f - f(\sqrt{-1}) \in \mathfrak{M}_0$ stands in violation of the fundamental lemma unless $f(\omega) \equiv f(\sqrt{-1})$.

Step 3. $\mathfrak{M}_1 = 0$ since $1/6$ is not expressible as the fundamental lemma requires.

Step 4. $\mathfrak{M}_2 = \mathbb{C}g_2$. The fundamental lemma states that $2 = 2m_\rho + 3m_{\sqrt{-1}} + 6m_\infty + 6m$ for nonvanishing forms of weight 4, so $m_\rho = 1$ while $m_{\sqrt{-1}}$, m_∞, and m vanish. This means that every such form f has a simple root at ρ and no others in the fundamental cell, even at infinity. Now observe that g_2 is such a form and that $f/g_2 \in \mathfrak{M}_0 = \mathbb{C}$, by step 2.

Step 5. $\mathfrak{M}_3 = \mathbb{C}g_3$. The proof is similar; in particular, g_3 has a simple root at $\sqrt{-1}$ and no others in the cell, even at infinity.

Aside 1. The vanishing of $g_2(\rho)$ was already seen in example 2.9.2 (triangular lattice) and that of $g_3(\sqrt{-1})$ in example 2.9.1 (square lattice). The values $j(\rho) = 0$ and $j(\sqrt{-1}) = 1$ follow; see Section 12 for other special values.

Step 6. $\mathfrak{M}_4 = \mathbb{C}g_2^2$. The fundamental lemma states that $4 = 2m_\rho + 3m_{\sqrt{-1}} + 6m_\infty + 6m$ for nonvanishing forms of weight 8, so $m_\rho = 2$ while $m_{\sqrt{-1}}$, m_∞, and m vanish. The form g_2^2 fits this bill, with the result that $f/g_2^2 \in \mathfrak{M}_0$ is constant.

Step 7. $\mathfrak{M}_5 = \mathbb{C}g_2g_3$ is proved in the same way.

Aside 2. To complete the description of the spaces $\mathfrak{M}_n$ it is necessary to know that the forms g_2 and g_3 are **algebraically independent**, that is, they do not satisfy any polynomial in two variables.

Proof. Let the irreducible polynomial $P \in \mathbb{C}[x, y]$ be satisfied by $x = g_2(\omega)$ and $y = g_3(\omega)$ and apply the modular substitution $[ab/cd] \in \Gamma_1$ to ω. Then $P((c\omega + d)^4 g_2, (c\omega + d)^6 g_3) \equiv 0$, independently of c and d, which may be chosen at will from $\mathbb{Z}$. (Why this freedom?) In particular, $P((c\sqrt{-1} + d)^4 g_2(\sqrt{-1}), 0) = 0$, so $P(x, 0) \equiv 0$ and y divides $P(x, y)$, contradicting the irreducibility of P.

Step 8. The independent forms $g_2^a g_3^b$ of fixed weight $2n = 4a + 6b$ span $\mathfrak{M}_n$, and $\dim \mathfrak{M}_n = [n/6]$ or $[n/6] + 1$ according as $n \equiv 1 \bmod 6$ or not.

Proof. The independence is the content of aside 2. Now Δ is of weight 12. It cannot vanish in the open half-plane, the numbers e_1, e_2, e_3 being distinct, so the fundamental lemma requires it to vanish simply at ∞. This is preparatory to the remark that

$$0 \to \mathfrak{M}_{n-6} \to \mathfrak{M}_n \to \mathbb{C} \to 0$$

is an exact sequence, meaning that the image of one arrow is the kernel of the next. The first arrow is injection, the next is multiplication by Δ, and the third is evaluation at ∞. The exactness is obvious: Multiplication by Δ has trivial kernel, and $f/\Delta \in \mathfrak{M}_{n-6}$ for forms $f \in \mathfrak{M}_n$ that vanish at ∞. This means that $\mathfrak{M}_n = \Delta\mathfrak{M}_{n-6} + \mathbb{C}\times$ any form of weight $2n$ not vanishing at ∞, and now you turn the crank:

$$\begin{aligned}
\mathfrak{M}_6 &= \Delta\mathfrak{M}_0 + \mathbb{C}g_3^2 &&= \text{the span of } g_2^3 \text{ and } g_3^2, \text{ of dimension 2,}\\
\mathfrak{M}_7 &= \Delta\mathfrak{M}_1 + \mathbb{C}g_2^3 g_3 &&= \text{the span of } g_2^3 g_3, \text{ of dimension 1,}\\
\mathfrak{M}_8 &= \Delta\mathfrak{M}_2 + \mathbb{C}g_2^4 &&= \text{the span of } g_2^4 \text{ and } g_2 g_3^2, \text{ of dimension 2,}
\end{aligned}$$

and so on. The discussion is finished.

Exercise 1. Prove that $\mathfrak{M}_n$ is the same as the class of all symmetric homogeneous polynomials of degree n in e_1, e_2, e_3. *Hint*: The es behave like forms of weight 2 except that they permute.

Exercise 2. Table 3.9.1 states that the null values $\vartheta_2(0)$, $\vartheta_3(0)$, $\vartheta_4(0)$ behave as modular forms of weight 1/2, or very nearly so; in fact, $\vartheta_2^8(0)\vartheta_3^8(0)\vartheta_4^8(0) = 2^{-4}\pi^{-1/2}\Delta$ is a form of weight 12. Find other combinations of null values that are honest forms. *Sample*: What is $\vartheta_2^8(0) + \vartheta_3^8(0) + \vartheta_4^8(0)$? or $\vartheta_2^8(0)\vartheta_3^4(0) + \vartheta_2^4(0)\,\vartheta_3^8(0) + \vartheta_2^8(0)\,\vartheta_4^4(0) + \vartheta_3^4(0)\,\vartheta_4^8(0) - \vartheta_3^8(0)\,\vartheta_4^4(0) - \vartheta_2^4(0)\vartheta_4^8(0)$?

Mumford [1983] gives a general account of such identities.

Exercise 3. Prove that $f = (1/2)(\log\Delta)'$ solves $f''' - 2ff'' + 3(f')^2 = 0$; see Chazy [1911]. Ablowitz and Clarkson [1991] explain the connection with solitons; compare ex. 2.10.3. *Hint*: $f''' - 2ff'' + 3(f')^2$ is a modular form of weight 8. What is its value at ∞?

Exercise 4. Check that the discriminant Δ satisfies

$$\Delta^3\Delta''' - 5(\Delta')^2\Delta''' - (3/2)\Delta^2(\Delta'')^2 + 12\Delta(\Delta')^2\Delta'' - (13/2)(\Delta')^4 = 0.$$

Exercise 5. One final example: The null value $\vartheta = \vartheta_3(0|\omega)$ satisfies

$$\begin{aligned}
&\left(\vartheta^2\vartheta''' - 15\vartheta\vartheta'\vartheta'' + 30(\vartheta')^3\right)^2 + 32\left(\vartheta\vartheta'' - 3(\vartheta')^2\right)^3\\
&\quad = \vartheta^{10}\left(\vartheta\vartheta'' - 3(\vartheta')^2\right)^2.
\end{aligned}$$

Why? Ehrenpreis [1994] explains how Jacobi [1847] discovered this remarkable fact.

4.8 Eisenstein Sums

Eisenstein was the favorite pupil of Gauss. He died at age 29. Weil [1975] gives a delightful account of his work. The nth **Eisenstein sum** is

$$E_n(\omega) = \sum_{\mathbb{Z}^2 - 0} (j + k\omega)^{-2n} \text{ divided by } \sum_{\mathbb{Z}-0} j^{-2n}.$$

It converges for $n \geq 2$ and is a modular form of weight $2n$, taking the value 1 at the cusp $\sqrt{-1}\infty$. The special sums $E_2 = (3/4)\pi^{-4} g_2$ and $E_3 = (27/8)\pi^{-6} g_3$ are already familiar. The constants may be checked from Table 4.8.1 which displays the first six Bernoulli numbers $B_n = (2n)!/(2\pi)^{2n} \sum_{\mathbb{Z}-0} j^{-2n}$.

Identities. The finite dimensionality of the spaces $\mathfrak{M}_n$ $(n \geq 0)$ necessitates the existence of numerous identities between Eisenstein sums, much as for theta null values; indeed, ex. 7.2 points to a common reason. The next item is preparatory to three samples of this phenomenon.

Exercise 1. Check that $\wp(x) = x^{-2} + \sum_{n=1}^{\infty} \sum_{\mathbb{Z}^2-0} (j + k\omega)^{-2n-2}(2n+1)x^{2n}$, in which the $\wp$-function is formed for the lattice $\mathbb{L} = \mathbb{Z} \oplus \omega\mathbb{Z}$; compare ex. 2.8.7.

Now feed this expansion into the differential equation for $\wp$ and compare like powers of x to produce $E_4 = E_2^2$, $E_5 = E_2 E_3$, and $E_6 = (2 \cdot 5^3/691)E_3^2 + (3^2 \cdot 7^2/691)E_2^3$, and more.

Exercise 2. Do it with the aid of the table.

Arithmetic Consequences. Both modular forms and null values can be expressed as sums of whole or fractional powers of $q = e^{2\pi\sqrt{-1}\omega}$, and it is a well-attested fact that the coefficients of the several powers have number-theoretic significance; see Koblitz [1993] and Sarnak [1990] for an overview. The Eisenstein sum is illustrative to begin with:

$$E_n(\omega) = 1 + (-1)^n 4n B_n^{-1} \sum_{m=1}^{\infty} \sigma_{2n-1}(m) q^m,$$

Table 4.8.1. *Bernoulli numbers*

B_1	B_2	B_3	B_4	B_5	B_6
$\frac{1}{6}$	$\frac{1}{30}$	$\frac{1}{42}$	$\frac{1}{30}$	$\frac{5}{66}$	$\frac{691}{2730}$

in which $\sigma_p(m)$ is the sum of the pth powers of the divisors of the whole number m. This follows from the elementary identity

$$\frac{1}{4\pi^2}\sum_{\mathbb{Z}}(j+\omega)^{-2} = [2\sin\pi\omega]^{-2} = -(q^{1/2}-q^{-1/2})^{-2} = -q\frac{d}{dq}(1-q)^{-1}$$
$$= -\sum_{m=1}^{\infty} mq^m.$$

Exercise 3. How? *Hint*: $(j+\omega)^{-2}$ is differentiated $2n-2$ times.

The stated identities between Eisenstein sums can now be converted into surprising arithmetic identities.

Sample 1. $E_4 = E_2^2$, $E_2 = 1+240\sum\sigma_3(m)q^m$, $E_4 = 1+480\sum\sigma_7(m)q^m$, so

$$1+480\sum_{m=1}^{\infty}\sigma_7(m)q^m = 1+480\sum_{m=1}^{\infty}\sigma_3(m)q^m + 240^2\sum_{m=1}^{\infty}\sigma_3\circ\sigma_3(m)q^m,$$

in which $\sigma\circ\sigma(m)$ is short for $\sum_{n=1}^{m}\sigma(m-n)\sigma(n)$. This produces $\sigma_7 = \sigma_3 + 120\sigma_3\circ\sigma_3$; in particular, $\sigma_7(m) \equiv \sigma_3(m) \bmod 2^3\cdot 3\cdot 5$.

Sample 2. Start with the expansion

$$\Delta = (1728)^{-1}(2\pi)^{12}\left(E_2^3 - E_3^2\right) = (2\pi)^{12}\sum_{m=1}^{\infty}\tau(m)q^m,$$

the coefficients of the sum being the Ramanujan numbers of sample 4 in Section 3.5. Now $E_3 = 1-504\sum\sigma_5(m)q^m$ and $E_6 = 1+(65520/691)\sum\sigma_{11}(m)q^m$ produce $65520\sum\sigma_{11}(m)q^m = 2^6\cdot 3^5\cdot 7^2\sum\tau(m)q^m$, up to a sum of whole powers of q with coefficients from $691\mathbb{Z}$. This is the same as saying $\tau(m)\equiv\sigma_{11}(m) \bmod 691$ in view of $65520 = 691\times 94+566$ and $2^6\cdot 3^5\cdot 7^2 = 691\times 1102+566$, 566 and 691 being coprime. Is that clever or what? Let's try it out: $\tau(4) = -1472$ (see ex. 3.5.5), $\sigma_{11}(4) = 1+2^{11}+2^{22} = 419653$, and $419653+1472 = 4197825 = 3^5\cdot 5^2\cdot 691$. Swinnerton and Dyer [1988] present many other congruences satisfied by the Ramanujan numbers.

4.9 Absolute Invariants

The goal of this section is to prove what was promised in Section 1: The absolute invariant $h(=j$ or $k^2)$ for the groups Γ of levels 1 and 2 merits its name in that (a) it maps the fundamental cell, completed by adjunction of cusps, 1:1 onto the projective line, and (b) the field $\mathbf{K}(\Gamma)$ of modular functions is one and the same as $\mathbb{C}(h)$.

First Level. Take the (shaded) half of the fundamental cell for $\Gamma_1 = PSL(2, \mathbb{Z})$ seen in Fig. 4.9; let f be the Riemann map of this region onto the upper

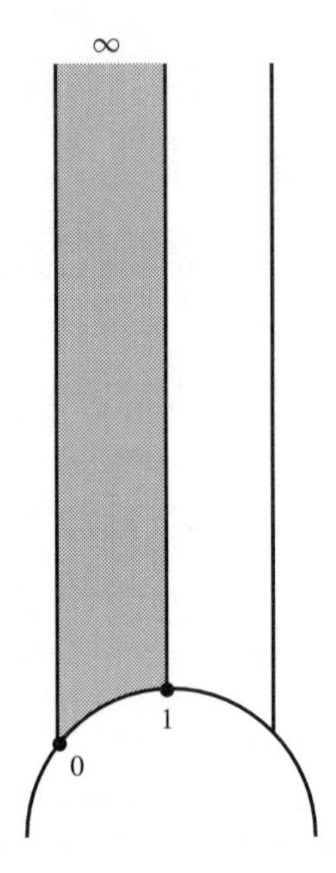

Figure 4.9. The fundamental region.

half-plane $\mathbb{H}$, standardized by $f(\sqrt{-1}\infty) = \infty$, $f(\rho = e^{2\pi\sqrt{-1}/3}) = 0$, and $f(\sqrt{-1}) = 1$; and extend it by reflection, after the manner of Section 6, to the whole tessellation of $\mathbb{H}$ as depicted in Fig. 4.10. The extended function is invariant under Γ_1, taking the value 0 sixfold at $\omega = 0$, and the value 1 twice at $\omega = \sqrt{-1}$. It is of period 1 and so may be expanded in the familiar manner as $\sum_{\mathbb{Z}} \hat{f}(n) q^n$, and from its 1:1 character on $\mathfrak{F}_1$ you conclude that the sum breaks off at $n = -1$, which is to say that f has a simple pole at $\sqrt{-1}\infty$.

Exercise 1. Check all that. *Hint*: $1/f$ vanishes at $q = 0$ and is 1:1 in the punctured disk $0 < |q| < 1$.

Now Δ is a modular form of weight 12 vanishing at ∞. It follows that Δf is also of weight 12; as such, it can be expressed as $a g_2^3 + b\Delta$, $\mathfrak{M}_6$ being of dimension 2, and from the vanishing of g_2 and f at ρ, you see that $b = 0$; similarly, $a = 1$ follows from $g_3(\sqrt{-1}) = 0$ and $f(\sqrt{-1}) = 1$ in view of

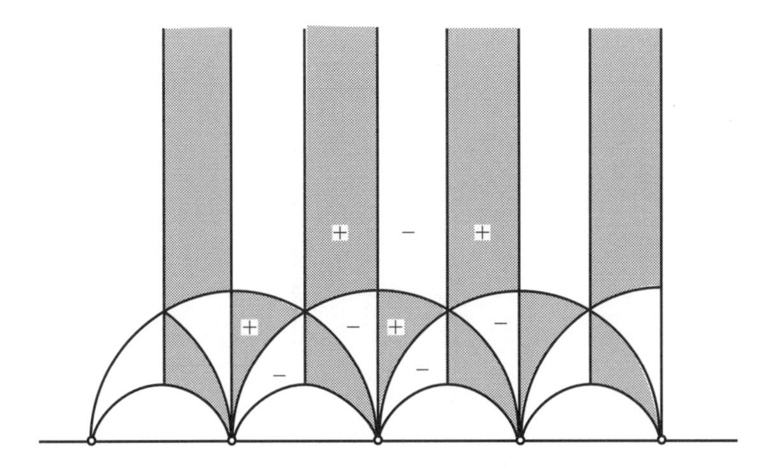

Figure 4.10. The tessellation.

$f\Delta(\sqrt{-1}) = ag_2^3(\sqrt{-1}) = a\Delta(\sqrt{-1})$ and the fact that $\Delta(\sqrt{-1})$ does not vanish. The upshot is that f is one and the same as the (so-called) absolute invariant $j = g_2^3/\Delta$; in particular, *j takes on every complex value just once in the fundamental cell $\mathfrak{F}_1$.* This is the justification of method 2 employed for the problem of inversion in Section 2.12.

Exercise 2. The fact that j takes each complex value c just once in the fundamental cell can be proved more directly by integration of $(2\pi\sqrt{-1})^{-1}d\log(j-c)$ about the contour seen in Fig. 4.11. The contributions of 1 and 4 cancel, and 5 produces a contribution of 1 or nearly so. What about $2+3$? Give the details.

Now comes the identification $\mathbf{K}(\Gamma_1) = \mathbb{C}(j)$, which justifies the name *absolute invariant.* j identifies the completed cell $\mathfrak{F}_1 + \sqrt{-1}\infty$ with the projective line. The map is not conformal at the three points ρ, $\sqrt{-1}$, and ∞: At ρ, it behaves like the cubic parameter $q_\rho(\omega) = (\omega-\rho)^3$, opening up the corner into a semicircle; at $\sqrt{-1}$, it behaves like $1+$ a constant multiple of $q_{\sqrt{-1}}(\omega) = (\omega-\sqrt{-1})^2$, producing a full disk; at the cusp $\sqrt{-1}\infty$, it behaves like $q_\infty(\omega) = (1728)^{-1}q^{-1}$, so here you also see a full disk after adjunction of $\sqrt{-1}\infty$ $(q = 0)$. Keeping this in mind, it is easy to see that any modular function $f \in \mathbf{K}(\Gamma_1)$ is of rational character in j: (1) It is of cubic character at ρ since $\omega \mapsto -1 - 1/\omega$ fixes this point and is of period 3; (2) it is of quadratic character at $\sqrt{-1}$ since $\omega \mapsto -1/\omega$ fixes that point and is of period 2; (3) it is of rational character at $\sqrt{-1}\infty$ since its expansion $\sum \hat{f}(n)q^n$ breaks off at some power $n = m > -\infty$. But such a function may be viewed as a function of rational character on the projective line $\mathbb{P}^1 = j(\mathfrak{F}_1 + \sqrt{-1}\infty)$, which is to say $\mathbf{K}(\Gamma_1) = \mathbb{C}(j)$, as advertised.

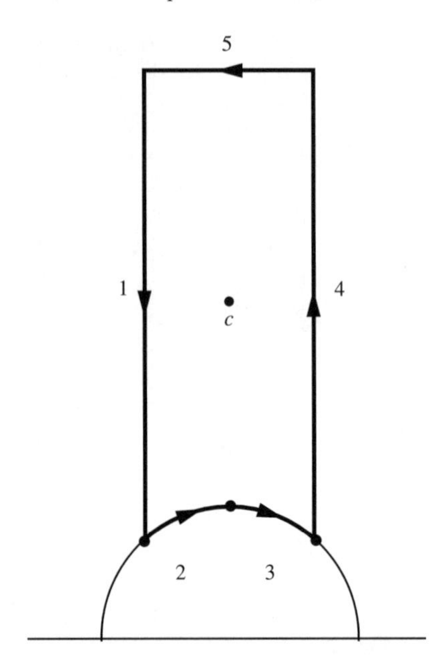

Figure 4.11. The contour.

Exercise 3. Check the details.

Exercise 4. $j'(\omega)$ behaves like but is not a form of weight 2. Check that Δ is a constant multiple of $j^{-4}(j-1)^{-3}(j')^6$. What is the constant?

Second Level. The same considerations apply to the Jacobi modulus k^2; it is the Riemann map of the shaded half of the fundamental cell $\mathfrak{F}_2$, standardized as in Fig. 4.12, but let us take a different route, for variety. The absolute invariant $h(x) = (4/27)(x^2 - x + 1)^3 x^{-2}(1-x)^{-2}$ of the group of anharmonic ratios is recalled from Section 6, and $h(k^2)$ is computed by hand with the help of $k^2 = (e_2 - e_3)/(e_1 - e_3)$, recapitulating ex. 2.12.3:

$$\begin{aligned} h(k^2) &= \frac{4}{27}\frac{(k^4 - k^2 + 1)^3}{k^4(1-k^2)^2} \\ &= \frac{64}{27}\frac{(e_1^2 + e_2^2 + e_3^2 - e_1e_2 - e_2e_3 - e_3e_1)^3}{16(e_1 - e_2)^2(e_2 - e_3)^2(e_3 - e_1)^2} \\ &= \frac{64}{27}\frac{3^3}{4^3}\frac{g_2^3}{\Delta} \\ &= j. \end{aligned}$$

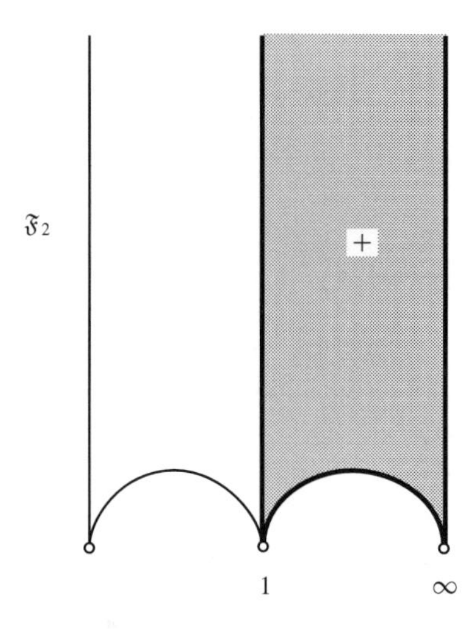

Figure 4.12. The Riemann map for $\mathfrak{F}_2$.

Now j takes on each complex value once in $\mathfrak{F}_1$ and so also once in each of the six copies of $\mathfrak{F}_1$ that make up $\mathfrak{F}_2$, as seen in Fig. 4.12. Fix a point $\omega \in \mathfrak{F}_1$, let $a = j(\omega)$, and pick $b \neq 0, \infty$ so that $a = h(b)$. Then $k^2(\omega)$ takes on the value of one of the anharmonic ratios of b. It follows that k^2 takes on each of those anharmonic ratios just once in $\mathfrak{F}_2$ because j takes its values sixfold in $\mathfrak{F}_2$ and h is 6:1. The upshot is that k^2 maps $\mathfrak{F}_2$ simply onto the twice-punctured plane $\mathbb{C} - 0 - 1$. It remains to identify k^2 with the Riemann map of Fig. 4.12 and to discuss its role as *absolute invariant* of Γ_2. k^2 is real for imaginary period ratios; this propagates to the whole boundary of $\mathfrak{F}_2$ by the rules of Table 4.2.1. Now

$$\omega = \frac{\sqrt{-1}K'}{2K} = \frac{\sqrt{-1}}{2}\frac{\int_1^{1/k}[(x^2-1)(1-k^2x^2)]^{-1/2}dx}{\int_0^1[(1-x^2)(1-k^2x^2)]^{-1/2}dx}$$

tends to $\sqrt{-1}\infty$ as $k^2 \downarrow 0$ and to 0 as $k^2 \uparrow 1$ via real values, so the vertical line joining $\sqrt{-1}\infty$ to $\sqrt{-1}\bullet 0$ is mapped to [0, 1], and a further application of Table 4.2.1 shows that the right-hand semicircle of Fig. 4.12 is mapped to $[1, \infty)$ while the right-hand vertical line is mapped to $(\infty, 0]$; in short, the shaded region is mapped (necessarily 1:1) onto the upper half-plane. The identification is complete.

The modular functions of Γ_2 are required to be of rational character in the special parameters $q_0(\omega) = e^{-\pi\sqrt{-1}/\omega}$ at 0, $q_1(\omega) = e^{\pi\sqrt{-1}/(1-\omega)}$ at 1, and $q_\infty(\omega) = e^{\pi\sqrt{-1}\omega}$ at ∞. It is to be proved that k^2 conforms to these rules and that any other modular function is a rational function of k^2. In fact, to leading order you find:

At $\sqrt{-1}\infty$, $j = h(k^2) = 4/27k^{-4}$ in view of $k(\sqrt{-1}\infty) = 0$, so $k^4 = 16^2 q_\infty^2$ and $k^2 = 16q_\infty$, k^2 being positive for imaginary ω.
At 0, $k^2 = 1 - k^2(-1/\omega) = 1 - 16q_0$, by Table 4.2.1 and the expansion at $\sqrt{-1}\infty$.
At 1, $k^2(\omega) = k^{-2}(\omega/(\omega - 1)) = -1/q_1$, by similar reasoning.

The rest will be plain: k^2 identifies the completed cell $\mathfrak{F}_2 + 0 + 1 + \sqrt{-1}\infty$ with the projective line, and every invariant function of Γ_2, of the proper rational character at the cusps 0, 1, $\sqrt{-1}\infty$, is a function of rational character on that line and so a rational function of its parameter k^2. In short, $\mathbf{K}(\Gamma_2) = \mathbb{C}(k^2)$.

Exercise 5. Get the next term in the expansion of k^2 at $\sqrt{-1}\infty$.

A Topological Bonus. The identification of k^2 as a map of the open upper half-plane onto the punctured plane $\mathbb{C} - 0 - 1$ has an unexpected bonus. The half-plane is simply connected and so can be viewed as the universal cover of the punctured plane, as proved in Section 1.14 already, but only by appeal to the full force of the Koebe–Poincaré theorem. The modulus k^2 is the projection and Γ_2, interpreted as the stabilizer of k^2 in Γ_1, is the covering group. The latter is naturally isomorphic to the fundamental group of the punctured plane as explained in Section 1.14, and since the fundamental group is free on two letters, so is Γ_2. This confirms ex. 4.3, the twice-punctured plane and the thrice-punctured sphere being one and the same thing.

Exercise 6. Check the details.

Picard's Little Theorem [1879]. A second bonus is the classical proof of the fact that *a nonconstant integral function takes on every complex value c with at most one exception.*

Proof. Let the integral function f omit the values a and b. Then $(b-a)^{-1}(f-a)$ omits 0 and 1, so it is permissible to take $a = 0$ and $b = 1$. Now use f to map a small disk to the punctured plane $\mathbb{C} - 0 - 1$, lift the map to the half-plane via any branch of the inverse function of k^2, and map the lift into the unit disk via $\omega \mapsto (\omega - \sqrt{-1})/(\omega + \sqrt{-1})$; see Fig. 4.13. The composite map may be continued without obstruction along paths in the left-hand plane. This produces a single-valued function in that plane by the monodromy theorem of Section 1.11, and as its values are confined to the disk, so it must be constant.

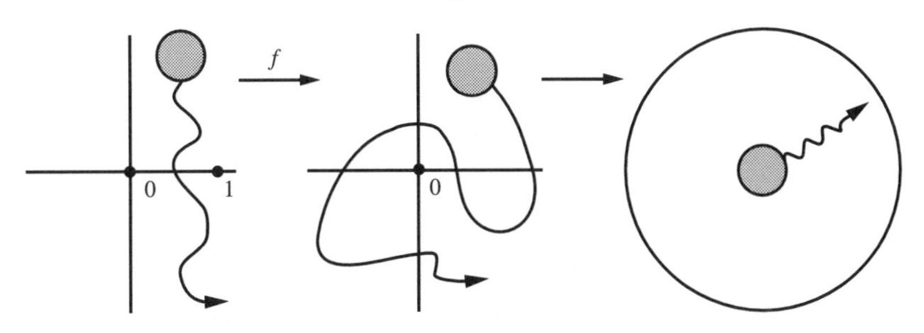

Figure 4.13. Picard's theorem.

But k^2 is not constant, nor is the map from the half-plane to the disk, and the only way out is for f to be constant.

Here the modulus is merely window dressing: All you need to know is that the half-plane is the universal cover of the twice-punctured plane, as in Section 1.14. That seems simpler than using k^2, but relies on the deep Koebe–Poincaré theorem. Ahlfors [1973] presents a beautiful elementary proof, employing neither coverings nor transcendental aids such as k^2; see also Nevanlinna [1970: 248–9].

The discussion of other arithmetic subgroups of $PSL(2, \mathbb{Z})$ and their absolute invariants is postponed until Section 16.

4.10 Triangle Functions

This section sketches the pretty discovery of Schwarz [1872] concerning absolute invariants and the hypergeometric equation:

$$\frac{d^2y}{dx^2} + \left[\frac{1-a}{x} + \frac{1-b}{x-1}\right]\frac{dy}{dx} + \frac{(1-a-b)^2 - c^2}{4x(x-1)}y = 0.$$

Ford [1972: 289–309] and Sansone and Gerretsen [1969(2): 407–506] are recommended for details. Let $\mathbf{y}$ be the ratio of two independent solutions taken at the point $x \neq 0, 1, \infty$ and continued in the punctured plane. Coming back to the starting point after a circuit about one or more punctures subjects $\mathbf{y}$ to a substitution from $PSL(2, \mathbb{C})$. These form a group Γ, the **monodromy group** of the equation. Now let $\mathbf{x}$ be the independent variable viewed as a function of $\mathbf{y}$; up to trivialities, it is a single-valued function if and only if each of the numbers a, b, c is the reciprocal of an integer $1, 2, 3, \ldots$ (∞ included), in which case $\mathbf{x}$ is the **absolute invariant** of the group Γ. The several branches of the function $\mathbf{y}$ map the upper and lower half-planes 1:1 onto a family of

Table 4.10.1. *Case 1:* $a+b+c=1$

a	1/3	1/2	1/2	1/2
b	1/3	1/4	1/3	1/2
c	1/3	1/4	1/6	0

nonoverlapping triangular regions, bounded by circular arcs making the internal angles $a\pi$, $b\pi$, $c\pi$, and any adjacent pair of triangles is mapped by $\mathbf{x}$ 1: 1 onto the plane. That is why $\mathbf{x}$ is the absolute invariant. There are three cases, distinguished by the value of the sum $a+b+c$.

Case 1. $a+b+c=1$. The triangles are straight-sided and cover the whole plane. The possibilities are displayed in Table 4.10.1. The corresponding functions $\mathbf{x}$ are exemplified by $\wp'$, $\wp^2$, $(\wp')^2$, and $\sin x$.

Exercise 1. Check this identification.

Case 2. $a+b+c>1$. Now there are only a finite number of triangles covering the sphere. The groups are those of the Platonic solids discussed in Section 1.7.

Case 3. $a+b+c<1$. This time the triangles cover the half-plane $\mathbb{H}$ and there is an infinite number of cases exemplified by the modular groups of first and second level: For the first level, $a=1/2, b=1/3, c=0$, and $\mathbf{x}=j$; for the second, $a=b=c=0$ and $\mathbf{x}=k^2$. The case $a=1/2, b=1/4, c=1/5$ will make an appearance in Section 5.7. Figure 4.14 depicts the situation for the group of first level with the half-plane replaced by the unit disk.

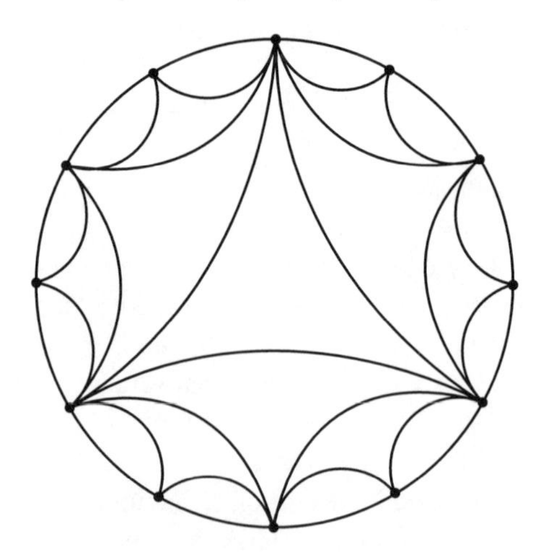

Figure 4.14. Triangulation by Γ_1.

4.11 The Modular Equation of Level 2

Think of two tori $\mathbf{X} = \mathbb{C}/\mathbb{L}$, one a double cover of the other, with period ratios ω and 2ω. Then the corresponding absolute invariants $j(\omega)$ and $j(2\omega)$ satisfy a (universal) irreducible polynomial equation: $P(j(\omega), j(2\omega)) = 0$. This is the **modular equation of level 2** to be derived here. More complicated modular equations, of so-called higher level, relate $j(n\omega)$ to $j(\omega)$ for $n \geq 3$; see Section 13.

Warning. The absolute invariant j is now multiplied by $1728 = 2^6 \cdot 3^3$ so that $j(\omega) = q^{-1} + \cdots$ at the cusp $\sqrt{-1}\infty$. This spoils the old normalization $j(\sqrt{-1}) = 1$, but simplifies the present discussion. *This convention will be in force from now on.*

The reason that a modular equation must be present at level 2 is simple; its explicit determination is not. The discussion is divided into five steps.

Step 1. Let $\omega \mapsto \omega' = [ab/cd]\omega$ be a modular substitution of level 2. Then

$$j(2\omega') = j\left[\frac{a \cdot 2\omega + 2b}{(c/2) \cdot 2\omega + d}\right] = j\left[\frac{a' \cdot 2\omega + b'}{c' \cdot 2\omega + d'}\right]$$

with $a' = a, b' = 2b, c' = c/2, d' = d \in \mathbb{Z}$, and $a'd' - b'c' = ad - bd = 1$ so that $[a'b'/c'd']$ is a modular substitution of level 1. This means that the right-hand side is just $j(2\omega)$, that is, this function is invariant under Γ_2. The same is true of $j(\omega)$ itself, so both functions belong to $\mathbf{K}(\Gamma_2) = \mathbb{C}(k^2)$. The existence of the modular equation is plain from that: Any two elements of a rational function field are algebraic, one over the other.

Step 2. Note that $j(2\omega)$ is not invariant under Γ_1; indeed, the six cosets of the quotient Γ_1/Γ_2 listed in Table 4.3.1 act upon it to produce three distinct values:

$$\begin{aligned}
j \circ 2[10/01]\omega &= j(2\omega), && \\
j \circ 2[0-1/10]\omega &= j(-2/\omega) &&= j(\omega/2), \\
j \circ 2[10/-11]\omega &= j(2\omega/(1-\omega)) &&= j(\omega/2+1/2), \\
j \circ 2[1-1/01]\omega &= j(2\omega-2) &&= j(2\omega), \\
j \circ 2[01/-11]\omega &= j(2/(1-\omega)) &&= j(\omega/2+1/2), \\
j \circ 2[1-1/10]\omega &= j((2\omega-2)/\omega) &&= j(\omega/2).
\end{aligned}$$

Exercise 1. Why is line 3 true?

Step 3. Introduce the field polynomial of $j(2\omega)$ over the ground field $\mathbf{K}(\Gamma_1) = \mathbb{C}(j)$. The action of Γ_1 is to permute the three values $j_1(\omega) = j(2\omega)$,

$j_2(\omega) = j(\omega/2)$, $j_3(\omega) = j(\omega/2 + 1/2)$, so their elementary symmetric functions

$$\sigma_1 = j_1 + j_2 + j_3, \quad \sigma_2 = j_1 j_2 + j_2 j_3 + j_3 j_1, \quad \text{and} \quad \sigma_3 = j_1 j_2 j_3$$

are modular functions of Γ_1, and this in the technical sense, that they are of rational character in the upper half-plane and also at the cusp $\sqrt{-1}\infty$; for example,

$$\sigma_1 = j(2\omega) + j(\omega/2) + j(\omega/2 + 1/2) = q^{-2} + q^{-1/2} - q^{-1/2}$$

to leading order at the cusp, so it must be a polynomial of degree 2 in j with top coefficient $+1$.

Exercise 2. Use the same method to check that σ_2 and σ_3 are polynomials in j of degree ≤ 2 and $= 3$, respectively, the top coefficient of σ_3 being -1.

The upshot is that the polynomial

$$\begin{aligned} F_2(x) &= [x - j(2\omega)][x - j(\omega/2)][x - j(\omega/2 + 1/2)] \\ &= x^3 - \sigma_1 x^2 + \sigma_2 x - \sigma_3 \end{aligned}$$

is of class $\mathbb{C}[j][x]$. It is even irreducible because Γ_1 permutes its roots transitively, so it must be the field polynomial of $j(2\omega)$ over the ground field $\mathbb{C}(j)$, alias the modular equation of level 2. To find the explicit form of F_2 requires detailed computations at the cusp. The absolute invariant has the expansion

$$\begin{aligned} j(\omega) = q^{-1} &+ 744 + 196884q + 21493760q^2 + 864299970q^3 \\ &+ 20245856256q^4 + \cdots; \end{aligned}$$

see Zuckerman [1939]. From that you may derive, with tears, the full details of $F_2(x)$:

$$\begin{aligned} F_2(x) = x^3 &- (j^2 - 2^4 \cdot 3 \cdot 31\ j + 2^4 \cdot 3^4 \cdot 5^3)\, x^2 \\ &+ (2^4 \cdot 3 \cdot 31\ j^2 + 3^4 \cdot 5^3 \cdot 4027\ j + 2^8 \cdot 3^7 \cdot 5^6)\, x \\ &+ (j^3 - 2^4 \cdot 3^4 \cdot 5^3\ j^2 + 2^8 \cdot 3^7 \cdot 5^6\ j - 2^{12} \cdot 3^9 \cdot 5^9). \end{aligned}$$

Step 4. This step determines σ_1. To the necessary accuracy at the cusp,

$$j(2\omega) = q^{-2} + 744, \quad j(\omega/2) = q^{-1/2} + 744, \quad j(\omega/2 + 1/2) = -q^{-1/2} + 744,$$

so $\sigma_1 = q^{-2} + 3 \cdot 744$ and from

$$j^2 = q^{-2} + 2 \cdot 744\, q^{-1} + (744)^2 + 2 \cdot 196884$$

you see that

$$\begin{aligned}\sigma_1 &= j^2 - 2\cdot 744\, j + (744)^2 - 2\cdot 196884 + 3\cdot 744 \\ &= j^2 - 2^4\cdot 3\cdot 31\, j + 2^4\cdot 3^4\cdot 5^3,\end{aligned}$$

as advertised.

Step 5. F_2 is now written as a polynomial in two variables $F_2(x, j(\omega))$. It vanishes for $x = j(2\omega)$, so $F_2(j(\omega), j(\omega/2)) \equiv 0$ upon division of ω by 2, that is, $j(\omega/2)$ is a root of $F_2(j(\omega), x) = 0$. But $F_2(x, j(\omega))$ is likewise the field polynomial of $j(\omega/2)$, and since both $F_2(x, j(\omega))$ and $F_2(j(\omega), x)$ have top coefficient $+1$, by ex. 2, they must be the same: $F_2(x, j(\omega)) = F_2(j(\omega), x)$. This suffices to determine σ_2 and σ_3 in part:

$$\sigma_2 = 2^4\cdot 3\cdot 31 j^2 + \cdots \quad \text{and} \quad \sigma_3 = -j^3 + 2^4\cdot 3^4\cdot 5^3 j^2 + \cdots.$$

The rest of the computation requires more terms in the expansion of j at the cusp.

Exercise 3. Check the expansion up to three terms using the expression of $j = 1728 g_2^3/\Delta$ by means of null values from Section 3.3. *Warning*: The present parameter q is the old q^2.

Exercise 4. Finish the computation of σ_2 and σ_3.

4.12 Landen's Transformation

The modular equation of level 2 between $j(2\omega)$ and $j(\omega)$ implies the existence of a similar relation between the Jacobian moduli $k^2(2\omega)$ and $k^2(\omega)$. This is just the Gauss–Landen transformation of Section 2.3:

$$k(2\omega) = \frac{1-k'}{1+k'} = \frac{1-\sqrt{1-k^2(\omega)}}{1+\sqrt{1-k^2(\omega)}}.$$

The function $k^2(\omega)$ is different from 0 and 1 in the upper half-plane, so $k(\omega)$ and $k'(\omega) = \sqrt{1-k^2(\omega)}$ are taken to be single-valued there, though, unlike k^2, the function $k(\omega)$ *is not a modular function* of Γ_2, as is clear on general grounds and will be confirmed subsequently. Landen's rule will now be rederived in the modular style, for variety.

Proof. $k^2(\omega) = 16 q^{1/2} + \cdots$ at the cusp $\sqrt{-1}\infty$; see Section 9 under *second level*, keeping in mind that $q^{1/2} = q_\infty$. A single-valued root is now determined by the requirement $k(\omega) = 4q^{1/4} + \cdots$, which is to say $k = \vartheta_2^2(0)/\vartheta_3^2(0)$; similarly, $k'(\omega) = \vartheta_4^2(0)/\vartheta_3^2(0) = 1 - 8q^{1/2} + \cdots$ at the cusp; and it is easy to see from Table 4.2.1 that

$$k(\omega + 1) = \sqrt{-1}k(\omega)/k'(\omega) \quad \text{and} \quad k(-1/\omega) = k'(\omega);$$

in particular,

$$k'(\omega + 1) = \sqrt{1 + k^2(\omega)/[k'(\omega)]^2} = 1/k'(\omega),$$
$$k(\omega + 2) = \sqrt{-1}k(\omega + 1)/k'(\omega + 1) = -k(\omega).$$

Landen's rule states that the function

$$f(\omega) = k(2\omega)\frac{1 + k'(\omega)}{1 - k'(\omega)}$$

is identically 1. The verification is made in three easy steps.

Step 1. f is invariant under the substitution $[11/01]\colon \omega \mapsto \omega + 1$ and is turned upside down by the substitution $[10/21]\colon \omega \mapsto \omega/(2\omega + 1)$ in view of

$$k\left(\frac{\omega}{2\omega + 1}\right) = k'(-1 - 1/2\omega) = 1/k'(-1/2\omega) = 1/k(2\omega)$$

and

$$k'\left(\frac{\omega}{2\omega + 1}\right) = k(-2 - 1/\omega) = -k(-1/\omega) = -k'(\omega).$$

Step 2. f is of rational character at the cusps 0, 1, $\sqrt{-1}\infty$ of the fundamental cell $\mathfrak{F}_2$ relative to the parameters of Section 9; for example, at $\sqrt{-1}\infty$, the parameter is $q_\infty = e^{\pi\sqrt{-1}\omega} = q^{1/2}$, $k(2\omega) = 4q_\infty + \cdots$, and $k'(\omega) = 1 - 8q_\infty + \cdots$, so $f(\omega) = 4q_\infty \times 2 \times (8q_\infty)^{-1} = 1$ to leading order.

Exercise 1. Check that $f = 1$ to leading order at the other two cusps 0 and 1.

Step 3. The modified function $f + 1/f$ is now seen to be a pole-free modular function of second level; as such, it can only be constant ($= 2$). The rest will be obvious.

Exercise 2. Think over step 3.

Special values. Landen's transformation can be expressed in the form of a modular equation for $x = k(2\omega)$ and $y = k(\omega)$ with $F_2(x, y) = (1+x)^2(1-y^2) - (1-x)^2$; its simplicity compared to the modular equation of level 2 for j, expounded in Section 11, is striking. The same disparity presents itself at higher levels: The arithmetic character of k^2 is simpler than that of j; see Yui [1978] and also Kaltofen and Yui [1984]. This has the particular consequence that to compute special values of j it is often easier to find k^2 and then to use the relation

$$j = \left[1728 \times \frac{4}{27} = 2^8\right] \times (k^4 - k^2 + 1)^3 / k^4(1-k^2)^2,$$

noted in Section 4.9; see Section 6.9 for the computation of $j(\sqrt{-5}) = 2^6 \cdot 5 \times (5^2 \cdot 79 + 2^3 \cdot 3 \cdot 7\sqrt{5})$ in this style. Here are four simpler instances to be checked by you.

Exercise 3. Use the known value $j\left(\frac{1}{2}(-1+\sqrt{-3})\right) = 0$ to deduce $k^2\left(\frac{1}{2}(-1+\sqrt{-3})\right) = \frac{1}{2}(1-\sqrt{-3})$, $k^2(\sqrt{-3}) = \frac{1}{4}(2-\sqrt{3})$, and $j(\sqrt{-3}) = 2^4 \cdot 3^3 \cdot 5^3 = 54000$; compare example 2.17.3.

Exercise 4. $k = (\sqrt{2}-1)/(\sqrt{2}+1)$ is the modulus of the square lattice (see example 2.17.1). Deduce $j(2\sqrt{-1}) = 2^3 \cdot 3^3 \cdot 11^3 = 287496$ from that.

Exercise 5. Check that the pairs $j(\sqrt{-1}) = 2^6 \cdot 3^3 = 1728$, $j(2\sqrt{-1}) = 2^3 \cdot 3^3 \cdot 11^3$ and $j((-1+\sqrt{-3})/2) = 0$, $j(\sqrt{-3}) = 2^4 \cdot 3^3 \cdot 5^3$ satisfy the modular equation of level 2.

Exercise 6. $j(\sqrt{-2}) = 2^6 \cdot 5^3$ is found from $k(\sqrt{-2}) = \sqrt{2}-1$; see example 2.17.2. What is $j(2\sqrt{-2})$? *Answer*: $2^9(1-\sqrt{2})^6$.

The number $j(\omega)$ has a deep arithmetic significance when, as in the present examples, ω is a quadratic irrationality. That is the subject of Chapter 6 on imaginary quadratic number fields.

4.13 Modular Equations of Higher Levels

The same type of thing can be done for any integral substitution $[ab/cd]$ with fixed nonvanishing determinant $ad - bc = n > 0$ in place of $[20/01]$: $\omega \mapsto 2\omega$. The corresponding absolute invariants $x = j(\omega')$ and $y = j(\omega)$ satisfy a (universal) irreducible polynomial $F_n \in \mathbb{C}[x, y]$: $F_n(j(\omega'), j(\omega)) = 0$. This is the **modular equation of level n**. The geometric background will be treated first.

Covering Tori. Let $\mathbf{X}' = \mathbb{C}/\mathbb{L}'$ be an n-fold cover of $\mathbf{X} = \mathbb{C}/\mathbb{L}$ with compatible complex structure. The question is: What is the relation of the corresponding absolute invariants $j(\mathbf{X}')$ and $j(\mathbf{X})$? The projection $\mathbf{X}' \to \mathbf{X}$ lifts to a conformal map of the universal cover $\mathbb{C}$, as in Section 2.6. Now the lattice $\mathbb{L}'$ appears as a sublattice of index n in $\mathbb{L}$. The primitive periods of $\mathbb{L}'$ are related to those of $\mathbb{L}$ by

$$\omega_2' = a\omega_2 + b\omega_1,$$

$$\omega_1' = c\omega_2 + d\omega_1$$

with $a, b, c, d \in \mathbb{Z}$ and $ad - bc = n$, this number being the ratio of the areas of the associated fundamental cells. The presence of a common divisor m in a, b, c, d indicates the presence of a trivial intermediate cover of $\mathbf{X}$ by a conformally equivalent torus $\mathbb{C}/m\mathbb{L}$ with the same absolute invariant as $\mathbf{X}$, so it is no loss of generality to suppose that a, b, c, d are coprime. The period ratio $\omega' = \omega_2'/\omega_1'$ is related to the period ratio $\omega = \omega_2/\omega_1$ by the (fractional) substitution $[ab/cd]$, so $j(\omega') = j\left((a\omega + b)/(c\omega + d)\right)$ is the absolute invariant of $\mathbf{X}'$. The modular equation of level n expresses its relation to the absolute invariant $j(\omega)$ of $\mathbf{X}$. The special case $\omega' = n\omega$ is a variant of the problem of division of periods mentioned in Section 2.17.

Correspondences. The substitution $[ab/cd] \in GL(2, \mathbb{Z})$ with coprime $a, b, c, d \in \mathbb{Z}$ and determinant $ad - bc = n$ is a **modular correspondence of level n**. The derivation of the modular equation requires a preliminary discussion of these. Γ_1 acts upon them by *left* multiplication; in this way, any modular correspondence may be brought into one of the inequivalent **reduced forms** $[ab/0d]$ with coprime $a, b, d \in \mathbb{Z}$, $ad = n$, and $0 \le b < d$. For example, if $n = p$ is prime, then the reduced correspondences are $p + 1$ in number: They are $[p0/01]$ and $[1q/0p]$ for $0 \le q \le p$.

Proof. The value of c is first reduced to 0 by application of a left-hand modular substitution: if $c = 0$, do nothing; if $a = 0$, apply $[01/-10]$ to make $c = 0$; if neither a nor c is 0, let q be their highest common divisor and use $[ij/kl]$ with coprime $k = -c/q, l = a/q$, and such $i, j \in \mathbb{Z}$ as make $il - jk = 1$. Now $c = 0$ in every case, $ad - bc = n, d > 0$ can be insisted upon as it is only ± 1 times the substitution that counts, and b may be reduced modulo d to obtain $0 \le b < d$ by use of $[1j/01]$. The reduction is finished.

Exercise 1. Check that the reduced forms are inequivalent under the (left) action of Γ_1.

Γ_1 also acts upon the modular correspondences from the *right* producing a transitive permutation of the reduced forms.

Proof. It is to be shown that the special reduced correspondence $[n0/01]$ can be brought into coincidence with any other reduced form $[ab/0d]$ by the action of Γ_1 from *both* sides. To spell it out,

$$\begin{pmatrix} n & 0 \\ 0 & 1 \end{pmatrix} \cdot \begin{pmatrix} i & j \\ k & l \end{pmatrix} = \begin{pmatrix} ni & nj \\ k & l \end{pmatrix}$$

can be made to match

$$\begin{pmatrix} i' & j' \\ k' & l' \end{pmatrix} \cdot \begin{pmatrix} a & b \\ 0 & d \end{pmatrix} = \begin{pmatrix} i'a & i'b + j'd \\ k'a & k'b + l'd \end{pmatrix}$$

by choice of the modular substitutions $[ij/kl]$ and $[i'j'/k'l']$. This requires

$$i = \frac{i'a}{n} = \frac{i'}{d}, \quad j = \frac{i'b + j'd}{n} = \frac{ib + j'}{a}, \quad k = k'a, \text{ and } \quad l = k'b + l'd.$$

Let i be the part of d coprime to a and pick j coprime to i. This determines $i' = id$ and $j' = -ib + ja$, and these are coprime: If not, they have a common prime p and either p divides i and so also ja (but not j), violating the fact that i is coprime to a, or else p divides d (but not i) and so also first a and then b, violating the fact that a, b, d are coprime. But if i' and j' are coprime, then $i'l' - j'k' = 1$ can be achieved by choice of k' and l'. This determines $k = k'a$ and $l = k'b + l'd$; in particular,

$$il - jk = \frac{i'}{d}(k'b + l'd) - \frac{(i'b + j'd)}{n}k'a = i'l' - j'k' = 1$$

at no extra cost. That does the trick.

The Modular Equation. The (fractional) action of Γ_1 on the period ratio is now seen to permute the **reduced invariants** $j(\omega') = j(d^{-1}(a\omega + b))$ with coprime $a, b, d \in \mathbb{Z}$, $ad = n$, and $0 \le b < d$, with the result that the product

$$F_n(x) = \prod \left[x - j\left(\frac{a\omega + b}{d}\right) \right],$$

taken over the reduced correspondences, is invariant under Γ_1. The fact is that $F_n \in \mathbb{C}[j][x]$ much as for $n = 2$: Indeed, the coefficients of the several powers of x are (1) invariant functions of Γ_1, (2) pole-free in the open half-plane, and (3) of rational character in $q = e^{2\pi\sqrt{-1}\omega}$ at the cusp $\sqrt{-1}\infty$, being of rational character in $q^{1/n}$ and of period 1 in ω, and as such, they must be polynomials in the absolute invariant j itself. Besides, the action of Γ_1 effects a transitive permutation of the reduced invariants so F_n is irreducible over the ground field $\mathbf{K}(\Gamma_1) = \mathbb{C}(j)$. F_n is now seen to be the common field polynomial, over the ground field $\mathbb{C}(j)$, of all absolute invariants $j(\omega')$ of level n. In short, $F_n(x, j(\omega)) = 0$ is the modular equation of level n.

Exercise 2. The degree of F_n is $p+1$ if $n = p$ is prime. What is it for general n? *Hint*: Do it for a prime power to get the idea. *Answer*: $n\Pi(1 + p^{-1})$ with p running over the prime divisors of n.

Exercise 3. The vanishing of $F_n(j(\omega'), j(\omega))$ expresses the fact that $j(\omega')$ is one of the reduced invariants. Prove that this relation is reciprocal, that is, $F_n(j(\omega), j(\omega'))$ also vanishes. Conclude that $F_n(x, y) = F_n(y, x)$ as for $n = 2$. *Hint*: $F_n(j(n\omega), j(\omega)) = 0$ holds equally for ω and ω/n.

Exercise 4. The group $\Gamma_0(n)$ of modular substitutions $[ij/kl]$ with k divisible by n is the stabilizer of $j(n\omega)$ in the full modular group. Check it. What is the relation of the (left) quotient $\Gamma_0(n)\backslash\Gamma_1$ to the reduced correspondences of level n?

The explicit computation of F_2 was already not simple. The case $n \geq 3$ can only be worse and is not attempted here; but see Fricke [1928 (2): 371–459] for $n = 2, 4, \ldots, 32, 3, 9, 27$, and more, and Yui [1978] for $n = 11$. The coefficients of F_n grow rapidly with n. The biggest of their absolute values is the height h_n of F_n. Cohen [1984] proved that

$$\log\, h_n = 6n \prod_{p|n} \left(1 + \frac{1}{p}\right) \times \left[\log\, n - 2\sum_{p|n} \frac{\log\, p}{p} + O(1)\right];$$

in particular, $\log\, h_{2^n} \sim 9 \log\, 2 \times n \cdot 2^n$ so the biggest coefficient is comparable to $2^{9n \cdot 2^n}$.

4.14 Jacobi's Modular Equation

Abel [1827] and Jacobi [1829] studied the higher modular equations relating $k^2(p\omega)$ to $k^2(\omega)$ for primes $p \geq 3$. The case $p = 2$ is covered by Landen's transformation. The special case $p = 5$ is the key to Hermite's solution of the quintic expounded in Chapter 5. Jacobi [1881–91 (1): 29–48] treated the eighth roots $\sqrt[4]{k}(p\omega)$ and $\sqrt[4]{k}(\omega)$ for $p = 3, 5$. Sohnke [1836] did $p = 7, 11, 13, 17, 19$. Hannah [1928] presents a list of the modular equations known up to that time. Berndt [1991] discusses the modular equations that appear in Ramanujan's work. The present account, for $p \geq 3$, is adapted from Fricke [1922 (2): 495–502].

Step 1. k^2 does not vanish in the open upper half-plane $\mathbb{H}$ and so has a single-valued eighth root specified by its development at the cusp $\sqrt{-1}\infty$: $\sqrt[4]{k}(\omega) = \sqrt{2}q^{1/16} + \cdots$ with $q = e^{2\pi\sqrt{-1}\omega}$, as is the custom now. The present task is to compute its stability group $\mathfrak{K}_1$ in the modular group of second level Γ_2. It is to be proved that $\sqrt[4]{k}$ obeys the rule

$$\sqrt[4]{k}\left(\frac{a\omega+b}{c\omega+d}\right) = (-1)^{[a^2-1+ab]/8}\sqrt[4]{k}(\omega).$$

This permits the identification of $\mathfrak{K}_1$ as the substitutions of second level with $a^2 + ab \equiv 1 \bmod 16$; in particular, $\mathfrak{K}_1$ contains the group Γ_{16} of modular substitutions congruent to the identity mod 16.

Exercise 1. Check that $[ab/cd] \mapsto (-1)^{[a^2-1+ab]/8}$ is a character of Γ_2.

Proof of the substitution rule. By ex.1, it suffices to check the rule for the generators $[12/01]$ and $[10/21]$ of Γ_2. The first fixes the cusp $\sqrt{-1}\infty$ and $k(\omega+2) = -k(\omega)$, as in Section 12, so $\sqrt[4]{k}(\omega+2) = e^{\pi\sqrt{-1}/4}\sqrt[4]{k}(\omega)$, in agreement with $a^2 - 1 + ab = 2$. The second fixes the cusp 0 and so also the value $\sqrt[4]{k(0)}$, in agreement with $a^2 - 1 + ab = 0$.

Step 2. This step determines the stability group $\mathfrak{K}_p \subset \Gamma_2$ of $\sqrt[4]{k}(p\omega)$. A substitution $[ab/cd]$ of second level fixes this function if and only if

$$\omega \mapsto \frac{a\omega + pb}{(c/p)\omega + d}$$

fixes $\sqrt[4]{k}(\omega)$ itself. This requires $c/p \in \mathbb{Z}$ and $a^2 - 1 + abp \equiv 0 \bmod 16$, which may be reduced to $a^2 - 1 + ab \equiv 0 \bmod 16$ since, in the presence of the odd whole number a, either condition implies $b \equiv 0 \bmod 8$, p being odd. In short, $\mathfrak{K}_p$ is the group $\Gamma_0(p) \cap \mathfrak{K}_1$ of substitutions $[ab/cd]$, with $c \equiv 0 \bmod p$, stabilizing $\sqrt[4]{k}$.

Exercise 2. The requirement $c/p \in \mathbb{Z}$ was passed over rapidly. Think it through.

Step 3. Note that $\mathfrak{K}_1$ acts upon $\left(\frac{2}{p}\right)\sqrt[4]{k}(p\omega)$ to produce the $p+1$ distinct forms

$$\left(\frac{2}{p}\right)\sqrt[4]{k}(p\omega) \quad \text{and} \quad \sqrt[4]{k}\left(\frac{\omega+16q}{p}\right) \quad \text{for} \quad 0 \le q < p,$$

in which $\left(\frac{2}{p}\right) = (-1)^{(p^2-1)/8}$ is the quadratic residue symbol of ex. 3.10.2.

Proof. $\mathfrak{K}_p$ reproduces the first form. Now take $[ab/cd]$ belonging to $\mathfrak{K}_1$ but not to $\mathfrak{K}_p$, that is, with $a^2 - 1 + ab \equiv 0 \bmod 16$ but c indivisible by p. Then $16c$ is prime to p so $d \equiv 16cq \bmod p$ with $0 \le q < p$; also

$$p\frac{a\omega+b}{c\omega+d} = \frac{ap\dfrac{\omega+16q}{p} + b - 16aq}{c\dfrac{\omega+16q}{p} + \dfrac{d-16cq}{p}} = \frac{a'\omega'+b'}{c'\omega'+d'}$$

with a self-evident notation, and $[a'b'/c'd'] \in \Gamma_2$ by inspection. The upshot is

$$\left(\frac{2}{p}\right)\sqrt[4]{k}\left(p\frac{a\omega+b}{c\omega+d}\right) = (-1)^{[p^2-1+a^2p^2-1+abp]/8}\sqrt[4]{k}\left(\frac{\omega+16q}{p}\right),$$

and it is easy to see that the power of -1 is trivial: $a^2 - 1 + abp \equiv 0 \bmod 16$ and p is odd.

Exercise 3. Check that every $q = 0, 1, 2, \dots, p-1$ actually occurs.

Step 4. The type of polynomial familiar from Sections 11 and 13 is now introduced:

$$F_p(x) = \left[x - \left(\frac{2}{p}\right)\sqrt[4]{k}(p\omega)\right]\prod_{0\le q<p}\left[x - \sqrt[4]{k}\left(\frac{\omega+16q}{p}\right)\right].$$

Naturally, its coefficients are invariant under $\mathfrak{K}_1$. Now $\mathfrak{K}_1 \subset \Gamma_2$ is of index 8 in the latter because Γ_2 stabilizes k^2 while $\mathfrak{K}_1$ stabilizes its eighth root $\sqrt[4]{k}$. This means that Γ_2 acts upon $\sqrt[4]{k}$ by multiplication by eighth roots of unity, and all these actually occur since $\sqrt[4]{k}(\omega+2) = e^{\pi\sqrt{-1}/4}\sqrt[4]{k}(\omega)$. Now it is plain that eight copies of the fundamental cell $\mathfrak{F}_2$ provide a fundamental cell for $\mathfrak{K}_1$ and that $\sqrt[4]{k}$ takes on every complex value just once in that new cell, cusps included of course; in particular, $\sqrt[4]{k}$ is the absolute invariant of $\mathfrak{K}_1$, with the proper precautions as to the rational character of invariant functions at the cusps. It follows, as in Section 11, that the coefficients of F_p are *polynomials* in $\sqrt[4]{k}$ since they are (1) invariant functions of $\mathfrak{K}_1$, (2) of rational character at the cusps, and (3) pole-free where $\sqrt[4]{k}$ is finite. Now (1) is clear. The reasoning involved in (2) is typified by the discussion of the root sum for $p = 5$ at the cusp $\sqrt{-1}\infty$:

$$\sigma_1(\omega) = \left(\frac{2}{5}\right)\sqrt[4]{k}(5\omega) + \sqrt[4]{k}\left(\frac{\omega}{5}\right) + \cdots + \sqrt[4]{k}\left(\frac{\omega+64}{5}\right)$$

is of rational character in $q^{1/80}$ and of period 16 in ω, so it has to be of rational character in $\sqrt[4]{k} = \sqrt{2}q^{1/16} + \cdots$. The verification of (3) occupies the next exercise.

Exercise 4. $k^2(\omega) = \infty$ on the orbit of $\omega = 1$ under Γ_2. Prove that this comprises all ratios of odd coprime integers. Use this to finish the proof that the coefficients of F_p belong to $\mathbb{C}[\sqrt[4]{k}]$. *Hint*: The ratio e/f of odd coprime integers can be expressed as $(a+b)/(c+d)$ with $[ab/cd] \in \Gamma_2$ by choice of $a = e - b, d = f - c$, and $ef - 1 = bf + ce$ with $b, c \equiv 0 \bmod 2$.

The final point is that F_p is irreducible over the ground field $\mathbb{C}(\sqrt[4]{k})$ by the principle employed in Section 11, namely, the stabilizer $\mathfrak{K}_1$ of $\sqrt[4]{k}$ effects a transitive permutation of its roots. In short, *F_p is the field polynomial of $\sqrt[4]{k}(p\omega)$ over $\mathbb{C}(\sqrt[4]{k})$.*

Step 5. The more explicit notation $F_p(x) = F_p(x, \sqrt[4]{k}(\omega))$ is now introduced, and a number of useful supplementary rules are proved.

Rule 1. $F_p\left(x, \left(\frac{2}{p}\right) y\right) = \left(\frac{2}{p}\right) F_p(y, x)$.

Rule 2. $F_p(x, 1) = (x-1)^{p+1}$ or $(x-1)(x+1)^p$ according as $p \equiv \pm 1 \bmod 8$ or $p \equiv \pm 3 \bmod 8$.

Rule 3. $F_p(e^{\pi\sqrt{-1}p/4}x, e^{\pi\sqrt{-1}/4}y) = e^{\pi\sqrt{-1}(p+1)/4}F_p(x, y)$, that is, the term $x^a y^b$ figures in F_p if and only if $pa + b \equiv p + 1 \bmod 8$.

Proof of rule 1. This could be done as in Section 11 for j, but here another route is taken, for variety. The substitution $[ab/cd]$ with $a = -1, b = -16q, c = 0, d = -1$ stabilizes $\sqrt[4]{k}(p\omega)$, so

$$F_p\left(\left(\frac{2}{p}\right)\sqrt[4]{k}(p\omega - 16q), \sqrt[4]{k}(\omega)\right) = 0,$$

and this is changed into

$$F_p\left(\left(\frac{2}{p}\right)\sqrt[4]{k}(\omega), \sqrt[4]{k}\left(\frac{\omega + 16q}{p}\right)\right) = 0$$

by the substitution $\omega \mapsto p^{-1}(\omega + 16q)$. The further substitution $[ab/cd]$ with $a = 1, b = 8p, c = 0, d = 1$ satisfies $a^2 - 1 + ab = 8p \equiv 8 \bmod 16$ and flips the signs of both $\sqrt[4]{k}(\omega)$ and $\sqrt[4]{k}\left(p^{-1}(\omega + 16q)\right)$, so

$$F_p\left(\sqrt[4]{k}(\omega), \left(\frac{2}{p}\right)\sqrt[4]{k}\left(\frac{\omega + 16q}{p}\right)\right) = 0$$

too, whether $\left(\frac{2}{p}\right) = +1$ or -1; similarly, the substitution $\omega \mapsto p\omega$ converts $F_p\left(\sqrt[4]{k}\left(\frac{\omega}{p}\right), \sqrt[4]{k}(\omega)\right) = 0$ into $F_p\left(\sqrt[4]{k}(\omega), \left(\frac{2}{p}\right) \times \left(\frac{2}{p}\right)\sqrt[4]{k}(p\omega)\right) = 0$. The upshot is that $F_p\left(\sqrt[4]{k}, \left(\frac{2}{p}\right)x\right)$ vanishes at the roots of $F_p(x, \sqrt[4]{k}) = 0$, and as the latter is the field polynomial, so it must divide the former. Now the degree of $F_p\left(\sqrt[4]{k}, \left(\frac{2}{p}\right)x\right)$ is precisely $p+1$, as follows by examination of the coefficient of $F_p\left(x, \left(\frac{2}{p}\right)\sqrt[4]{k}\right)$ of highest degree at the cusp 1:

$$\left(\frac{2}{p}\right)\sqrt[4]{k}(p\omega) \times \prod_{0\leq q<p} \sqrt[4]{k}\left(\frac{\omega+16q}{p}\right) = \left(\frac{2}{p}\right)[\sqrt[4]{k}(\omega)]^{p+1} + \cdots.$$

This may be checked in two steps. At the cusp 1, each factor to the left blows up since its argument belongs to the orbit of 1 under Γ_2 (see ex. 4), and the rule of step 1 together with the expansion $-e^{\pi\sqrt{-1}/(\omega-1)} + \cdots$ of $k^2(\omega)$ at the cusp (see Section 9) confirms that the whole product is of degree $p+1$ in $\sqrt[4]{k}$. $F_p\left(\sqrt[4]{k}, \left(\frac{2}{p}\right)x\right)$ and $F_p\left(x, \sqrt[4]{k}\right)$ are now seen to be proportional, and the numerical factor $\left(\frac{2}{p}\right)$ is elicited by use of the expansion $\sqrt[4]{k}(\omega) = \sqrt{2}e^{\pi\sqrt{-1}\omega/8} + \cdots$ at the cusp $\sqrt{-1}\infty$. The upshot is that $F_p\left(\sqrt[4]{k}, \left(\frac{2}{p}\right)x\right)$ is just $\left(\frac{2}{p}\right) \times F_p(x, \sqrt[4]{k})$, as stated.

Proof of rule 2. $\sqrt[4]{k}(0) = 1$, so $F_p(x, \sqrt[4]{k})$ reduces to $F_p(x, 1)$ at $\omega = 0$ and its roots are $\left(\frac{2}{p}\right)$ and $\sqrt[4]{k}(16q/p)$ for $0 \leq q < p$. Take $[ab/cd] \in \Gamma_2$ with $a \equiv p \bmod 8$, $b = 16q$, $c \equiv 0 \bmod 2$, and $d = p$; this can be done since $p^2 \equiv 1 \bmod 8$, as you will check. Then $[ab/cd]$ maps 0 to $b/d = 16q/p$, and $a^2 - 1 + ab \equiv p^2 - 1 \bmod 16$, so

$$\sqrt[4]{k}\left(\frac{16q}{p}\right) = (-1)^{[p^2-1]/8} = \left(\frac{2}{p}\right)$$

by the rule of step 1. The rest will be plain.

Exercise 5. Prove rule 3. *Hint*: Look at the action of [12/01] and use step 1.

Special Cases. For $p = 3$, 2 is a quadratic nonresidue, so the residue symbol $\left(\frac{2}{3}\right)$ is -1, and $F_3(y, x) = -F_3(x, -y)$ by rule 1. Now

$$F_3(x, y) = x^4 - \sigma_1 x^3 + \sigma_2 x^2 - \sigma_3 x + \sigma_4$$

is of degree $p + 1 = 4$ in x with coefficients of degree ≤ 4 from $\mathbb{C}[y]$, and the requirement of rule 3, that the only permissible terms $x^a y^b$ satisfy $3a + b \equiv 4 \bmod 8$, already provides a lot of information. For instance, σ_1 is a multiple of y^3 since $a = 3$ and $b \equiv -5 \equiv 3 \bmod 8$ requires b (≤ 4) to be 3;

$\sigma_2 = 0$ since $a = 2$ and $b \equiv -2 \equiv 6 \bmod 8$ cannot be < 4; σ_3 is a multiple of y since $a = 1$ and $b \equiv 1 \bmod 8$ requires $b = 1$; finally, $\sigma_4 = -y^4$ by rule 1. The result is that

$$F_3(x, y) = x^4 + ax^3y^3 + bxy - y^4$$

with constant a and b. Rule 2 does the rest:

$$F_3(x, 1) = (x - 1)(x + 1)^3 = x^4 + 2x^3 - 2x - 1,$$

so $a = 2$ and $b = -2$, the final result being

$$F_3(x, y) = x^4 + 2x^3y^3 - 2xy - y^4.$$

Exercise 6. Compute F_5 in the same style. *Answer*:

$$F_5(x, y) = x^6 - 4x^5y^5 + 5x^4y^2 - 5x^2y^4 + 4xy - y^6.$$

Exercise 7. Jacobi [1829] and Legendre [1825: 26–7] obtained an alternative form of the modular equation of third level:

$$\sqrt{k(\omega)k(3\omega)} + \sqrt{k'(\omega)k'(3\omega)} = 1.$$

Check it. *Hint*: $F_3(x, y)F_3(x, -y) = (1 - x^2y^2)^4 - (1 - x^8)(1 - y^8)$. Do the same for level 5 to obtain

$$F_5(x, y)F_5(x, -y) = (x^2 - y^2)^6 - 16x^2y^2(1 - x^8)^4 \cdot (1 - y^8)^4.$$

Exercise 8. Jacobi and Legendre's form of the modular equation at level 3 is equivalent to the identity of null values:

$$\vartheta_3(0 \mid \omega)\vartheta_3(0 \mid 3\omega) = \vartheta_2(0 \mid \omega)\vartheta_2(0 \mid 3\omega) + \vartheta_4(0 \mid \omega)\vartheta_4(0 \mid 3\omega);$$

compare Section 3.9 under "Landen's transformation." Check that this identity is the same as

$$2\sum q^{m^2+3n^2} = \sum q^{(a+1/2)^2+3(b+1/2)^2},$$

in which the left sum is taken over $(m, n) \in \mathbb{Z}^2$ of opposite parity and the right sum is taken over the whole of $\mathbb{Z}^2$. Give a direct proof. Hardy [1940: 218] explains this and more. *Hint*: $m + n = a + b + 1$ and $m - n = 2b + 1$.

4.15 Jacobi and Legendre's Derivation: Level 5

Jacobi [1829] and Legendre [1825] employed a very different derivation of the modular equation of level 5, based upon Jacobi's function sin amp. The procedure is outlined in five steps; compare Cayley [1895: 191–4] for this and for similar derivations at higher level.

Step 1. Let $\mathbf{x} = \mathrm{sn}(x, k(\omega/5))$ and $\mathbf{y} = \mathrm{sn}(x, k(\omega))$. The periods of $\mathbf{y}$ are $4K$ and $2\sqrt{-1}K$; the argument of $\mathbf{x}$ is adjusted so that its periods are $4K$ and $(1/5)\times 2\sqrt{-1}K'$. Then $\mathbf{x}$ takes on the value $+1$ with multiplicity 2 at the points

$$K + (n/5)\,2\sqrt{-1}K' \ (0 \le n < 5)$$

in the fundamental cell of $\mathbf{y}$; similarly, it takes the value -1 at the points $3K + (n/5)\,2\sqrt{-1}K'(0 \le n < 5)$. The values of $\mathbf{y}$ at these two series of points are, respectively, $+1$ and -1 taken with multiplicity 2 for $n = 0$; $\pm a$ taken with multiplicity 1 for $n = 1$ and 4, and $\pm b$ taken with multiplicity 1 for $n = 2$ and 3, the numbers $1, a, b$ being distinct and different from 0. The upshot is that

$$\frac{1-\mathbf{x}}{1+\mathbf{x}} = \frac{1-\mathbf{y}}{1+\mathbf{y}} \times \frac{(B\mathbf{y}^2 - A\mathbf{y} + 1)^2}{(B\mathbf{y}^2 + A\mathbf{y} + 1)^2}$$

with $A/B = a + b$ and $B = 1/ab$; indeed, the quotient of the two expressions is constant by inspection of roots and poles, and the fact that $\mathbf{x} = \mathbf{y} = \infty$ at $\sqrt{-1}K'$ forces the constant to be 1. Now, it is easy to check that $\mathbf{x} = 0$ at $\mathbf{y} = 0$ and at the roots of

$$B^2\mathbf{x}^4 + (A^2 + 2AB + 2B)\mathbf{x}^2 + 1 + 2A = 0 \tag{1}$$

and that $\mathbf{x} = \infty$ at $\mathbf{y} = \infty$ and at the roots of

$$(B^2 + 2AB)\mathbf{y}^4 + (A^2 + 2A + 2B)\mathbf{y}^2 + 1 = 0. \tag{2}$$

Step 2. Translation by $\sqrt{-1}K$ changes $\mathbf{y}$ into $-[k(\omega)\mathbf{y}]^{-1}$, as can be seen from the value distribution of $\mathbf{y}$ and the differential equation for sin amp; similarly, $\mathbf{x}$ is changed into a multiple of $\mathbf{x}^{-1}$, so $\mathbf{y}$ is a root of (1) if and only if $-[k(\omega)\mathbf{y}]^{-1}$ is a root of (2). This provides two identities for $k = k(\omega)$ by comparison of like powers of $\mathbf{y}$:

$$k^2(A^2 + 2AB + 2B) = B^2(A^2 + 2A + 2B) \tag{3}$$

and

$$k^4(1 + 2A) = B^2(B^2 + 2AB). \tag{4}$$

Step 3. The differential $dx = [(1-\mathbf{y}^2)(1-k^2(\omega)\mathbf{y}^2)]^{-1/2}d\mathbf{y}$ of the parameter x of the universal cover $\mathbb{C}$ is proportional to $dy = [(1-\mathbf{x}^2)(1-k^2(\omega/5)\mathbf{x}^2)]^{-1/2}d\mathbf{x}$: In fact, $dx = (1+2A)dy$ by evaluation of $\mathbf{x}/\mathbf{y}$ at $\mathbf{x} = \mathbf{y} = 0$, with the help of the formula for $(1-\mathbf{x})(1+\mathbf{x})^{-1}$ displayed in step 1. The slope $d\mathbf{x}/d\mathbf{y}$ can also be computed from that identity after the extraction of square roots:

$$\frac{d\mathbf{x}}{(\mathbf{x}+1)\sqrt{1-\mathbf{x}^2}}$$

$$= \left[\frac{1}{(\mathbf{y}+1)\sqrt{1-\mathbf{y}^2}}\frac{B\mathbf{y}^2 - A\mathbf{y}+1}{B\mathbf{y}^2 + A\mathbf{y}+1} - \sqrt{\frac{1-\mathbf{y}}{1+\mathbf{y}}}\frac{2A(B\mathbf{y}^2-1)}{(B\mathbf{y}^2+A\mathbf{y}+1)^2}\right] d\mathbf{y},$$

producing the further identity

$$(1+2A)\frac{k(\omega/5)}{k(\omega)} = 1 + \frac{2A}{B}, \tag{5}$$

by comparison of slopes at $\mathbf{x} = \mathbf{y} = \infty$.

Step 4. The last two expressions of step 3 are now combined to evaluate A and B in terms of $X = \sqrt[4]{k}(\omega)$ and $Y = \sqrt[4]{k}(\omega/5)$:

$$A = \frac{1}{2}\frac{Y}{X} \bullet \frac{Y^4 - X^4}{YX^3 - 1}, \qquad B = \frac{Y^5}{X}.$$

Step 5. Insert these values into (3); after some manipulation, it develops that

$$X^6 - 4X^5Y^5 + 5X^4Y^3 - 5X^2Y^4 + 4XY - Y^6 = 0,$$

which is just the modular equation of level 5.

Exercise 1. Derive the modular equation of level 3 in the same style. *Hints*: Form $\mathbf{x}$ and $\mathbf{y}$ as for level 5 but with the moduli $k(\omega/3)$ and $k(\omega)$. They are related by

$$\frac{1-\mathbf{x}}{1+\mathbf{x}} = \frac{1-\mathbf{y}}{1+\mathbf{y}} \bullet \left[\frac{1-c\mathbf{y}}{1+c\mathbf{y}}\right]^2$$

with $c \neq 1$. Deduce $k^2(\omega) = c^3(c+2)/(2c+1)$ and $k^2(\omega/3) = c\,[(c+2)/(2c+1)]^3$. Now eliminate c.

Exercise 2. Check the details for level 5. This is not so easy but is instructive.

4.16 Arithmetic Subgroups: Overview

The modular group Γ_1 contains a bewildering variety of subgroups Γ_0, such as the principal congruence subgroups Γ_n of level $n \geq 2$ specified by $[ab/cd] \equiv [10/01] \bmod n$, the group $\Gamma_0(n)$ of ex. 13.4 specified by $c \equiv 0 \bmod n$, and so forth. If Γ_0 is of finite index n in Γ_1, then it has a fundamental cell $\mathfrak{F}_0$ comprised of n copies of $\mathfrak{F}_1$, in the shape of a hyperbolic polygon with corners and/or cusps. Each of these special points may be equipped with a local parameter that opens up the neighborhood into a little disk, whereupon the quotient $\mathbb{H}/\Gamma_0$+cusps appears as a handlebody $\mathbf{X}_0$, and the class of invariant functions of Γ_0 having the proper rational character at corners and cusps appears as the associated function field $\mathbf{K}(\mathbf{X}_0)$. An inclusion of subgroups $\Gamma \subset \Gamma_0$ with finite index d means that the big curve $\mathbf{X} = \mathbb{H}/\Gamma$+cusps is a d-fold cover of the little curve $\mathbf{X}_0 = \mathbb{H}/\Gamma$+cusps and that the big field $\mathbf{K}(\mathbf{X})$ is an extension of the ground field $\mathbf{K}(\mathbf{X}_0)$ of degree d. This produces lots of intricate relations between invariant functions, exemplified by the modular equations for j and k^2. The rest of the section illustrates what happens next; see, for example, Serre [1973] and Shimura [1971] for more information.

Genus. The computation of the genus of $\mathbf{X}_0$ is illustrated by the group $\Gamma_0 = \Gamma_0(p)$ for prime $p \geq 2$. It is the stability group of $j(p\omega)$ in the full modular group and so of index $p+1$ in the latter. For the computation of the genus, it is simpler to deal with the conjugate group $[0-1/10]\Gamma_0(p)[0-1/10]$ specified by $b \equiv 0 \bmod p$. This is denoted by Γ_0 from now on. Now $\mathbf{X}_0 = \mathbb{H}/\Gamma_0+$ cusps may be viewed as the $p+1$ copies of $\mathfrak{F}_1 + \sqrt{-1}\infty$ produced by the substitutions $\omega \mapsto -1/\omega$ and $\omega \mapsto \omega + q$ $(0 \leq q < p)$, with edge identifications provided by Γ_0. It is ramified over a point of the projective line $\mathbf{X}_1 = \mathbb{H}/\Gamma_1 + \sqrt{-1}\infty$ if and only if Γ_0 identifies some or more points of the fiber $\omega_\infty = -1/\omega$, $\omega_q = \omega + q$ $(0 \leq q < p)$ over the projection $\omega \in \mathfrak{F}_1 + \sqrt{-1}\infty$ down below, and a count of such ramifications permits you to compute the genus of $\mathbf{X}_0$ by means of the Riemann–Hurwitz formula of Section 1.12: ramification index = 2 (sheet number + genus -1).

Exercise 1. $\mathbf{X}_0$ is ramified over $\omega \in \mathfrak{F}_1 + \sqrt{-1}\infty$ only if ω is fixed by some modular substitution, that is, only if $\omega = \infty, \rho = e^{2\pi\omega/3}$, or $\sqrt{-1}$. Why? Check that $\sqrt{-1}$ is fixed only by $[0-1/10]$ and ρ only by $[11/-10]$ and $[0-1/11]$.

Now for the count.

Step 1. The fiber over ∞ is $-1/\infty = 0$ together with p copies of ∞, so it produces a single ramification of index $p - 1$.

Step 2. The fiber over $\sqrt{-1}$ is $\omega_\infty = -1/\sqrt{-1} = \sqrt{-1}$ together with $\omega_q = \sqrt{-1} + q\,(0 \le q < p)$. Now the identification of ω_m and ω_n for $0 \le m < n < p$ requires

$$\begin{pmatrix} 1 & n \\ 0 & 1 \end{pmatrix} \sqrt{-1} = \begin{pmatrix} a & b \\ c & d \end{pmatrix} \begin{pmatrix} 1 & m \\ 0 & 1 \end{pmatrix} \sqrt{-1}$$

with $[ab/cd] \in \Gamma_0$, which is to say that

$$\begin{pmatrix} 1 & -n \\ 0 & 1 \end{pmatrix} \begin{pmatrix} a & b \\ c & d \end{pmatrix} \begin{pmatrix} 1 & m \\ 0 & 1 \end{pmatrix}$$

fixes $\sqrt{-1}$ and so coincides with $[0 - 1/10]$. Then

$$\begin{pmatrix} a & b \\ c & d \end{pmatrix} = \begin{pmatrix} 1 & n \\ 0 & 1 \end{pmatrix} \begin{pmatrix} 0 & -1 \\ 1 & 0 \end{pmatrix} \begin{pmatrix} 1 & -m \\ 0 & 1 \end{pmatrix} = \begin{pmatrix} * & -mn - 1 \\ * & * \end{pmatrix},$$

so you must have $mn \equiv -1 \bmod p$ and, in particular, $m \ne 0$. This shows that the coincidence $\omega_\infty = \omega_0$ represents a simple ramification of index 1 and that you have as many additional simple ramifications as there are pairs $1 \le m < n < p$ with $mn \equiv -1 \bmod p$, that is, $(1/2)(p - 1 - n_1)$, n_1 being the number of quadratic residues of $-1 \bmod p$, as you will check.

Step 3. The fiber over $\rho = e^{2\pi\sqrt{-1}/3}$ is $\omega_\infty = e^{\pi\sqrt{-1}/3}$ together with $\omega_q = e^{2\pi\sqrt{-1}/3} + q\,(0 \le q < p)$, and ω_∞ coincides with $-1/\omega_0 = \omega_1$. Now the identification of ω_m and ω_n for $0 \le m < n < p$ requires that

$$\begin{pmatrix} 0 & -n \\ 0 & 1 \end{pmatrix} \begin{pmatrix} a & b \\ c & d \end{pmatrix} \begin{pmatrix} 1 & m \\ 0 & 1 \end{pmatrix}$$

fixes ρ and so coincides with $[11/-10]$ or $[0-1/11]$, which is to say, either $mn - m \equiv -1 \bmod p$ or $mn - n \equiv -1 \bmod p$. The special values $m = 0$ or 1 produce a triple identification of ω_∞, ω_0, and ω_1, the rest of the fiber breaking up into singlets or triplets according as $q^2 - q \equiv -1 \bmod p$ or not. The moral is that the present ramifications are $1 + (1/3)(p - 2 - n_2)$ in number, of common index 2, n_2 being the number of solutions of $q^2 - q \equiv -1 \bmod p$.

Exercise 2. Check the details.

Step 4. Now the genus is presented. There are $p+1$ sheets, so

$$\begin{aligned} g &= -p + \frac{1}{2} \times \text{ the ramification index} \\ &= -p + \frac{1}{2}(p-1) + \frac{1}{2} + \frac{1}{4}(p-1-n_1) + 1 + \frac{1}{3}(p-2-n_2) \\ &= \frac{p+1}{12} - \frac{1}{4}n_1 - \frac{1}{3}n_2. \end{aligned}$$

Exercise 3. Check that $g = 0$ for $p = 2, 3, 5, 7, 13$, but not for $p = 11$. What is it then? What kind of function field is it now?

If $\mathbf{x}_0$ has genus 0, then $\mathbf{K}(\mathbf{X}_0)$ is a *rational* function field, which is to say that Γ_0 has an absolute invariant. This is not the case for $p = 11$.

Exercise 4. Prove that the absolute invariant of $\Gamma_0(2)$ is

$$j_2(\omega) = \Delta(2\omega)/\Delta(\omega) = 2^{-16}k^4(1-k^2)^{-1}.$$

Hint: j_2 is of degree 2 in k^2 and $\Gamma_2 \subset \Gamma_0(2)$ is of index 2 in the latter.

Exercise 5. The absolute invariant of $\Gamma_0(7)$ is

$$j_7(\omega) = [\Delta(7\omega)/\Delta(\omega)]^{1/6}.$$

Why? What is the general rule for $p = 2, 3, 5, 7, 13$? *Hint*: See sample 3.5.4 to understand the role of 24, which the numbers $p-1 = 1, 2, 3, 4, 6, 12$ divide. Note that $9-1=8$ and $25-1=24$ also divide 24. What does this do for $\Gamma_0(9)$ and $\Gamma_0(25)$?

Higher Genus. Fricke [1922: 371–459] worked out numerous examples of higher genus. The recipe he favors is illustrated by $p = 11$ for which the genus is 1; see Fricke [1928: 428–32] for more details and Birch [1973] for an entertaining supplement.

The story begins with the sum

$$f_1(\omega) = \sum_{\mathbb{Z}^2} q^{(m^2+mn+3n^2)}$$

which is of arithmetic interest in that it counts the number of representations of whole numbers by the quadratic form $m^2 + mn + 3n^2$ of discriminant -11. It obeys the rule

$$f_1\left(\frac{a\omega+b}{c\omega+d}\right) = \left(\frac{a}{11}\right)(c\omega+d)f_1(\omega) \quad \text{for} \quad [ab/cd] \in \Gamma_0(11);$$

in particular, f_1 is an improper form of $\Gamma_0(11)$ of weight $1/2$; the cautionary adjective improper alludes to a possible change of sign. The full proof of this rule is complicated, but here is a sample with $a = 1,\ b = 0, c = 11, d = 1$. Let $Q(\mathbf{n})$ be a positive quadratic form in $\mathbf{n} = (m, n) \in \mathbb{Z}^2$, Q^{-1} the form inverse to Q, and ϑ_Q the sum of $\exp(2\pi\sqrt{-1}\omega Q(\mathbf{n}))$ over $\mathbb{Z}^2$. The Poisson summation formula of Section 3.9, properly extended to dimension 2, yields the rule

$$\vartheta_Q(-1/\omega) = (-\det Q)^{-1/2}\vartheta_{Q^{-1}}(\omega).$$

Now $f_1 = \vartheta_Q$ for $Q(m,n) = m^2 + mn + 3n^2$, and $Q^{-1} = Q/11$, up to an exchange of variables, so

$$\begin{aligned} f_1\left(\frac{\omega}{11\omega+1}\right) &= -\frac{11\omega+1}{\omega\sqrt{-11}} \times f_1\left(-\frac{11\omega+1}{11\omega}\right) \\ &= \frac{-11\omega+1}{\omega\sqrt{-11}} \times f_1\left(\frac{-1}{11\omega}\right) \end{aligned}$$

in view of $f_1(\omega+1) = f_1(\omega)$. The proof is finished by a reprise:

$$f_1\left(\frac{-1}{11\omega}\right) = \frac{11\omega}{\sqrt{-11}} f_1(\omega) \quad \text{so} \quad f_1\left(\frac{\omega}{11\omega+1}\right) = (11\omega+1) f_1(\omega),$$

in conformity with the fact that 1 is a quadratic residue of 11.

Now the same rule applies to the sum

$$f_2(\omega) = \sum_m (-1)^n q^{m^2+mn+3n^2}, \qquad m \text{ odd}$$

and it is a fact, of the same variety as the fundamental lemma of Section 7, that such forms as f_1 and f_2 vanish to the total degree 1 in any fundamental cell plus cusps of $\Gamma_0(11)$. It follows that $\mathbf{x} = f_1^2/f_2^2$ is a single-valued function in the fundamental cell, of degree 2, having one root of that degree. Now $f_3 = 11^2 g_2(11\omega) - g_2(\omega)$ is a *proper* form of $\Gamma_0(11)$ of weight 2, and it is an instance of a general principle that it can be expressed in terms of the proper forms f_1^2 and f_2^2 of weight 1, much as in Section 7. The result is that f_3^2 is a constant multiple of

$$f_2^2(f_1^6 - 2^2 \cdot 5 f_1^4 f_2^2 + 2^3 \cdot 7 f_1^2 f_2^4 - 2^2 \cdot 11 f_2^6),$$

whence

$$\mathbf{y} = f_3/f_2^4 = \sqrt{\mathbf{x}(\mathbf{x}^3 - 20\mathbf{x}^2 + 56\mathbf{x} - 44)}$$

is a single-valued function in the fundamental cell, of degree 4. The fact that $\mathbf{x}$ is of degree 2 now implies that the map from the fundamental cell to the pair $\mathfrak{p} = (\mathbf{x}, \mathbf{y})$ is (mostly) 1:1, permitting the identification of $\mathbb{H}/\Gamma_0(11)+$

cusps with the elliptic curve $\mathbf{X}$: $\mathbf{y}^2 = \mathbf{x}(\mathbf{x}^3 - 20\mathbf{x}^2 + 56\mathbf{x} - 44)$, and of the field of invariant functions of $\Gamma_0(11)$ with $\mathbb{C}(\mathbf{x})[\sqrt{\mathbf{x}(\mathbf{x}^3 - 20\mathbf{x}^2 + 56\mathbf{x} - 44)}]$; in particular, the absolute invariant j of Γ_1 is rationally expressible (with tears) in terms of $\mathbf{x}$ and $\mathbf{y}$:

$$j = \frac{4(61\mathbf{x}^2 - 2^4 \cdot 23\mathbf{x} + 2^5 \cdot 11 - 2^2 \cdot 3 \cdot 5\mathbf{y})^3}{(\mathbf{x}^3 - 3 \cdot 7\mathbf{x}^2 + 2^3 \cdot 11\mathbf{x} + \mathbf{xy} - 11\mathbf{y})^2}.$$

Singular Values. The formula for j leads to numerical values of the absolute invariant at the ramifications of $\mathbf{X}$ over the projective line $\mathbb{P}^1 = \mathbf{x}(\mathbf{X})$. These so-called **singular values** occur at points where $\mathbf{y}$ vanishes and have to do with the imaginary quadratic field $\mathbb{Q}(\sqrt{-11})$, of which more to come in Section 6.9. The form f_3 vanishes at the points

$$\omega_1 = \frac{\sqrt{-1}}{\sqrt{11}}, \ \omega_2 = \frac{-1}{2} + \frac{\sqrt{-1}}{2\sqrt{11}}, \ \omega_3 = \frac{-1}{3} + \frac{\sqrt{-1}}{3\sqrt{11}}, \ \omega_4 = \frac{-1}{4} + \frac{\sqrt{-1}}{4\sqrt{11}},$$

as you may check by writing

$$g_2(11\omega) = (11\omega \cdot c + d)^{-4} g_2\left(\frac{11\omega \cdot a + b}{11\omega \cdot c + d}\right)$$

and requiring that $(c \cdot 11\omega + d)^4 = 11^2$ and that $[ab/cd]$ fix ω. Γ_1 carries these points to inequivalent points of $\mathfrak{F}_1$:[2]

$$\sqrt{-11}, \quad \frac{-1+\sqrt{-11}}{2}, \quad \frac{-1+\sqrt{-11}}{3}, \ \frac{1+\sqrt{-11}}{3},$$

so they are all the more inequivalent under $\Gamma_0(11)$ and represent all four roots of $f_3 = 0$ per cell. Now $\mathbf{x}$ cannot vanish at ω_1 since f_1 is real and positive for purely imaginary ω. To see where $\mathbf{x}$ *does* vanish, note first that $f_1(\omega)$ and $\omega^{-1} f_1(-1/11\omega)$ obey the same transformation rule under $\Gamma_0(11)$: $\left(\frac{a}{11}\right) = \left(\frac{d}{11}\right)$ since $ad \equiv 1 \bmod 11$, so their ratio is a single-valued function on $\mathbf{X}$, of degree 1, and can only be a (nonvanishing) constant. Now the reality of f_1 just cited requires that its roots be symmetric with regard to the imaginary axis, so if $\mathbf{x} = 0$ at $\omega_3 = -1/3 + \sqrt{-1}/3\sqrt{11}$, then it also vanishes at $\omega_3^+ = +1/3 + \sqrt{-1}/3\sqrt{11}$ and at $-1/\omega_3^+ = \omega_4$, which is impossible because $\mathbf{x}$ vanishes at just one place per cell. The same reasoning excludes ω_4, the upshot being that $\mathbf{x} = 0$ only at ω_2, and you obtain the value

$$j(\omega_2) = j\left(\frac{1}{2}(-1 + \sqrt{-11})\right) = -2^{15}$$

[2] $-\frac{1}{2} + \frac{\sqrt{-1}}{2\sqrt{11}} = \frac{-6}{11+\sqrt{-11}} \mapsto \frac{11+\sqrt{-11}}{6} \mapsto \frac{-1+\sqrt{-11}}{6} = \frac{-2}{1+\sqrt{-11}} \mapsto \frac{1+\sqrt{-11}}{2}$, for example.

by inserting $\mathbf{y} = \sqrt{-44}\mathbf{x}$ and $\mathbf{x} = 0+$ into the expression for j. The other three roots of $f_3 = 0$ come from $\mathbf{x}^3 - 20\mathbf{x}^2 + 56\mathbf{x} - 44 = 0$. The corresponding absolute invariants

$$j(\sqrt{-11}),\quad j\left(\frac{1}{3}(-1+\sqrt{-11})\right),\quad j\left(\frac{1}{3}(1+\sqrt{-11})\right)$$

are obtained by substituting those roots into the expression for j with $\mathbf{y} = 0$. Fricke, that indefatigable calculator, reports that the results are *etwas umständlich* (= a little involved). It would be prudent to believe him.

5

Ikosaeder and the Quintic

The title refers to Klein's little book *Ikosaeder* [1884] on the extraction of roots of the general equation of degree 5. Ruffini [1799] had indicated that this cannot be done by radicals alone, but it was Abel [1826] who found the first real proof; see Ayoub [1982] for complete details and Rosen [1995] for Abel's original proof. Hermite [1859] discovered how it *can* be done with the help of Jacobi's modulus k^2 and the modular equation of level 5 relating $\sqrt[4]{k}(5\omega)$ to $\sqrt[4]{k}(\omega)$. Kronecker [1858] had a variant; see also Brioschi [1858]. The icosahedral group A_5 of Section 1.7 is the Galois group of the general quintic and plays a central role, whence *Ikosaeder*. The present chapter tells the story from the start; compare Briot and Bouquet [1875: 654–60] and, for more information, see Klein [1884], Cole's instructive review [1887], and Green's modern geometrical account [1978] described in Section 7.

5.1 Solvability of Equations of Degree ≤ 4

The general equation of degree 2 is $P_2(x) = x^2 - c_1 x + c_2 = 0$. It is solvable by radicals as every schoolchild knows: The discriminant is $d = c_1^2 - 4c_2$ and the roots are $x = (1/2)(c_1 \pm \sqrt{d})$. Diophantus (250) knew this, but it was not until the sixteenth century that the general equation of degree 3 was solved by Tartaglia [1546] and Cardano [1545]. Ore [1965: 53–108] tells the story of their disreputable rivalry. The general equation $P_3(x) = x^3 - c_1 x^2 + c_2 x - c_3 = 0$ is reduced to $P_3(x) = x^3 + ax + b = 0$ by the substitution $x \mapsto x + c_1/3$ and the three roots are expressed as

$$x = \rho \sqrt[3]{-b/2 + \sqrt[2]{d}} + \rho^2 \sqrt[3]{-b/2 - \sqrt[2]{d}}$$

with ρ running over the three cube roots of unity $1, e^{2\pi\sqrt{-1}/3}, e^{4\pi\sqrt{-1}/3}$ and $d = a^3/27 + b^2/4 =$ the discriminant of the reduced cubic. Ferrara (1545) solved the general equation of degree 4: $P_4(x) = x^4 - c_1x^3 + c_2x^2 - c_3x + c_4 = 0$. The reduction to $c_1 = 0$ is achieved by the substitution $x \mapsto x + c_1/4$. The reduced equation is equivalent to $(x^2 + c_0)^2 = yx^2 + c_3x + c_0^2 - c_4$ with $c_0 = (1/2)(c_2 + y)$ and undetermined y. The discriminant of the right side is proportional to a cubic in y and can be made to vanish with the help of Cardano's rule. Then the right side has a double root x_0, and $x^2 + c_0 = \pm\sqrt{y}(x - x_0)$ may be solved by the extraction of one more square root. Now it stops: *It is impossible to extract the roots of the general equation of degree 5 or more by radicals alone*. The proof will be recalled in a moment.

5.2 Galois Groups Revisited

The principal tool for the study of polynomial equations is the Galois group. Birkhoff and Maclane [1948: 409–22], Stewart [1973], and Stillwell [1994] are recommended for proofs and more information; see also Ore [1957], Kollros's little biography of Galois [1949], and Bell [1937: 362–77] for historical details.

Galois Groups. Fix a **ground field** $\mathbf{K}_0$ such as the rational numbers $\mathbb{Q}$, or $\mathbb{Q}(\sqrt{5})$, or $\mathbb{C}(\sqrt[4]{k})$, and let $P(x) = x^d - c_1x^{d-1} + \cdots \pm c_d \in \mathbf{K}_0[x]$ be an irreducible polynomial of degree d with necessarily simple roots $x_1, \ldots, x_d$. The enlarged field $\mathbf{K} = \mathbf{K}_0(x_1, \ldots, x_d)$ obtained by their adjunction to $\mathbf{K}_0$ is the **splitting field** of P over $\mathbf{K}_0$; it is of dimension (degree) $d = [\mathbf{K}: \mathbf{K}_0]$ over the ground field. The **Galois group** $\mathfrak{G}$ of $\mathbf{K}$, or of P, over $\mathbf{K}_0$ is the class of automorphisms of the larger field $\mathbf{K}$ fixing the smaller field $\mathbf{K}_0$, and it is a fact that it fixes nothing more: Any element of $\mathbf{K}$ that is fixed by $\mathfrak{G}$ lies in $\mathbf{K}_0$. The old-fashioned hands-on view is that $\mathfrak{G}$ is the group of permutations of the d letters $x_1, \ldots, x_d$ distinguished by two competing features: (1) The value of any rational function $r(x_1, \ldots, x_d) \in \mathbf{K}_0(x_1, \ldots, x_d)$ that is fixed by $\mathfrak{G}$ lies in $\mathbf{K}_0$, and (2) any such function with values in $\mathbf{K}_0$ is so fixed. Feature (1) forces $\mathfrak{G}$ to be big, (2) forces it to be small, and the two together specify it unambiguously. The degree $[\mathbf{K}: \mathbf{K}_0] = d$ is likewise the number of substitutions in $\mathfrak{G}$; that is, it is the order $|\mathfrak{G}|$ of the group. It is a general principle that a faithful pairing exists between subgroups $\mathfrak{G}_p$ of $\mathfrak{G}$ and intermediate fields $\mathbf{K}_0 \subset \mathbf{K}_1 \subset \mathbf{K}$; see Fig. 5.1. The index $[\mathfrak{G}: \mathfrak{G}_1]$ of $\mathfrak{G}_1$ in $\mathfrak{G}$ is the degree $[\mathbf{K}_1: \mathbf{K}_0] = d$ of $\mathbf{K}_1$ over the ground field. $\mathbf{K}_1$ is a splitting field over $\mathbf{K}_0$ if and only if $\mathfrak{G}_1$ is an invariant subgroup of $\mathfrak{G}$; in this case, $\mathfrak{G}/\mathfrak{G}_1$ is the Galois group of $\mathbf{K}_1$ over $\mathbf{K}_0$, as the figure indicates.

Exercise 1. Let $\Gamma_0 \subset PSL(2, \mathbb{C})$ be any of the Platonic groups of Section 1.7 and j its absolute invariant. Check that Γ_0 is the Galois group of the equation

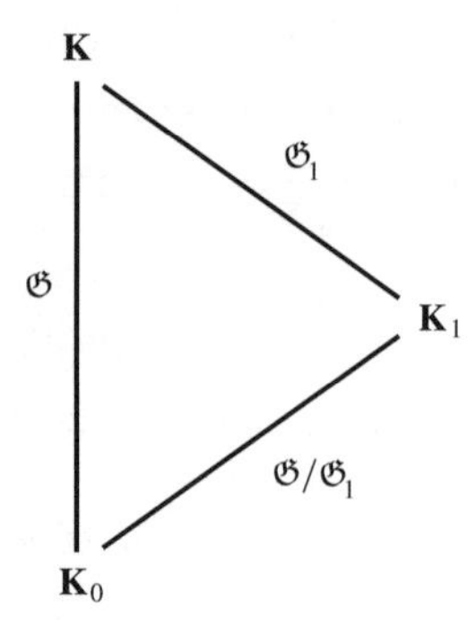

Figure 5.1. Intermediate fields.

$j(x) - y = 0$ over the ground field $\mathbf{K}_0 = \mathbb{C}(y)$. *Hint*: The totality of roots of $j(x) = y$ can be (rationally) produced from a single one. How?

Solving by Radicals. Let p be a rational prime and let $\mathbf{K}_0$ contain the pth roots of unity. Then $\mathfrak{G}$ is cyclic of order p if and only if $\mathbf{K}$ is obtained from $\mathbf{K}_0$ by the adjunction of a single radical $\sqrt[p]{x_0}$ of an element $x_0 \in \mathbf{K}_0$. This can be elevated to a general principle over any ground field $\mathbf{K}_0$ of this type: The equation $P(x) = 0$ is **solvable by radicals** in the sense that $\mathbf{K}$ is produced from $\mathbf{K}_0$ by successive adjunctions of roots $\sqrt[n]{x}$ if and only if $\mathfrak{G}$ is **solvable** in the sense that it contains a descending chain of subgroups $\mathfrak{G} \supset \mathfrak{G}_1 \supset \mathfrak{G}_2 \supset \cdots$ terminating at the identity, each invariant in the one before, with cyclic quotient.

The general equation of degree d is

$$P_d(x) = (x - x_1) \cdots (x - x_d) = x^d - c_1 x^{d-1} + c_2 x^{d-2} - \cdots \pm c_d = 0$$

with the independent indeterminates $x_1, \ldots, x_d$ as its roots; naturally, it splits in $\mathbf{K} = \mathbb{C}(x_1, \ldots, x_d)$ but not in the ground field $\mathbb{C}(c_1, \ldots, c_d)$ over which the game is played, $d = 1$ excepted. The quantities $c_1, \ldots, c_d$, exemplified by

$$c_1 = x_1 + x_2 + x_3, \quad c_2 = x_1 x_2 + x_2 x_3 + x_3 x_1, \quad c_3 = x_1 x_2 x_3$$

for $d = 3$, are the **elementary symmetric functions** of the roots, and it is a fact that every symmetric polynomial in the roots is a polynomial in these. It follows that the Galois group of P_d over $\mathbf{K}_0$ is the full symmetric group S_d. This is solvable if $d \le 4$ but not if $d = 5$ or more; in fact, S_5 contains just one intermediate invariant subgroup: the icosahedral group A_5 of permutations of parity $+1$ of Section 1.7, and this is simple, that is, it has no intermediate invariant subgroups at all, whence the discovery of Abel [1826] that *it is impossible to extract the roots of the general equation of degree 5 by radicals alone*.

Naturally, the solvability of special equations is not precluded; for example, if $c_1, \ldots, c_5$ take special values in $\mathbb{Q}$, then the Galois group of P_5 over *that* field can perfectly well be solvable, as for $x^5 = 1$.

Exercise 2. S_3 is solvable in two steps with quotients of order 2 and 3. Check it by means of Cardano's formula. *Hint*: The cube roots are determined so their product is $-a/3$.

Exercise 3. S_4 is solvable in four steps with quotients of order 2, 3, 2, 2. This means that $P_4(x) = 0$ can be solved by the adjunction of two square roots after extracting the roots of a cubic. Check it and compare what Ferrara actually did.

Exercise 4. The general cubic can be reduced to $4x^3 - 3x + y = 0$ by elementary substitutions. This is the problem of antiquity of trisection of angles: $x = \sin(\omega/3)$, $y = \sin(\omega)$. It is incapable of solution by ruler and compass. Why? Note for future reference that the other two roots of the cubic are $x_2 = \sin(\omega/3 + 2\pi/3)$ and $x_3 = \sin(\omega/3 + 4\pi/3)$.

Exercise 5. Look up the proof that A_5 and, indeed, all the higher alternating groups are simple. Stewart [1973: 129–31] has it.

5.3 The Galois Group of Level 5

The ideas of Section 2 are nicely illustrated by the modular equation of level 5 computed in ex. 4.14.6 and Section 4.15:

$$F_5(x, y) = x^6 - y^6 + 5x^2y^2(x^2 - y^2) - 4xy(x^4y^4 - 1) = 0.$$

It is to be proved that *the Galois group $\mathfrak{G}$ of F_5 over the ground field* $\mathbf{K}_0 = \mathbb{C}(y)$ *is just the ubiquitous icosahedral group A_5.*

Step 1. Let $y = \sqrt[4]{k}(\omega)$ so that the roots of $F_5(x, y) = 0$ can be expressed as in Section 4.14:

$$x_\infty = -\sqrt[4]{k}(5\omega) \text{ and } x_n = \sqrt[4]{k}\left(\frac{\omega}{5} + \frac{16n}{5}\right) \; (0 \le n < 5).$$

The group $\mathfrak{K}_1$ of modular substitutions $[ab/cd]$ of second level with $a^2 - 1 + ab \equiv 0 \bmod 16$ stabilizes y, while its subgroup $\mathfrak{K}_5$, distinguished by $c \equiv 0 \bmod 5$, stabilizes the **principal root** x_∞. The congruence $a^2 - 1 + ab \equiv 0 \bmod 16$ is the same as $a^2 - 1 + 5ab \equiv 0 \bmod 16$: a being odd, either implies that 8 divides b. Now the joint stabilizer of $x_\infty, x_0, x_1, x_2, x_3, x_4$ in $\mathfrak{K}_1$ is easily computed: Besides the condition $c \equiv 0 \bmod 5$, also $b \equiv 0 \bmod 5$ is required for the stabilization of x_0, and $d \equiv a \equiv 1 \bmod 5$ for the rest; in short, the joint stabilizer is just the invariant subgroup $\mathfrak{K}_1 \cap \Gamma_5$ of $\mathfrak{K}_1$, as you will check.

Step 2. Now the quotient group $\mathfrak{K}_1/\mathfrak{K}_1 \cap \Gamma_5$ acts faithfully on the splitting field $\mathbf{K} = \mathbf{K}_0(x_\infty, x_0, \ldots, x_4)$ of F_5, and the fact that $\sqrt[4]{k}$ is the absolute invariant of $\mathfrak{K}_1$, as explained in Section 4.14, permits you to identify the quotient as the Galois group $\mathfrak{G}$ of F_5 over $\mathbf{K}_0$. The point is that $x \in \mathbf{K}$ is fixed by $\mathfrak{K}_1$ if and only if it is a rational function of $\sqrt[4]{k} = y$.

Exercise 1. Check the last statement.

Step 3. A bit of the structure of $\mathfrak{G}$ is already apparent: $\mathfrak{K}_1 \supset \mathfrak{K}_5 \supset \mathfrak{K}_1 \cap \Gamma_5$ and $[\mathfrak{K}_1 : \mathfrak{K}_5] = 6$, as noted in Section 4.14, so

$$\begin{aligned}|\mathfrak{G}| &= 6 \times [\mathfrak{K}_5 : \mathfrak{K}_1 \cap \Gamma_5] \\ &= 6 \times \text{the number of distinct permutations of} \\ &\quad\ x_0, \ldots, x_4 \text{ produced by } \mathfrak{K}_5.\end{aligned}$$

The count of those permutations is done by hand. The substitution $[ab/cd] \in \mathfrak{K}_5$ sends x_m to x_n $(0 \le m < n < 5)$ if and only if

$$\sqrt[4]{k}\left(\frac{1}{5}\frac{a\omega + b}{c\omega + d} + \frac{16m}{5}\right) = \sqrt[4]{k}\left(\frac{a'(\frac{\omega}{5} + \frac{16n}{5}) + b'}{c'(\frac{\omega}{5} + \frac{16n}{5}) + d'}\right)$$

with $[a'b'/c'd'] \in \mathfrak{K}_5$. To spell it out,

$$\begin{aligned}a' &= a + 16mc \equiv 1 \bmod 2, \\ b' &= \frac{1}{5}[b + 16md - (a + 16mc)16n], \\ c' &= 5c \equiv 0 \bmod 2, \\ d' &= d - 16nc \equiv 1 \bmod 2,\end{aligned}$$

and for the inclusion $[a'b'/c'd'] \in \Gamma_2$, you need $b' \equiv 0 \bmod 2$, which is to say that 5 divides $b + 16md - 16na$, seeing as 5 divides c. This can be achieved by one and only one choice of $0 \le n < 5$, a being prime to 5 in view of $ad - bc = 1$ and $c \equiv 0 \bmod 5$. Then $[a'b'/c'd'] \in \Gamma_2$ as well; in fact $a^2 - 1 + ab \equiv 0 \bmod 16$ implies that 8 divides b, a being odd, so $b' \equiv \frac{1}{5}[b + 16md - 16na] \equiv b \bmod 16$, and $(a')^2 - 1 - a'b' \equiv a^2 - 1 + ab \equiv 0 \bmod 16$. The upshot is that the image of the root x_m under $[ab/cd] \in \mathfrak{K}_5$ coincides with the root x_n determined by

$n = n(ad - bc) \equiv nad \equiv 16nad \equiv bd + 16md^2 \equiv bd + md^2 \bmod 5$, whence the rule of permutation:

$$m \mapsto n \equiv bd + md^2 \bmod 5.$$

Exercise 2. Check that the rule produces just the 10 permutations of parity $+1$ listed here:

$$(01234),\quad (12340),\quad (23401),\quad (34012),\quad (40123),$$

$$(04321),\quad (43210),\quad (32104),\quad (21043),\quad (10432).$$

Hint: The first row is produced by $[ab/cd]$ with $a = 1, b = 16, c = 0, d = 1$, the second by $a = 3, b = -40, c = 10, d = -133$, and there are no other possibilities. Why?

The upshot is that $\mathfrak{G}$ is of order $6 \times 10 = 60$. This is enough to identify it as a copy of A_5, as will be seen in one more step.[1]

Step 4. $\mathfrak{G} \subset S_6$ is automatic and it is plain that either $\mathfrak{G} \subset A_6$ or else it contains an equal number of elements of parity $+1$ and -1, 30 of each. Now the stabilizer $\mathfrak{G}_\infty = \mathfrak{K}_5/\mathfrak{K}_1 \cap \Gamma_5$ of x_∞ in $\mathfrak{G}$ is of order 10 and contains the substitution $e = (\infty 12340) \in A_6$; see ex. 2. Let $f \in \mathfrak{G}$ move ∞ to 0, say. Then the six substitutions $1, f, ef, e^2 f, e^3 f, e^4 f$ map ∞ to $\infty, 0, 1, 2, 3, 4$ and so represent the six cosets of $\mathfrak{G}_\infty$ in $\mathfrak{G}$, with the contradictory implication that $\mathfrak{G}$ contains at least $50 = 5 \times |\mathfrak{G}_\infty|$ odd substitutions if f is odd (parity -1). This confirms $\mathfrak{G} \subset A_6$. Now $[A_6 : \mathfrak{G}] = 360/60 = 6$, and A_6 acts upon the six cosets of $\mathfrak{G}$ in A_6, viewed as six new letters. The simplicity of A_6 ensures that the action is faithful, and $\mathfrak{G}$ in its role of permuter of the new letters appears as the stability group of the identity coset. This identifies $\mathfrak{G}$ as a copy of A_5.

Depressing the Degree. The existence of elements of degree 5 in the splitting field $\mathbf{K}$ of F_5 over $\mathbf{K}_0$ is read off Fig. 5.2. The icosahedral group A_5 contains the tetrahedral group A_4, so $\mathbf{K}$ contains an intermediate field $\mathbf{K}_1$ of degree $[A_5 : A_4] = 60/12 = 5$ over $\mathbf{K}_0$, and this may be viewed as a simple extension of $\mathbf{K}_0$ produced by the adjunction of a single element x_0 *of degree* 5, by a general principle of field theory enunciated in Section 1.4.

Exercise 3. $\mathbf{K}$ contains the splitting field of x_0; in fact, it IS the splitting field. Why?

[1] T. Bickel and S. Whiteside helped with this.

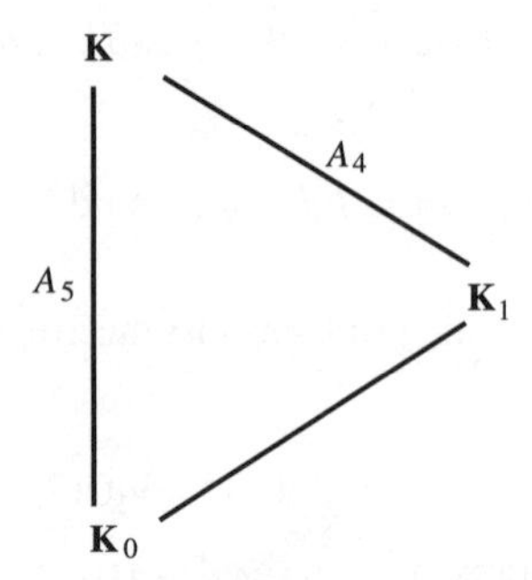

Figure 5.2. An element of degree 5.

Galois knew this;[2] indeed, he proved that the degree of the modular equation is *depressible* in this sense not only for $p = 5$ from 6 to 5, but also for $p = 7$ from 8 to 7, and for $p = 11$ from 12 to 11, but not for any larger prime. Galois identified the group of the modular equation of level n as $PGL(2, \mathbf{F}_p)$ = the group of 2×2 invertible matrices with coefficients from the finite field of p elements. The subgroup $PSL(2, \mathbf{F}_p)$ is of index 2 and it contains a subgroup of order p only for $p = 5, 7$, and 11. Galois stated, and Betti [1851] proved, that this is precisely what is needed to depress the degree.

5.4 An Element of Degree 5

The element $\mathfrak{x}_0 = (x_\infty - x_0)(x_1 - x_4)(x_2 - x_3)y$ fulfills the prediction of Galois: *It is of degree* 5 *over the ground field* $\mathbf{K}_0 = \mathbb{C}(y)$, as will be confirmed next.

Exercise 1. The 10 permutations of ex. 3.2 representing the action of $\mathfrak{K}_5$ on x_m $(0 \le n < 5)$ produce four variants of $\mathfrak{x}_0$:

$$\mathfrak{x}_1 = (x_\infty - x_1)(x_2 - x_0)(x_3 - x_4)y,$$

$$\mathfrak{x}_2 = (x_\infty - x_2)(x_1 - x_3)(x_0 - x_4)y,$$

$$\mathfrak{x}_3 = (x_\infty - x_3)(x_2 - x_4)(x_1 - x_0)y,$$

$$\mathfrak{x}_4 = (x_\infty - x_4)(x_0 - x_3)(x_1 - x_2)y.$$

Check this.

Now the point at issue is whether the full action of $\mathfrak{K}_1 \supset \mathfrak{K}_5$ produces further variants of this quantity. The answer is *no*. $[\mathfrak{K}_1 \colon \mathfrak{K}_5] = 6$ so there are six cases to test.

[2] Galois [1832], *Lettre á Auguste Chevalier*.

Sample. [10/21] belongs to $\mathfrak{K}_1$ but not to $\mathfrak{K}_5$. It acts upon $x_\infty, x_0, \ldots, x_4$ via the permutation $(302\infty41)$ and so maps $\mathfrak{x}_0$ to $\mathfrak{x}_4$. First

$$x_\infty \mapsto -\sqrt[4]{k}\left(\frac{5\omega}{2\omega+1}\right) = -\sqrt[4]{k}\left(\frac{5(\omega/5+48/5)-48}{2(\omega/5+48/5)-19}\right)$$

$$= \sqrt[4]{k}\left(\frac{\omega}{5}+\frac{48}{5}\right) = x_3,$$

since the substitution $[a\ b/c\ d]$ with $a = 5,\ b = -48,\ c = 2,\ d = -19$ is a modular substitution of second level ($-5 \cdot 19 + 48 \cdot 2 = 1$), and $5^2 - 1 - 5 \cdot 48 \equiv 8 \bmod 16$ so the rule of Section 4.14, step 1 applies with multiplier -1. Next

$$x_0 \mapsto \sqrt[4]{k}\left(\frac{1}{5}\frac{\omega}{2\omega+1}\right) = \sqrt[4]{k}\left(\frac{\omega/5}{10(\omega/5)+1}\right) = \sqrt[4]{k}(\omega/5) = x_0,$$

since the substitution $[a\ b/c\ d]$ with $a = 1,\ b = 0,\ c = 10,\ d = 1$ is of second level and $1^2 - 1 + 0 \equiv 0 \bmod 16$. Third,

$$x_2 \mapsto \sqrt[4]{k}\left(\frac{1}{5}\frac{\omega}{2\omega+1}+\frac{32}{5}\right) = \sqrt[4]{k}\left(\frac{13\cdot(5\omega)+32}{2\cdot(5\omega)+5}\right) = -\sqrt[4]{k}(5\omega) = x_\infty,$$

since the substitution $[a\ b/c\ d]$ with $a = 13,\ b = 32,\ c = 2,\ d = 5$ is of second level and $13^2 - 1 + 13 \cdot 32 \equiv 8 \bmod 16$; similarly, x_1, x_3, x_4 map to x_2, x_4, x_1 in that order:

$$x_1 \mapsto \sqrt[4]{k}\left(\frac{1}{5}\frac{\omega}{2\omega+1}+\frac{16}{5}\right) = \sqrt[4]{k}\left(\frac{33(\omega/5+32/5)-208}{10(\omega/5+32/5)-63}\right) = x_2,$$

$$x_3 \mapsto \sqrt[4]{k}\left(\frac{1}{5}\frac{\omega}{2\omega+1}+\frac{48}{5}\right) = \sqrt[4]{k}\left(\frac{97(\omega/5+64/5)-1232}{10(\omega/5+64/5)-127}\right) = x_4,$$

$$x_4 \mapsto \sqrt[4]{k}\left(\frac{1}{5}\frac{\omega}{2\omega+1}+\frac{64}{5}\right) = \sqrt[4]{k}\left(\frac{129(\omega/5+16/5)-400}{10(\omega/5+16/5)-31}\right) = x_1.$$

The verification is finished.

Exercise 2. Check that [1 0/2 1] and $\mathfrak{K}_5$ generate $\mathfrak{K}_1$. *Hint*: Multiplication to the left by $[a\ b/c\ d]$ with $a = 1,\ b = 0,\ c = 2n,\ d = 1$; maps $\mathfrak{K}_5$ into itself only if $n \equiv 0 \bmod 5$; also, $[\mathfrak{K}_1 : \mathfrak{K}_5] = 6$.

This finishes the proof that $\mathfrak{K}_1$ produces only the four variants of $\mathfrak{x}_0$ stated in ex. 1. It follows that $F(x) = (x - \mathfrak{x}_0) \cdots (x - \mathfrak{x}_4)$ belongs to $\mathbf{K}_0[x]$, $\sqrt[4]{k} = y$ being the absolute invariant of $\mathfrak{K}_1$. $F(x) = 0$ is the **depressed equation**.

Exercise 3. Use the expansion $\sqrt[4]{k} = \sqrt{2}q^{1/8} + \cdots$ at the cusp $\sqrt{-1}\infty$ to deduce that $\mathfrak{x}_n$ is a fixed multiple of $2^{6/5}\sqrt{5}y^{8/5}e^{6\pi\sqrt{-1}n/5} + \cdots$ for $n = 0, 1, 2, 3, 4$. Here $q = e^{\pi\sqrt{-1}\omega}$. *Hint*: The number $\sqrt{5}$ appears in the disguise $\epsilon(1-\epsilon)^2(1+\epsilon)$ with $\epsilon = e^{2\pi\sqrt{-1}/5}$.

The exercise shows that the roots $\mathfrak{x}_0, \ldots, \mathfrak{x}_4$ of $F(x) = 0$ are distinct and, as $\mathfrak{K}_1$ permutes them transitively, so F is the field polynomial of $\mathfrak{x}_0$; in particular, $\mathfrak{x}_0$ is of degree 5 over $\mathbf{K}_0$.

5.5 Hermite on the Depressed Equation

Hermite [1859] computed the field polynomial F of $\mathfrak{x}_0$ over $\mathbf{K}_0 = \mathbb{C}(\sqrt[4]{k})$:

$$F(x) = x^5 - 2^4 \cdot 5^3 k^2(1-k^2)^2 x - 2^6 \cdot 5^{5/2} k^2 (1-k^2)^2(1+k^2).$$

The details are spelled out in five steps.

Step 1. Employ the modular equation $F_5(x, y) = 0$ to check that the field polynomial of $x_\infty y^{-5}$ belongs to $\mathbb{C}[y^{-8}][x]$ and to conclude that the same is true of $\mathfrak{x}_0' = \mathfrak{x}_0 y^{-16}$, as follows from a look at the elementary symmetric functions of $\mathfrak{x}_0'$ and its variants $\mathfrak{x}_n'$ $(1 \le n \le 4)$ formed as in ex. 4.1.

Exercise 1. Do it.

Step 2. It is an advantage of $\mathfrak{x}_0'$ over $\mathfrak{x}_0$ that it is infinite only for $y = 0$, that is, only at the cusp $\sqrt{-1}\infty$, where you have the simple expansion $y = \sqrt{2}q^{1/8} + \cdots$.

Exercise 2. Confirm the statement of step 2 at $\omega = 1/5m$, for example. *Hint*: $k^2(\omega) = 2^{-4}e^{\pi\sqrt{-1}/(\omega-1)} + \cdots$ at $\omega = 1$ and [10/41] maps 1 to 1/5.

Step 3. Write the field polynomial of $\mathfrak{x}_0'$ in the form $x^5 - c_1x^4 + \cdots - c_5$ with coefficients $c_1, \ldots, c_5$ from $\mathbb{C}[y^{-8}] = \mathbb{C}[k^{-2}]$ and use the expansions of $\mathfrak{x}_n'(0 \le n < 5)$ to show that their degrees do not exceed 1, 3, 5, 7, 9, respectively.

Exercise 3. Do it.

Step 4. The substitution [10/11] changes

$$x_\infty, x_0, x_1, x_2, x_3, x_4 \quad \text{into} \quad 1/x_1, 1/x_0, 1/x_3, 1/x_4, 1/x_2, 1/x_\infty,$$

as you may confirm by explicit computation; for example, x_∞ is mapped to

$$\begin{aligned}
-\sqrt[4]{k}\left(\frac{5\omega}{\omega+1}\right) &= -\sqrt[4]{k}\left(\begin{pmatrix}5 & -16\\ 6 & -19\end{pmatrix}\begin{pmatrix}10\\ -11\end{pmatrix}\left(\frac{\omega}{5}+\frac{16}{5}\right)\right)\\
&= \sqrt[4]{k}\left(\begin{pmatrix}10\\ -11\end{pmatrix}\left(\frac{\omega}{5}+\frac{16}{5}\right)\right)\\
&= \frac{1}{\sqrt[4]{k}}\left(\frac{\omega}{5}+\frac{16}{5}\right)\\
&= 1/x_1,
\end{aligned}$$

$11^2 - 1 + 11 \cdot 16 \equiv 8 \bmod 16$ being employed in line 1 and the anharmonic ratio $k^2\left([10/-11]\omega\right) = k^{-2}(\omega)$ in line 3. Now $x_\infty x_0 \cdots x_4 = -y^6$ by the modular equation. It follows that, under the substitution $[10/11]$, the quantities $\mathfrak{x}'_n (0 \le n < 5)$ are permuted and multiplied by y^{24}, while y itself is turned upside down. This is possible only if $c_n(k^{-2}) = c_n(k^2)k^{-6n}$ for $1 \le n \le 5$, whence

$$\begin{aligned}
c_1 &= 0,\\
c_2 &= c_2' k^{-6},\\
c_3 &= c_3'(k^{-8} + k^{-10}),\\
c_4 &= c_4'(k^{-10} + k^{-14}) + c_4'' k^{-12},\\
c_5 &= c_4'(k^{-10} + k^{-18}) + c_5''(k^{-14} + k^{-16})
\end{aligned}$$

with *constant* primed coefficients, by the restrictions on the degrees of $c_1, \ldots, c_5$ noted in step 3.

Step 5. Rule 2 of Section 4.14 provides further information: If $\omega = 0$ $(y = 1)$, then the roots $x_\infty, x_0, \ldots, x_4$ of the modular equation are all -1, with one exception, namely, $+1$. This means that $\mathfrak{x}'_0, \ldots, \mathfrak{x}'_4$ all vanish and forces

$$\begin{aligned}
c_2 = 0, \quad c_3 = 0, \quad c_4 &= c_4'(k^{-10} + k^{-14} - 2k^{-12}),\\
c_5 &= c_5'(k^{-12} + k^{-18} - k^{-14} - k^{-16}).
\end{aligned}$$

The field polynomial of $\mathfrak{x}_0$, itself, is now expressed as

$$F(x) = x^5 + c_4' k^2 (1-k^2)^2 x + c_5' k^2 (1-k^2)^2 (1+k^2),$$

and c_5' is evaluated as $-2^6 \cdot 5^{5/2}$ from the expansion $\mathfrak{x}_0 = 2^{6/5} 5^{1/2} k^{2/5} + \cdots$ obtained in ex. 4.9.4. The value $c_4' = -2^4 \cdot 5^3$ is now obtained from the more accurate expansion

$$\mathfrak{x}_0 = 2^{6/5}5^{1/2}k^{2/5}[1 + 2^{-4/5}k^{2/5}] + \cdots.$$

Exercise 4. Check that. *Hint*: $\sqrt[4]{k} = \sqrt{2}q^{1/8}(1-q) + \cdots$ at $\sqrt{-1}\infty$ with $q = e^{\pi\sqrt{-1}\omega}$; see Section 4.9 for help.

5.6 Hermite on the Quintic

Jerrard (1834) reduced the general equation of degree 5, from

$$P_5(x) = (x - x_1)\cdots(x - x_5) = x^5 - c_1x^4 + c_2x^3 - c_3x^2 + c_4x - c_5 = 0,$$

to the simpler form $x^5 - x + c = 0$ by adjunction of radicals to the ground field $\mathbf{K}_0 = \mathbb{C}(c_1, \ldots, c_5)$, and this reduced version can be matched to the depressed equation of Section 5.5 by the substitutions

$$x \mapsto 2{\cdot}5^{3/4}\sqrt{k(1-k^2)}x, \quad c \mapsto \frac{-2{\cdot}5^{-5/4}(1+k^2)}{\sqrt{k(1-k^2)}}.$$

The modulus k^2 is a root of $2^4(1+k^2)^4 = 5^5c^4k^2(1-k^2)^2$, which is of degree 4 in k^2 and so can be solved by radicals over the ground field $\mathbb{C}(c)$. A root of $x^5 - x + c = 0$ is now obtained in the form $2{\cdot}5^{3/4}\sqrt{k(1-k^2)}\,\mathfrak{x}_0$ and is used to depress the degree of *this* equation from 5 to 4. The other four roots

$$2{\cdot}5^{3/4}\sqrt{k(1-k^2)}\,\mathfrak{x}_n \quad (n = 1, 2, 3, 4)$$

may now be extracted by radicals. This is Hermite's recipe [1859], expressing the roots of the general equation of degree 5 in terms of the eighth roots of the moduli

$$k^2(\omega),\ k^2(5\omega),\ k^2(\omega/5 + 16n/5) \quad (0 < n < 5)$$

after adjunction of the radicals. In this way, *the solution of the general equation of degree 5 is made to depend upon the equations for the division of periods of the elliptic functions*, as they used to say. This recipe is to be compared to the trigonometric solution of the cubic noted in ex. 2.4.

The present scheme, of adjunction of new **radicals** formed from the level 2 modular function k^2, fails for $d \geq 6$, but Jordan [1870: 370–84] showed that invariant functions of higher arithmetic subgroups will serve; compare Lindemann [1884, 1892] and also Umemura's article in Mumford [1984: 261–72], where the roots are expressed by means of Riemann's theta function.

Reduced Quintics. Here is what Jerrard did. Let $k_0, \ldots, k_4$ be new indeterminates and compute the powers $\mathfrak{x}^n$ $(0 \le n \le 5)$ of $\mathfrak{x} = k_0 + k_1 x + \cdots + k_4 x^4$, eliminating x^5 by means of $P_5(x) = 0$. The six reduced powers so introduced are polynomials in x, of degree ≤ 4, with coefficients from $\mathbf{K}_1 = \mathbf{K}_0(k_0, \ldots, k_4)$, and as the class of such polynomials is of dimension 5 over $\mathbf{K}_1$, so $\mathfrak{x}$ is a root of a polynomial $Q_5(\mathfrak{x}) = \mathfrak{x}^5 - c_1 \mathfrak{x}^4 + \cdots - c_5$ from $\mathbf{K}_1[\mathfrak{x}]$; in particular, every root of $P_5(x) = 0$ produces a distinct root of $Q_5(\mathfrak{x}) = 0$, and these are all a roots of the latter. The coefficients

$$c_1 = \mathfrak{x}_1 + \cdots + \mathfrak{x}_5 = 5k_0 + k_1(x_1 + \cdots + x_5) + \cdots + k_4(x_1^4 + \cdots + x_5^4),$$

$$c_2 = \mathfrak{x}_1\mathfrak{x}_2 + \cdots + \mathfrak{x}_4\mathfrak{x}_5, \ldots, \qquad c_5 = \mathfrak{x}_1 \cdots \mathfrak{x}_5$$

are homogeneous forms in $k_0, \ldots, k_4$ of degrees $1, 2, \ldots, 5$ with coefficients from $\mathbb{C}[c_1, \ldots, c_5]$. The plan is to make c_1, c_2, and c_3 (but not c_4 and c_5) vanish by choice of $k_0, \ldots, k_4$ in a radical extension of $\mathbf{K}_0 = \mathbb{C}(c_1, \ldots, c_5)$.

Step 1. c_1 is made to vanish by choice of $k_0 = -(1/5)[k_1(x_1 + \cdots + x_5) + \cdots + k_4(x_1^4 + \cdots + x_5^4)]$ without extension of $\mathbf{K}_0$.

Step 2. c_2 is now a form of degree 2 in $k_1, \ldots, k_4$; as such, it is a sum $f_1^2 + f_2^2 + f_3^2 + f_4^2$ of four squares of forms of degree 1 with coefficients in a radical extension $\mathbf{K}_2$ of $\mathbf{K}_0$, and is made to vanish by taking $f_1 = \sqrt{-1}\, f_2$ and $f_3 = \sqrt{-1}\, f_4$, at the cost of eliminating two of the indeterminates.

Step 3. c_3 is now a cubic form in the two surviving indeterminates and is made to vanish in a cubic extension $\mathbf{K}_3$ of $\mathbf{K}_2$ by fixing one of them and solving for the other in such a way that k_1, k_2, k_3, k_4 do not all vanish. $Q_5(\mathfrak{x})$ is now reduced to $\mathfrak{x}^5 + c_4\mathfrak{x} - c_5$ with $c_4, c_5 \in \mathbf{K}_3$, and the splitting field of P_5 over $\mathbf{K}_0$ is seen to be included in the radical extension of the splitting field of Q_5 over $\mathbf{K}_3$ obtained by solving $\mathfrak{x} = k_0 + \cdots + k_4 x^4$ for x in terms of $\mathfrak{x}$. The elements c_4 and c_5 do not vanish since $P_5(x)$ is not solvable by radicals. This permits a final reduction to $Q_5(\mathfrak{x}) = \mathfrak{x}^5 - \mathfrak{x} + c$ with nonvanishing c in a radical extension of $\mathbf{K}_0$.

5.7 A Geometric View

It is instructive to look at the whole matter from a geometric point of view to see why the roots of the general equation $P_5(x) = 0$ cannot help but be expressible by means of modular functions. Green's pretty account [1978] is followed here. Klein [1884] proceeded differently, via the so-called icosahedral equation, of which more will follow.

The Root Curve. The equation is taken in the reduced form $P_5(x) = x^5 + c_4x - c_5 = 0$ with $(c_4, c_5) \in \mathbb{C}^2 - 0$, so the root vector $(x_1, \dots, x_5)$ belongs to $\mathbb{C}^5 - 0$. The latter is viewed projectively, so that the locus $c_1 = c_2 = c_3 = 0$ appears as a projective curve $\mathbf{X}$ in $\mathbb{P}^4$, equipped with a projection $(x_1, \dots, x_5) \mapsto c_4^5/c_5^4$ onto the projective line $\mathbb{P}^1$, two families of roots being identified in $\mathbb{P}^4$ if and only if they have the same projection. $\mathbf{X}$ is a smooth curve, that is, the gradients of the defining relations $c_1 = c_2 = c_3 = 0$ are independent on $\mathbf{X}$: In fact, the gradients are $\imath = (1, \dots, 1)$, $c_1\imath - (x_1, \dots, x_5)$, $c_2\imath - c_1(x_1, \dots, x_5) + (x_1^2, \dots, x_5^2)$, and a dependence among them implies that the roots $x_1, \dots, x_5$ satisfy an equation of degree 2, which cannot be maintained in the face of $c_1 = c_2 = c_3 = 0$ and $(c_4, c_5) \in \mathbb{C}^2 - 0$, as you will check. $\mathbf{X}$ is the **root curve** of $P_5(x) = 0$; compare Fricke [1928 (2): 121] who alludes to the same object. Now the symmetric group S_5 acts upon $\mathbf{X}$ by permutation of the letters $x_1, \dots, x_5$. Indeed, the projection is equivalent to the quotient map $\mathbf{X} \to \mathbf{X}/S_5$, so $\mathbf{X}$ is a $5! = 120$-sheeted ramified cover of $\mathbb{P}^1$, and it is easy to see that it has

24 ramifications of degree 5 over 0,

30 ramifications of degree 4 over ∞,

60 ramifications of degree 2 over $-5^5/4^4$.

Proof. A ramification occurs if a nontrivial substitution of S_5 maps $(x_1, \dots, x_5)$ into a multiple of itself. The multiplier ω must be a fourth or a fifth root of unity, so three cases present themselves.

Case 1. The multiplier is a fifth but *not* a fourth root of unity. Then the roots are proportional to $x = \omega^n$ $(0 \le n < 5)$. These may be permuted in 120 ways and identified in $\mathbb{P}^4$ in families of five by the multipliers $1, \omega, \omega^2, \omega^3, \omega^4$, leaving 24 ramified points of $\mathbf{X}$, of degree 5 apiece, over the projection $c_4^5/c_5^4 = 0$.

Case 2. The multiplier is a fourth but *not* a fifth root of unity. Then $\omega = \pm\sqrt{-1}$ implies that $(x_1, \dots, x_5)$ is proportional to $(0, \sqrt{-1}, -1, -\sqrt{-1}, +1)$, up to permutations. The 120 possible permutations are identified in $\mathbb{P}^4$ in families of four by the action of the multipliers $1, \sqrt{-1}, -1, -\sqrt{-1}$, leaving 30 ramified points of $\mathbf{X}$, of degree 4 apiece, lying over the projection ∞. The multiplier could also have been $\omega = -1$, but that leads to the vanishing of all the roots, which is not permitted.

Case 3. The multiplier is a fourth *and* a fifth root of unity, that is, $\omega = 1$. Then $P_5(x) = x^5 + c_4x - c_5 = 0$ has a double root and $P_5'(x) = 5x^4 + c_4 = 0$

requires the projection to be $-5^4/4^4$. The other roots are simple, so there are $\binom{5}{2} \times 3! = 60$ such ramified points of $\mathbf{X}$, of degree 2 apiece, as you will check. The genus of $\mathbf{X}$ is now computed via the Riemann–Hurwitz formula:

$$\text{ramification index } = 24 \cdot 4 + 30 \cdot 3 + 60 = 246 = 2 \times (120 \text{ sheets } + g - 1),$$

with the result that $g = 4$; in particular, the universal cover of $\mathbf{X}$ is the open upper half-plane $\mathbb{H}$. This means that the roots $x_1, \ldots, x_5$ can be expressed, up to a multiplier, by means of functions on the half-plane invariant under the covering group of $\mathbf{X}$. It is just this possibility that Hermite exploited.

Exercise 1. Make the same computation for degrees $d \leq 4$. The reduced form is $x^2 - c_1 x + c_2$ for $d = 2$, $x^3 + c_2 x - c_3$ for $d = 3$, and $x^4 - c_3 x + c_4$ for $d = 4$, the projections being c_1^2/c_2, c_2^3/c_3^2, c_3^4/c_4^3. $\mathbf{X}$ is now a projective line, as you will verify by counting ramifications, with the implication that genus 0 has to do with solvability by radicals. *Answers*: For $d = 2$, $\mathbf{X}$ has 2 ramifications of degree 2 over 4 and 0; for $d = 3$, 3 ramifications of degree 2 over $-3^3/2^2$ and 3 more over ∞, plus 2 ramifications of degree 3 over 0; for $d = 4$, 12 ramifications of degree 2 over $4^4/3^3$, 8 of degree 3 over ∞, and 6 of degree 4 over 0.

The Stellated Dodecahedron. Kepler [1596] invented higher non-Platonic solids incapable of embedding in $\mathbb{R}^3$. It comes as an unexpected pleasure that the curve $\mathbf{X}$ is one of these; indeed, *wonders never cease*. The figure in question is the **stellated dodecahedron**, constructed as follows. Begin with the icosahedron of 20 triangular faces, 30 edges, and 12 corners, described in Section 1.7. Each corner has five neighbors which may be spanned by a pentagon; this done, throw the original icosahedron away. The result is a self-intersecting polyhedron $\mathbf{X}_0$ which may be provided with a complex structure by projecting its faces onto the circumscribed sphere $\mathbb{P}^1$, self-intersections being ignored as an artifact of the ambient $\mathbb{R}^3$. Figure 5.3 depicts a corner of the icosahedron and its five neighbors. The pentagon spanning 12345 is a face of $\mathbf{X}_0$; other faces span 103, 305, 502, 204, 401 (+ two adjacent corners), and these meet at 0, to produce a ramification of degree 2; moreover $\mathbf{X}_0$ covers $\mathbb{P}^1$ threefold via the spherical projection. The genus of $\mathbf{X}_0$ is now seen to be 4:

$$\text{ramification index} = 12 = 2 \times (3 \text{ sheets } + g - 1).$$

A second projection of $\mathbf{X}_0$ onto $\mathbb{P}^1$ now comes into play. The group of symmetries of the icosahedron is the alternating group A_5 comprising 60 substitutions: To wit, besides the identity, (1) $24 = 4 \times 6$ rotations by multiples of $2\pi/5$ about corners, (2) $20 = 2 \times 10$ rotations by multiples of $2\pi/3$ about the centers

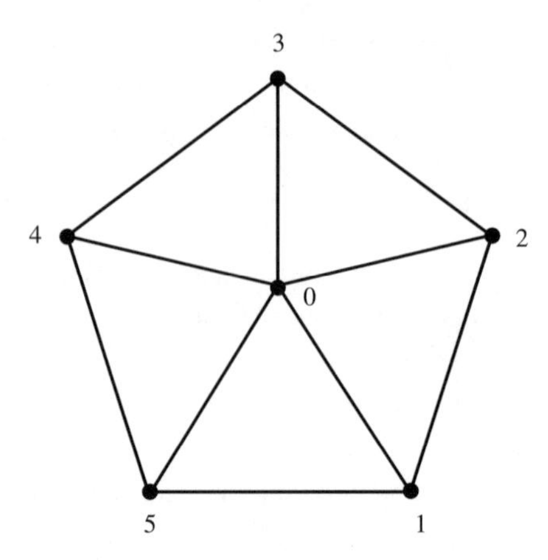

Figure 5.3. A corner of the icosahedron.

of faces, (3) 15 rotations by π about centers of edges, for a total count of $1+24+20+15=60$. A_5 acts upon $\mathbf{X}_0$ in a self-evident way. It remains to identify the quotient $\mathbf{X}_0/A_5$ as a projective line.

Proof. The appropriate symmetry of type 1 fixes the center of the pentagon 12345; it identifies nearby points in families of five, producing ramifications of $\mathbf{X}_0$ of degree 5 lying over two distinct points of the base $\mathbf{X}_0/A_5$. There are 12 points of each kind lying over the same two base points. A symmetry of type 2 effects a cyclic permutation of the three sheets of $\mathbf{X}_0$ lying over the associated icosahedral face; no ramification is produced. A symmetry of type 3 exchanges two of the sheets over the associated icosahedral edge and fixes the other. This accounts for 30 more ramifications of degree 2 lying over the third point of the base. The genus of the base is now computed by a variant of the Riemann–Hurwitz formula: The Euler characteristic of $\mathbf{X}_0$ is $2-2\times 4=-6$, while that of the base is $2-2g$. The curve $\mathbf{X}_0$ covers the base 60 times, so $-6=60\times(2-2g)-$ the ramification index as in Section 1.12, with the result that $120g=120+6-24\times 4-30=0$, that is, $g=0$; in short, $\mathbf{X}_0/A_5$ is a projective line, as advertised.

Now it is easy to see that the root curve $\mathbf{X}\subset\mathbb{P}^4$ of $x^5+c_4x-c_5=0$ is of the same character over its quotient by $A_5\subset S_5$, to wit, $\mathbf{X}/A_5$ is a projective line and $\mathbf{X}$ is ramified over three points e_1, e_2, e_3 of the base, having 12 ramifications of degree 5 over each of e_1 and e_2, and 30 more ramifications of degree 2 over e_3. This fact is now used to identify the root curve as a stellated dodecahedron.

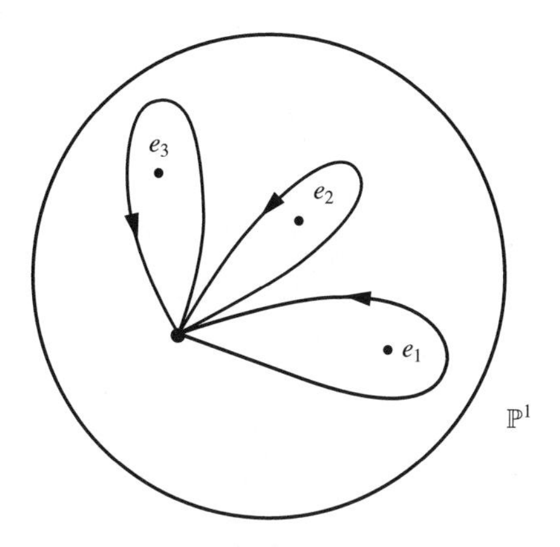

Figure 5.4. Generators of Γ.

Proof. A preliminary step is to make the branch points e_1, e_2, e_3 of the two curves coincide by application of a self-map of $\mathbb{P}^1$. Now fix three loops γ generating the fundamental group Γ of the thrice-punctured sphere $\mathbb{P}^1 - e_1 - e_2 - e_3$, as in Fig. 5.4, with $\gamma_1\gamma_2\gamma_3 = 1$. Any element of Γ determines a covering map of $\mathbf{X}$ over the punctured sphere, and this map can be identified with an element of A_5, so what you have is a representation π of Γ in A_5. The ramifications of $\mathbf{X}$ are mirrored in π by the fact that the permutations $\sigma_1 = \pi(\gamma_1), \sigma_2 = \pi(\gamma_2)$, and $\sigma_3 = \pi(\gamma_3)$ have orders $5, 4, 2$; conversely, it is plain that the conformal type of $\mathbf{X}$ is specified by π, so the identity of the two covers comes down to the fact that, up to conjugation inside A_5, Γ has only *one* such representation, as will now be confirmed. The fact that $\sigma_3^2 = 1$ permits you to take $\sigma_3 = (1\ 3\ 2\ 5\ 4)$ by labeling of letters, products of 2-cycles being the only elements of A_5 of order 2; similarly, only 5-cycles have degree 5, so $\sigma_1^{-1} = \sigma_2\sigma_3$ and σ_2 are such, and you can relabel the letters so as to make $\sigma_2(1) = 2$. Then $\sigma_2(2) = 3$ implies $\sigma_2\sigma_3(3) = \sigma_2(2) = 3$, which is contradictory, as is $\sigma_2(2) = 1$, so $\sigma_2(2)$ is neither $1, 2$, nor 3 and can be taken as 4 by a further relabeling. This precludes $\sigma_2(4) = 5$ and, as $\sigma_2 \colon 1 \mapsto 2 \mapsto 4$ cannot now map 4 to any of $1, 2, 4$, or 5, so σ_2 being a 5-cycle must map $1 \mapsto 2 \mapsto 4 \mapsto 3 \mapsto 5 \mapsto 1$. The proof is finished.

The Universal Cover. The uniformization theorem of Koebe and Poincaré being notoriously nonconstructive, Green [1978: 238–9] elaborates upon the way in which the half-plane $\mathbb{H}$ covers the root curve $\mathbf{X}$ by means of the triangle functions of Section 4.10 in the case $a + b + c = 1/2 + 1/4 + 1/5 < 1$. This provides a tessellation of $\mathbb{H}$ by hyperbolic triangles having interior angles $\pi/2, \pi/4, \pi/5$, as seen in Fig. 5.5. The double reflections A, B, C in the sides

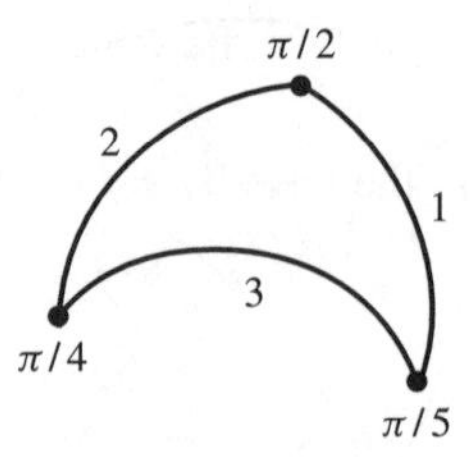

Figure 5.5. The hyperbolic triangle.

12, 23, 31 have periods 2, 4, 5, respectively, and $ABC = 1$. They generate a subgroup Γ of $PSL(2,\mathbb{R})$ preserving the tessellation, for which any two adjacent triangles form a fundamental cell; in particular, $\mathbb{H}/\Gamma$ is a copy of $\mathbb{P}^1$. Now the invariant subgroup $\Gamma_0 \subset \Gamma$ generated by $(AC^{-2}AC^2)^2$ is of index 120 in the latter; see Coxeter and Moser [1980: 137]. In fact, Γ/Γ_0 is a copy of S_5, and the projection of $\mathbb{H}/\Gamma_0$ to $\mathbb{P}^1 = \mathbb{H}/\Gamma$ has three branch points e_1, e_2, e_3 of degrees 2, 4, 5 corresponding to the orders of the cosets of A, B, C in Γ/Γ_0. This provides a representation in S_5 of the fundamental group of the punctured base, of the same character as that associated to the root curve $\mathbf{X}$ and its projection to $\mathbb{P}^1 = \mathbf{X}/S_5$ punctured at $e_1 = -5^5/4^4$, $e_2 = \infty$, $e_3 = 0$, and it is easy to see, as for the stellated dodecahedron, that there is, effectively, only one such representation. In short, $\mathbb{H}/\Gamma_0$ may be identified with the root curve. This leads to a more concrete view of the relation of $\mathbf{X}$ to $\mathbb{H}$, permitting the expression of the roots by modular forms of Γ_0 of weight -2; see Green [1978: 239–41] for details of this and further information.

Klein's method is described in Fricke [1928 (2): 57–181]. It depends upon the icosahedral invariant j_4 of Section 1.7. This is a map of degree 60 from the projective line $\mathbb{P}^1$ to itself, with covering group A_5, the latter being, at one and the same time, the symmetry group of the icosahedron and the Galois group of the icosahedral equation $j_4(x) = y$ over the ground field $\mathbf{K}_0 = \mathbb{C}(y)$; compare ex. 2.1. Now A_5 contains the tetrahedral group A_4, so the splitting field of $j_4(x) - y$ contains an element of degree 5 over the ground field, just as in Section 4. This means that *the roots of* $P_5(x) = 0$ *can be made to depend upon those of* $j_4(x) = y$. The latter can be expressed by means of modular functions: If j is the absolute invariant of the modular group Γ_1 and if $y = j(\omega)$, then, as Klein proved,

$$x = q^{-3/5}\frac{\vartheta_1(\omega|5\omega)}{\vartheta_1(2\omega|5\omega)} = q^{2/5} \times \frac{\sum_{\mathbb{Z}}(-1)^n q^{5n^2+3n}}{\sum_{\mathbb{Z}}(-1)^n q^{5n^2+n}} \quad \text{with} \quad q = e^{\pi\sqrt{-1}\omega}$$

is a root of the icosahedral equation; all the other roots may be obtained by applications to x of the substitutions effecting the icosahedral rotations of $\mathbb{P}^1$. A variant of the geometric picture is now obtained by modeling the covering projective line by $\mathbb{H}/\Gamma_5$ and the base by $\mathbb{H}/\Gamma_1$, the degree of the cover over the base being the index $[\Gamma_1\colon\Gamma_5] = 60$.

6

Imaginary Quadratic Number Fields

Kronecker (1853) tried to prove, and Weber [1886] succeeded in proving, that every finite extension of the rational number field $\mathbb{Q}$ with commutative Galois group is a subfield of some cyclotomic field obtained by adjoining to $\mathbb{Q}$ a root of unity; see Greenberg [1974] for a fairly simple proof. Kronecker (1860) also tried to describe the commutative extensions of the **imaginary quadratic field** $\mathbb{Q}(\sqrt{D})$, obtained by the adjunction to $\mathbb{Q}$ of the square root of a negative square-free number D, and perceived the relevance of certain "singular" values of the absolute invariant j. The deep impression this program made upon nineteenth-century mathematics is reflected in the romantic title by which it came to be known: **Kronecker's *Jugendtraum***,[1] and Weber's completion of it [1891] represents a remarkable feat. But it was objected that such transcendental tools as the absolute invariant were inappropriate; and, indeed, a purely algebraic understanding was temporarily retarded. Hilbert's class fields [1897] met the objection, providing a complete description of commutative extensions generally, avoiding transcendental elements; see, especially, Hasse [1980] and Cohn [1985] for more information. The next two articles review the basic facts about algebraic number fields. The portion of the *Jugendtraum* dealt with here is more fully explained in Section 3. The rest of the chapter is devoted to the proofs, following chiefly Fricke [1922 (2): 292–502] and Weber [1891: 413–91], with small simplifications made possible by more modest objectives. Deuring [1958] can be consulted for a streamlined account in the same (transcendental) style; see also Borel et al. [1966], Cox [1989], and, for a nice old-fashioned account, Matthews [1911]. Vladut [1991] reviews the whole subject and connects it to current number-theoretic questions.

[1] Kronecker [1895–1931(5): 455–7].

6.1 Algebraic Numbers

The basic facts about algebraic number fields can be found in a convenient form in Borevich and Shafarevich [1966], Hasse [1980], Pollard [1950], Hardy and Wright [1979], or Weyl [1940]; see also Stark [1970] for a nice introduction to the imaginary quadratic field $\mathbb{Q}(\sqrt{D})$. This field illustrates all the important points, so read the examples with care.

Number Fields. Let $\mathbf{K}$ be an extension of the **ground field** $\mathbb{Q}$ of finite degree $d = [\mathbf{K}:\mathbb{Q}]$. $\mathbf{K}$ is an **algebraic number field** if it is obtained from $\mathbb{Q}$ by a tower of intermediate fields. $\mathbf{K}$ can be obtained from $\mathbb{Q}$ by the adjunction of a single element r so that $\mathbf{K} = \mathbb{Q}(r)$. The irrationality r is of degree d over $\mathbb{Q}$ so every element of $\mathbf{K}$ is of the form $x = c_0 + c_1 r + \cdots + c_{d-1} r^{d-1}$ with $c_0, \ldots, c_{d-1} \in \mathbb{Q}$.

Example 1. $\mathbf{K} = \mathbb{Q}(\sqrt{-3}, \sqrt{5})$ is obtained from $\mathbb{Q}(\sqrt{-3})$ by adjunction of $\sqrt{5}$; it consists of all numbers of the form $x = c_0 + c_1\sqrt{-3} + c_2\sqrt{5} + c_3\sqrt{-15}$ with c_0, c_1, c_2, c_3 from $\mathbb{Q}$. It contains $r = \sqrt{-3} + \sqrt{5}$, $\sqrt{5} = -(1/16)(r^3 - 12r)$, and also $\sqrt{-3} = (1/16)(4r + r^3)$ in view of $r^3 = 12\sqrt{-3} - 4\sqrt{5}$, with the outcome that $\mathbf{K} = \mathbb{Q}(r)$. $\mathbf{K}$ is of degree 4 over $\mathbb{Q}$: in fact, the field polynomial of $r = \sqrt{-3} + \sqrt{5}$ is $x^4 - 4x^2 + 64$ and $\mathbf{K}$ is its splitting field.

Exercise 1. Give details.

Conjugates. The field polynomial of r is of degree d and so has d roots $r_1 (= r), r_2, \ldots, r_d$. The numbers

$$x_i = c_0 + c_1 r_i + \cdots + c_{d-1} r_i^{d-1} \ (i = 1, \ldots, d)$$

are the **conjugates** of $x = x_1 \in \mathbf{K}$. They are the roots of the field polynomial P of x, each repeated $m = d/n$ times, n being the degree of P. The product

$$x_1 \cdots x_d = N_{\mathbf{K}/\mathbb{Q}}(x)$$

is the **norm** of x; it is a rational number. The norm is multiplicative: $N(xy) = N(x)N(y)$.

Example 2. The conjugates of $r = \sqrt{-3} + \sqrt{5}$ in $\mathbb{Q}(\sqrt{-3}, \sqrt{5})$ are $r_1 = \sqrt{-3} + \sqrt{5}$, $r_2 = -\sqrt{-3} + \sqrt{5}$, $r_3 = \sqrt{-3} - \sqrt{5}$, and $r_4 = -\sqrt{-3} - \sqrt{5}$. The conjugates of $\sqrt{-3}$ are its conjugates in $\mathbb{Q}(\sqrt{-3})$ repeated twice. This gives $N_{\mathbf{K}/\mathbb{Q}}(\sqrt{-3} + \sqrt{5}) = 64$ and $N_{\mathbf{K}/\mathbb{Q}}(\sqrt{-3}) = 9$.

Galois Group. $\mathbf{K} = \mathbb{Q}(r)$ need not be the splitting field of r but can always be so enlarged. Then the Galois group G permutes the conjugates of r inside the bigger field.

Example 3. $\mathbb{Q}(\sqrt{-3}, \sqrt{5})$ is a splitting field with commutative Galois group G comprised of the four permutations (1234), (2143), (4321), (3412) of the letters r_1, r_2, r_3, r_4 of example 2; it is the product of two cyclic groups of order 2. The Kronecker–Weber theorem is illustrated: $\mathbb{Q}(\sqrt{-3}, \sqrt{5}) \subset \mathbb{Q}(\omega)$ with $\omega = e^{2\pi\sqrt{-1}/15}$; in fact,

$$\sqrt{-3} = 2\omega^5 + 1 \text{ and } \sqrt{5} = 2(\omega^3 + \omega^{12}) - 1.$$

Exercise 2. Check the Galois group.

The fields $\mathbb{Q}(\sqrt{p})$, for prime p, afford further illustrations of the Kronecker–Weber theorem: $\pm\sqrt{p}$ is the value of a Gauss sum of Section 3.10, so $\mathbb{Q}(\sqrt{p})$ is contained in a cyclotomic field.

Algebraic Integers. The number $x \in \mathbf{K}$ is an **algebraic integer** if its field polynomial has coefficients from $\mathbb{Z}$ and top coefficient $+1$. This class of polynomials is denoted by $\mathbb{SZ}[x]$ from now on. The algebraic integers of $\mathbf{K}$ form a ring $\mathcal{O}(\mathbf{K})$ just like $\mathbb{Z}$. Naturally, $\mathbb{Z} = \mathcal{O}(\mathbb{Q})$. These are the **rational integers** if the distinction is required. $\mathbf{K}$ is the quotient field of $\mathcal{O}(\mathbf{K})$.

Example 4. The integers of the imaginary quadratic field $\mathbb{Q}(\sqrt{D})$ are $m+n\sqrt{D}$ with $m, n \in \mathbb{Z}$ if $D \not\equiv 1 \bmod 4$; $(1/2)(1+\sqrt{D})$ must be adjoined if $D \equiv 1 \bmod 4$, so the integers are now $(1/2)(m + n\sqrt{D})$ with $m, n \in \mathbb{Z}$ of like parity.

Proof. The algebraic integer $x \in \mathbb{Q}(\sqrt{D})$ is of the form $a + b\sqrt{D}$ with $a, b \in \mathbb{Q}$. The conjugate of x is $a - b\sqrt{D}$, so its integrality requires $2a \in \mathbb{Z}$ and $a^2 - b^2 D \in \mathbb{Z}$.

Case 1. $a \in \mathbb{Z}$ implies $b^2 D \in \mathbb{Z}$, and $b \in \mathbb{Z}$ follows from the fact that D is square-free.

Case 2. $a \in \mathbb{Z} + 1/2$ implies $b^2 D \in \mathbb{Z} + 1/4$. Let $b = i/j$ with coprime $i, j \in \mathbb{Z}$. Then $4Di^2 = 4nj^2 + j^2$ and j^2 divides $4i^2$, D being square-free. This forces $j = 2$, i odd, and $D \equiv 1 \bmod 4$. The rest will be plain.

Units. The integer $x \in \mathcal{O}(\mathbf{K})$ is a **unit** if $1/x$ is also integral.

Exercise 3. Check that $N_{\mathbf{K}/\mathbb{Q}}(x)$ is a rational integer if $x \in \mathcal{O}(\mathbf{K})$. Conclude that x is a unit if and only if $N_{\mathbf{K}/\mathbb{Q}}(x) = \pm 1$.

Example 5. The units of the imaginary quadratic field $\mathbb{Q}(\sqrt{D})$ are ± 1 always, with the addition of $\pm\sqrt{-1}$ if $D = -1$, and of $(1/2)(\pm 1 \pm \sqrt{-3})$ if $D = -3$.

Proof. $x = a + b\sqrt{D}$ is a unit if and only if

$$\frac{1}{x} = \frac{a - b\sqrt{D}}{a^2 - Db^2}$$

is also integral. Then

$$N_{\mathbf{K}/\mathbb{Q}}\left(\frac{1}{x}\right) = (a^2 - Db^2)^{-1}$$

must be a whole number, that is, $a^2 - Db^2 = +1$, D being negative, and if $D \not\equiv 1 \bmod 4$ then $a, b \in \mathbb{Z}$ and either $a = \pm 1$, $b = 0$, and $x = \pm 1$, or else $a = 0$, $b = \pm 1$, $D = -1$, and $x = \pm\sqrt{-1}$. Now take $D \equiv 1 \bmod 4$ and half-integral a and b. Then $b^2 \geq 1/4$ and $a^2 - b^2 D = 1$ imply $D = -3$, $a = \pm 1/2$, $b = \pm 1/2$, and $x = (-1 \pm \sqrt{-3})/2$, as promised.

Discriminant. $\mathcal{O}(\mathbf{K})$ has an **integral basis**. This means that it is possible to find integers $\omega_1, \dots, \omega_d \in \mathbf{K}$, $d = [K\colon Q]$ being the degree of K over Q, so that every integer ω is uniquely expressible as $\omega = n_1\omega_1 + \cdots + n_d\omega_d$ with rational integers $n_1, \dots, n_d$. Let ω_{ij} $(j \leq d)$ be the conjugates of ω_i for $i \leq d$. The square of the determinant $[\omega_{ij}\colon 1 \leq i, j \leq d]$ is a nonvanishing rational integer independent of the choice of the base. It is the **discriminant** of $\mathbf{K}$ over $\mathbb{Q}$.

Example 6. The numbers 1 and $(1/2)(1+\sqrt{D})$ form an integral basis of $\mathbb{Q}(\sqrt{D})$ if $D \equiv 1 \bmod 4$; otherwise, 1 and $\sqrt{D}$ will serve. The discriminant is D in the first case and $4D$ otherwise.

6.2 Primes and Ideal Numbers

The idea of a **prime integer** in $\mathbb{Z}$ is plain and the notion extends to the general field $\mathbf{K}$ in a natural way. But is it still true that *the integers of* $\mathbf{K}$ *can be uniquely factored into primes, as for* $\mathbb{Q}$? The answer is *no*!

Example 1. $\mathbb{Q}(\sqrt{-5})$ does not enjoy this property. The congruence $-5 \not\equiv 1 \bmod 4$ implies that the integers are $\mathbb{Z} \oplus \mathbb{Z}\sqrt{-5}$, and $21 = 3 \cdot 7 = (1 + 2\sqrt{-5})(1 - 2\sqrt{-5})$ are distinct splittings of 21 into prime integers of $\mathbb{Q}(\sqrt{-5})$, as you will check.

Ideals. Gauss [1801] cured this difficulty with the introduction of **integral ideals**: $\mathfrak{i} \subset \mathcal{O}(\mathbf{K})$ is an (integral) ideal if it is closed under addition and subtraction within itself and also under multiplication by integers from the *full* ring $\mathcal{O}(\mathbf{K})$. The integral multiples $\omega \times \mathcal{O}(\mathbf{K})$ of a fixed integer $\omega \in \mathbf{K}$ form the **principal ideal** $\mathfrak{i} = (\omega)$.

A nonprincipal ideal $\mathfrak{j}$, if any such exists, can always be expressed as the intersection of $\mathcal{O}(\mathbf{K})$ with the principal ideal $\mathfrak{i} = (\omega)$ of an integer ω in a higher field containing $\mathbf{K}$: $\mathfrak{j} = \mathcal{O}(\mathbf{K}) \bigcap (\omega)$. This explains the name, $\mathfrak{j}$ being, so to say, the expression *inside* $\mathbf{K}$ of the ideal number ω residing in the higher field.

Example 2. The set[2] $\mathfrak{j} = 3\mathbb{Z} \oplus (-1+\sqrt{-5})\mathbb{Z}$ is an ideal of $\mathbb{Q}(\sqrt{-5})$ since $\mathcal{O}(\mathbb{Q}(\sqrt{-5})) = \mathbb{Z} \oplus \sqrt{-5}\mathbb{Z}$ and

$$\sqrt{-5}[3a + (-1+\sqrt{-5})b] = 3(a-2b) + (-1+\sqrt{-5})(3a-b) \in \mathfrak{j}.$$

This is *not* a principal ideal since $\mathfrak{j} = (\omega)$ with $\omega = a + b\sqrt{-5}$ and $a, b \in \mathbb{Z}$ requires the existence of $x \in \mathcal{O}(\mathbf{K})$ such that $3 = x\omega$. But then the norm $N(\omega) = |a + b\sqrt{-5}|^2 = a^2 + 5b^2$ would divide the norm $3^2 = 9$ of $3 \in \mathfrak{j}$, and none of the possibilities $a^2 + 5b^2 = 1, 3, 9$ pans out: $a^2 + 5b^2 = 1$ forces $\omega = 1$, $a^2 + 5b^2 = 3$ is impossible, and $a^2 + 5b^2 = 9$ requires either $a = \pm 3$ and $b = 0$ or $a = \pm 2$ and $b = \pm 1$; so the only chance is $\omega = 2 \pm \sqrt{-5}$, and this is not successful either:

$$3 = (m + n\sqrt{-5})(2 \pm \sqrt{-5}) = (2m \mp 5n) + (2n \pm m)\sqrt{-5}$$

for some $m, n \in \mathbb{Z}$ requires $m = \mp 2n$ and $3 = \mp 9n$.

Now the smallest ideal $\mathfrak{j}^2$ containing all products of elements of $\mathfrak{j}$ is a principal ideal:

$$\begin{aligned}\mathfrak{j}^2 &= (3\mathbb{Z} + (-1+\sqrt{-5})\mathbb{Z}) \times (3\mathbb{Z} + (-1+\sqrt{-5})\mathbb{Z}) \\ &= 9\mathbb{Z} + (-3+3\sqrt{-5})\mathbb{Z} + (-4+2\sqrt{-5})\mathbb{Z} \\ &= (2+\sqrt{-5})(2-\sqrt{-5})\mathbb{Z} + 3(2+2\sqrt{-5})\mathbb{Z} + 2(2+\sqrt{-5})\mathbb{Z} \\ &= \text{ the principal ideal } (2+\sqrt{-5}).\end{aligned}$$

This prompts the introduction of the **ideal integer** $\omega = \sqrt{2+\sqrt{-5}}$ from outside $\mathbb{Q}(\sqrt{-5})$, with field polynomial $P(x) = x^4 - 4x^2 + 9 = (x^2-2)^2 + 5$.

[2] The notation is faulty: In this instance, $(-1+\sqrt{-5})$ is the *number*, not the ideal. The context will remove any ambiguity.

The principal ideal $\mathfrak{i} = (\omega)$ is formed in the ring of integers of the associated splitting field. Then $\mathfrak{i} \cap \mathcal{O}(\mathbb{Q}(\sqrt{-5}))$ is an integral ideal of $\mathbb{Q}(\sqrt{-5})$ containing

$$\sqrt{2+\sqrt{-5}} \times \sqrt{2-\sqrt{-5}} = 3$$

and

$$(2+\sqrt{-5})^2 - 3 = 4(-1+\sqrt{-5})$$

and so also the whole ideal $\mathfrak{j}$. The fact is that $\mathfrak{i} \cap \mathcal{O}(\mathbb{Q}(\sqrt{-5})) = \mathfrak{j}$. The proof is postponed.

Exercise 1. Let $\omega = c^{-1}(a+b\sqrt{D}) \in \mathbb{Q}(\sqrt{D})$ with coprime $a, b, c \in \mathbb{Z}$. Prove that $\mathbb{L} = \mathbb{Z} \oplus \omega\mathbb{Z}$ is closed under multiplication by $\mathcal{O}(\mathbf{K})$ for $D \not\equiv 1 \bmod 4$ if and only if $b = 1$ and $a^2 \equiv D \bmod c$, and for $D \equiv 1 \bmod 4$ if and only if $b = 1$ and $a \equiv 1 \bmod 2$, $c \equiv 0 \bmod 2$, and $a^2 \equiv D \bmod 2c$.

Divisibility of Ideals. The **product** $\mathfrak{i}\mathfrak{j}$ of two ideals is the smallest ideal containing all products ab with $a \in \mathfrak{i}$ and $b \in \mathfrak{j}$. The ideal $\mathfrak{j}$ **divides** another ideal $\mathfrak{i}$ if there exists a third ideal $\mathfrak{k}$ such that $\mathfrak{i} = \mathfrak{j}\mathfrak{k}$. This occurs if and only if $\mathfrak{i} \subset \mathfrak{j}$. The ideal $\mathfrak{p}$ is **prime** if the product of two integers from $\mathcal{O}(\mathbf{K})$ belongs to $\mathfrak{p}$ only if one factor already belongs to $\mathfrak{p}$; equivalently, $\mathfrak{p}$ is prime if it cannot be split into the product of two nontrivial ideals of $\mathcal{O}(\mathbf{K})$. It is immediate that every ideal splits into a finite product of primes, but it is not immediate that this happens in only one way, up to trivialities, as for the decomposition of a whole number into a product of rational primes. The computation of the prime ideals in $\mathbb{Q}(\sqrt{D})$ is postponed in favor of an illustration.

Example 2 continued. Now the proof that $\mathfrak{i} \cap \mathcal{O}(\mathbb{Q}(\sqrt{-5}))$ is the same as the ideal $\mathfrak{j} = 3\mathbb{Z} + (-1+\sqrt{-5})\mathbb{Z}$ is easy to finish. $\mathfrak{i} \cap \mathcal{O}(\mathbb{Q}(\sqrt{-5}))$ includes $\mathfrak{j}$ and so divides it. But $\mathfrak{j}$ is prime since a product of integers

$$\begin{aligned}&[a + b(-1+\sqrt{-5})][c + d(-1+\sqrt{-5})]\\&= (ac - 6bd) + (ad + bc - 2bd)(-1+\sqrt{-5})\end{aligned}$$

belongs to $\mathfrak{j}$ only if 3 divides ac, which is to say that one factor or the other already belongs to $\mathfrak{j}$.

Discriminants Continued. Like the ring $\mathcal{O}(\mathbf{K})$, any ideal $\mathfrak{i}$ of $\mathcal{O}(\mathbf{K})$ has an integral basis $\omega_1, \ldots, \omega_d$. The discriminant of $\mathfrak{i}$ is formed as for $\mathcal{O}(\mathbf{K})$; it is a nonvanishing rational integer, independent of the choice of the base, and

$$\left|\frac{\text{ideal discriminant}}{\text{field discriminant}}\right|^{\frac{1}{2}} = \text{the index } [\mathcal{O}(\mathbf{K})\colon \mathfrak{i}] \text{ of } \mathfrak{i} \text{ in } \mathcal{O}(\mathbf{K}).$$

This number is the **norm** $N_{\mathbf{K}/\mathbb{Q}}(\mathfrak{i})$ of the ideal. Norms are multiplicative: $N(\mathfrak{i}\mathfrak{j}) = N(\mathfrak{i})N(\mathfrak{j})$. For principal ideals $\mathfrak{j} = (\omega)$, $N(\mathfrak{j})$ is the absolute value of the norm $N_{\mathbf{K}/\mathbb{Q}}(\omega)$ of the number ω, introduced before. An ideal always contains its norm, that is, if $N(\mathfrak{j}) = n$, then $\mathfrak{j}$ divides the principal ideal (n).

Example 3. Let $\mathfrak{i}$ be an ideal of $\mathbb{Q}(\sqrt{-5})$ with integral basis $\omega_1 = a + b\sqrt{-5}$, $\omega_2 = c + d\sqrt{-5}$ so that $\mathfrak{i}$ is the lattice $\mathbb{Z}\omega_1 \oplus \mathbb{Z}\omega_2$. Its discriminant is

$$\begin{vmatrix} a + b\sqrt{-5} & a - b\sqrt{-5} \\ c + d\sqrt{-5} & c - d\sqrt{-5} \end{vmatrix}^2 = (ad - bc)^2 \times (-20).$$

-20 is the field discriminant and $|ad - bc|$ is the quotient of the area of the fundamental cell of $\mathfrak{i}$ by the area of the fundamental cell of $\mathbb{Z} \oplus \sqrt{-5}\mathbb{Z}$. The connection with the index of $\mathfrak{i}$ in $\mathcal{O}(\mathbb{Q}(\sqrt{-5}))$ is geometrically plain from that: The latter is just the number of times the fundamental cell of $\mathcal{O}(\mathbb{Q}(\sqrt{-5}))$ fits into the larger fundamental cell of $\mathfrak{i}$.

Prime Ideals: Finale. A prime ideal $\mathfrak{p}$ in a general number field $\mathbf{K}$ divides (i.e., contains) just one rational prime p, so a complete list of prime ideals of $\mathcal{O}(\mathbf{K})$ can be obtained by factoring the rational primes p, or more accurately, the principal ideals (p), into prime ideals in $\mathcal{O}(\mathbf{K})$. Naturally, (p) could be prime in $\mathcal{O}(\mathbf{K})$; if not, it splits into two or more prime ideal pieces. The prime number p is **ramified** in $\mathbf{K}$ if the splitting involves repeated prime ideals. Dedekind [1877] proved that this takes place if and only if p divides the field discriminant. This is illustrated in the next example. The same ideas and terminology apply to the splitting of prime ideals of $\mathbb{Q}(\sqrt{-3})$, say, in a higher field such as $\mathbb{Q}(\sqrt{-3}, \sqrt{5})$.

Exercise 2. Check that every prime ideal contains a single rational prime.

Example 4. The prime ideals of $\mathbb{Q}(\sqrt{D})$ are to be computed according to the plan just proposed. The quadratic residue symbol of Section 3.10 is used to distinguish three cases: $\left(\frac{D}{p}\right) = \pm 1$ or 0, according as D is or is not a quadratic residue of p, or p divides D.

Let the prime ideal $\mathfrak{p}$ contain the rational prime p. The notation $\mathfrak{p} = [p, x + \sqrt{D}]$ means that p and $x + \sqrt{D}$ constitute an integral basis of $\mathfrak{p}$; with this notation, $\bar{\mathfrak{p}}$ is the ideal $[p, x - \sqrt{D}]$ comprised of the conjugates of the elements of $\mathfrak{p}$. The results are as follows.

Case 1. $D \not\equiv 1 \bmod 4$.

$\left(\frac{D}{p}\right) = +1$: $(p) = \mathfrak{p}\bar{\mathfrak{p}}$, $\mathfrak{p} = [p, x + \sqrt{D}]$, $x^2 \equiv D \bmod p$, $1 \le x < p$. $p = 2$ is exceptional: $\mathfrak{p} = \bar{\mathfrak{p}}$, that is, p is ramified, only in this case.
$\left(\frac{D}{p}\right) = -1$: $(p) = \mathfrak{p}$, that is, (p) is prime as it stands.
$\left(\frac{D}{p}\right) = 0$: $(p) = \mathfrak{p}^2$, $\mathfrak{p} = [p, \sqrt{D}]$, that is, $x = 0$ in line 1.

Case 2. $D \equiv 1 \bmod 4$.

$\left(\frac{D}{p}\right) = +1$: $(p) = \mathfrak{p}\bar{\mathfrak{p}}$, $\mathfrak{p} = [p, (1/2)(x + \sqrt{D})]$, $x^2 \equiv D \bmod p$, $1 \le x < p$, and x is odd, with one exception: If $p = 2$ and if $D \equiv 5 \pmod 8$, then (2) is prime as it stands.
$\left(\frac{D}{p}\right) = -1$: $(p) = \mathfrak{p}$ as before.
$\left(\frac{D}{p}\right) = 0$: $(p) = \mathfrak{p}^2$, $\mathfrak{p} = [p, (p + \sqrt{D})/2]$, that is, $x = p$ in line 1.

Sample proof. $D \equiv 1 \bmod 4$, $\left(\frac{D}{p}\right) = +1$, $p \neq 2$. This is broken into four little steps.

Step 1. D is a quadratic residue of p so $x^2 \equiv D \bmod p$ can be solved with odd $x < p$ since $x^2 \equiv (p - x)^2$ and x and $p - x$ have opposite parity.

Step 2. $\mathfrak{p} = [p, (1/2)(x + \sqrt{D})]$ is an ideal by ex. 1 applied to the lattice $[1, (1/2p)(x + \sqrt{D})]$: It is enough to note that $x^2 \equiv D \bmod p$ can be improved to $x^2 \equiv D \bmod 4p$ in view of $D \equiv 1 \bmod 4$ and $x \equiv 1 \bmod 2$.

Step 3. $\mathfrak{p}$ is prime. In fact, if $a + b\sqrt{D}$ and $c + d\sqrt{D}$ are integers from $\mathbb{Q}(\sqrt{D})$ with $a, b, c, d \in \mathbb{Z}/2$ and product

$$ac + bdD - x(ad + bc) + 2(ad + bc)\frac{x + \sqrt{D}}{2} \equiv m + n\frac{x + \sqrt{D}}{2}$$

in $\mathfrak{p}$, then $m \in \mathbb{Z}$ is divisible by p, and the same is true either of $a - bx$ or of $c - dx$ in view of

$$\begin{aligned}(a - bx)(c - dx) &= ac + bdD - x(ad + bc) + bd(x^2 - D) \\ &= m + bd(x^2 - D)\end{aligned}$$

and the divisibility of $x^2 - D$ by $4p$ used in step 2. But if p divides $a - bx$, say, then

$$a + b\sqrt{D} = a - bx + b(x + \sqrt{D}) \in \mathbb{Z}p \oplus \mathbb{Z}\frac{x+\sqrt{D}}{2} = \mathfrak{p}.$$

Step 4. Check that $\mathfrak{p}\bar{\mathfrak{p}} = (p)$. The product is spanned by rational integral multiples of p^2, $\frac{p}{2}(x \pm \sqrt{D})$, and $\frac{1}{4}(x^2 - D)$, so, with a self-explanatory notation,

$$\begin{aligned}
\mathfrak{p}\bar{\mathfrak{p}} &= \left[p^2, p\frac{x-\sqrt{D}}{2}, p\frac{x+\sqrt{D}}{2}, \frac{x^2-D}{4}\right] \\
&= \left[p^2, px, p\frac{x+\sqrt{D}}{2}, \frac{x^2-D}{4}\right] \\
&= \left[p, p\frac{x+\sqrt{D}}{2}, \frac{x^2-D}{4}\right] \text{ (x being coprime to p since $x^2 \equiv D \bmod 4p$)} \\
&= \left[p, p\frac{x+\sqrt{D}}{2}\right] \\
&= p\left[1, \frac{x+\sqrt{D}}{2}\right] \\
&= p\left[1, \frac{1+\sqrt{D}}{2}\right] \text{ (because $x \equiv 1 \bmod 2$)} \\
&= p \times \mathcal{O}(\mathbb{Q}(\sqrt{D})) \\
&= (p).
\end{aligned}$$

Ideal Classes. Two integral ideals $\mathfrak{i}$ and $\mathfrak{j}$ of the general number field $\mathbf{K}$ belong to the same **class** if $y\mathfrak{i} = x\mathfrak{j}$ for suitable integers x and y from $\mathcal{O}(\mathbf{K})$; in particular, every principal ideal belongs to the class of $(1) = \mathcal{O}(\mathbf{K})$. The formation of classes is respected by products; also, to each ideal $\mathfrak{i}$ corresponds a reciprocal ideal $\mathfrak{j}$ such that $\mathfrak{i}\mathfrak{j}$ belongs to the class of (1). In short, the classes form a multiplicative group $C(\mathbf{K})$. This is the **class group** of $\mathbf{K}$; it is of finite order $=$ the **class number** $h(\mathbf{K})$. Class number $h(\mathbf{K}) = 1$ is equivalent to saying that every integral ideal of $\mathbf{K}$ is principal or, what is the same, that unique factorization into prime integers prevails in $\mathcal{O}(\mathbf{K})$.

Exercise 3. Prove this statement.

Example 5. The finiteness of the class number $h(D)$ of $\mathbb{Q}(\sqrt{D})$ is easily proved. The period ratio of any ideal is of the form $c^{-1}(a + b\sqrt{D})$ with $a, b \in \mathbb{Z}$ and may be taken in the fundamental cell of $PSL(2, \mathbb{Z})$. Then $-1/2 \le a/c < 1/2$ and $c^{-2}(a^2 - D) \ge 1$, so $a^2 \le c^2/4$ and $3c^2/4 \le -D$. This limits the number of classes.

Example 6. The actual determination of the class number of the imaginary quadratic field $\mathbb{Q}(\sqrt{D})$ is a supremely elusive problem first studied by Gauss [1801]. For example, it is already a deep fact that $h(D) = 1$ for $D = -1, -2, -3, -7, -11, -19, -43, -67$ and -163; and for *no other value of* D; see Baker [1971a,b], Heegner [1952], Stark [1967, 1969], and especially Goldfeld [1985], where the whole problem is reviewed. The cases $D = -1, -2, -3, -7, -11$ can be decided by the simple

Criterion. The class number is 1 if, for every element $x \in \mathbf{K}$, integral or not, it is possible to find an integer $y \in \mathcal{O}(\mathbf{K})$ with $N_{\mathbf{K}/\mathbb{Q}}(x - y) < 1$.

Proof. Let x and y be nonvanishing integers of $\mathbf{K}$ and pick $k = k_1$ from $\mathcal{O}(\mathbf{K})$ so that $N(x/y - k_1) < 1$. Then $x = k_1 y + y_1$ with integral remainder y_1 of norm strictly less than $N(y)$. The procedure can be repeated if $y_1 \neq 0$ to produce integral k_2 so that $y = k_2 y_1 + y_2$ with integral remainder y_2 of still smaller norm. The procedure stops when the remainder vanishes: $y_{n-1} = k_n y_n$. The factor y_n is a highest common divisor of x and y. In fact, it is a common divisor and it divides every other such divisor in $\mathcal{O}(\mathbf{K})$; plainly, it is unique up to units. It follows, just as for $\mathbb{Z}$, that every ideal is principal: The ideal $\mathfrak{i}$ has an integral base ω_1, ω_2 and coincides with the principal ideal generated by the highest common divisor of these two numbers.

The criterion is geometrically transparent for $\mathbf{K} = \mathbb{Q}(\sqrt{D})$: It says that, *in the fundamental cell of the lattice* $\mathcal{O}(\mathbf{K})$, *every point is at distance less than 1 from a corner*. If $D \not\equiv 1 \bmod 4$, then $\mathcal{O}(\mathbf{K}) = [1, \sqrt{D}]$, and the criterion requires

$$(1/2)|1 + \sqrt{D}| = (1/2)\sqrt{1 - D} < 1,$$

so $D = -1$ or -2 is allowed but not $D \ge -3$. If $D \equiv 1 \bmod 4$, then $\mathcal{O}(\mathbf{K}) = [1, (1/2)(1 + \sqrt{D})]$, $D \le -3$, and now the criterion requires

$$\frac{1}{2}\left|1 - \frac{1}{2}(1 + \sqrt{D})\right| = \frac{1}{4}\sqrt{1 - D} < 1$$

so $D = -3, -7$, or -11 are permitted but not $D \le -15$.

Bonus. Let $D = -1, -2, -3, -7$, or -11. The rational prime p splits into two ideal factors in $\mathbb{Q}(\sqrt{D})$ if $\left(\frac{D}{p}\right) = +1$ or 0. These factors are principal ideals, so there is an integer $a + b\sqrt{D}$ of $\mathbb{Q}(\sqrt{D})$ with norm $a^2 - b^2 D = p$. This produces a number of classical results for odd primes $p \geq 3$:

Example 7. $D = -1$. Then $\left(\frac{-1}{p}\right) = (-1)^{[(p-1)/2]^2}$ by Section 3.10, so $\left(\frac{-1}{p}\right) = +1$ if and only if $p \equiv 1 \bmod 4$, confirming the theorem of Fermat [1894] that p is the sum of two squares just in this case; compare Section 3.6 and the discussion of quadratic forms in Section 4.5.

Example 8. $D = -2$. Now $\left(\frac{-2}{p}\right) = \left(\frac{-1}{p}\right)\left(\frac{2}{p}\right) = -1$, by ex. 3.10.2, so $\left(\frac{-2}{p}\right) = +1$ if and only if $p \equiv 1$ or $3 \bmod 8$, and $p = a^2 + 2b^2$ with $a, b \in \mathbb{Z}$ just in this case. This is due to Dirichlet.

Exercise 4. What happens for $D = -3, -7, -11$? *Answer*: For $D = -3$, $a^2 + 3b^2$ represents primes $p \equiv 1 \bmod 3$ as well as $p = 3$; for $D = -7$, $a^2 + 7b^2$ represents primes $p \equiv 1, 2, 4 \bmod 7$ as well as $p = 7$; for $D = -11$, $a^2 + 11b^2$ represents primes $p \equiv 1, 3, 5, 9 \bmod 11$ as well as $p = 11$.

The other numbers $D = -19, -43, -67, -163$ producing class number $h(D) = 1$ can also be dealt with in a perfectly elementary manner. The period ratio of an ideal of $\mathbb{Q}(\sqrt{D})$ is of the form $c^{-1}(a + \sqrt{D})$ with either $a^2 \equiv D \bmod c$ or $a \equiv 1 \bmod 2$, $c \equiv 0 \bmod 2$ and $a^2 \equiv D \bmod 2c$ according as $D \not\equiv 1 \bmod 4$ or $D \equiv 1 \bmod 4$, by ex. 1. The further limitations $a^2 \leq c^2/4$ and $3c^2/4 \leq -D$ may be imposed by placing the ratio in the fundamental cell of $PSL(2, \mathbb{Z})$, as noted in example 5. Now $D = -19 \equiv 1 \bmod 4$, and $c = 2$ and 4 are the only even numbers with $c^2 \leq 4 \cdot 19/3$: $c = 2$ permits only $a = -1$, -1 being odd and $(-1)^2 + 19 \equiv 0 \bmod 4$, producing the period ratio $(-1 - \sqrt{-19})/2$, while $c = 4$ is impossible since it permits only $a = \pm 1$ and $(\pm 1)^2 + 19 = 20 \not\equiv 0 \bmod 8$. This identifies the class number of $\mathbb{Q}(\sqrt{-19})$ as 1.

Exercise 5. Check that $h(D) = 1$ for $D = -43, -67, -163$ in the same style.

Exercise 6. Compute the class number of $\mathbb{Q}(\sqrt{-57})$. *Answer*: 3.

More on Class Numbers. Gauss [1801: article 179] expressed the class number of $\mathbb{Q}(\sqrt{D})$ as

$$h(D) = \frac{2}{\pi}\sqrt{|D|}\prod\left[1 - \frac{1}{p}\left(\frac{|D|}{p}\right)\right]^{-1},$$

the product being extended over all rational primes p. Dirichlet [1894: 212–83] found other remarkable expressions for class numbers in terms of quadratic residues; for example, if $-D = p$ is a prime number of the form $4n + 3$ and if n_+ and n_- are the numbers of quadratic residues and nonresidues mod p between 1 and $(p - 1)/2$, then $h = n_+ - n_-$ or $(1/3)(n_+ - n_-)$ according as $p \equiv 7$ or $p \equiv 3 \bmod 8$. Gauss [1801] conjectured and Siegel [1936] proved that $h(D) = |D|^{1/2+o(1)}$ for large D. The proof is based upon the beautiful identity of Dirichlet [1894]

$$h(D) = \sqrt{|D|}\frac{1}{\pi}\sum_{n=1}^{\infty}\left(\frac{D}{n}\right)n^{-1}.$$

Oesterle [1985] has obtained the effective bound

$$h(D) > \frac{1}{7000}(\log|D|)\prod\left(1 - \frac{[2\sqrt{d}]}{d+1}\right),$$

where the product runs over the proper divisors of D. This may be used to identify all integers $D < 0$ with a given class number, up to a finite amount of computation. Cox [1989], Ireland and Rosen [1990], and Zagier [1984] are recommended for more information about this question; see also Borevich and Shafarevich [1966] for this and for the class numbers of more general number fields.

6.3 Class Invariants and Kronecker's *Jugendtraum*

Let $\mathfrak{i}$ be an ideal of the ring of integers $\mathcal{O}(\mathbf{K})$ of an imaginary quadratic field $\mathbf{K} = \mathbb{Q}(\sqrt{D})$. Then $\mathbf{X} = \mathbb{C}/\mathfrak{i}$ is an elliptic curve, and it is plain that two such curves $\mathbb{C}/\mathfrak{i}$ and $\mathbb{C}/\mathfrak{j}$ are conformally equivalent if and only if the underlying ideals $\mathfrak{i}$ and $\mathfrak{j}$ belong to the same class $\mathbf{k}$.

The common value of the absolute invariants of $\mathbb{C}/\mathfrak{i}$ and $\mathbb{C}/\mathfrak{j}$ is the **class invariant** $j(\mathbf{k})$. Distinct classes have distinct class invariants; moreover, the class invariants of one imaginary quadratic field are distinct from those of any other.

Exercise 1. Check all that. *Hint*: D is square-free.

Exercise 2. $\mathbf{X} = \mathbb{C}/\mathbb{L}$ admits nontrivial complex multiplication by $m \in \mathbb{Z}$ only if its period ratio belongs to an imaginary quadratic field $\mathbf{K} = \mathbb{Q}(\sqrt{D})$; compare Section 2.6. Then the ring of all complex multiplications of $\mathbf{X}$ is part of $\mathcal{O}(\mathbf{K})$. Why? It is the full ring $\mathcal{O}(\mathbf{K})$ if and only if the absolute invariant of $\mathbf{X}$ is a class invariant of $\mathbf{K}$. Check these statements.

Kronecker's Jugendtraum. The class group $C(\mathbf{K})$ of $\mathbf{K} = \mathbb{Q}(\sqrt{D})$ is comprised of h classes $\mathbf{k}$, each with a different class invariant $j(\mathbf{k})$. The chief facts about these numbers are sufficiently striking to stand without extra comment.

Item 1. The class invariants $j(\mathbf{k}_1), \ldots, j(\mathbf{k}_h)$ are algebraic integers; in fact, they comprise the full set of roots of a polynomial of degree h from[3] $\mathbb{SZ}[x]$ which is irreducible, not only over $\mathbb{Q}$, but over $\mathbf{K} = \mathbb{Q}(\sqrt{D})$ as well.

Item 2. The splitting field $\mathbf{K}_1$ obtained by the adjunction of the class invariants to $\mathbf{K} = \mathbb{Q}(\sqrt{D})$ is commutative, meaning that its Galois group $\mathfrak{G}$ over the ground field $\mathbf{K}$ is such; in fact, $\mathfrak{G}$ is a copy of the class group $C(\mathbf{K})$, the rule associating the class $\mathbf{k}$ to the corresponding substitution $\sigma_{\mathbf{k}} \in \mathfrak{G}$ being $\sigma_{\mathbf{k}}\colon j(\mathbf{k}') \mapsto j(\mathbf{k}^{-1}\mathbf{k}')$.

Item 3. $\mathbf{K}_1$ is the absolute class field of $\mathbf{K} = \mathbb{Q}(\sqrt{D})$, meaning that it is the biggest commutative extension of the latter in which prime ideals of the little field do not ramify but merely split into distinct prime factors in the big field.

The rest of the chapter is devoted to items 1 and 2 following Fricke [1928] and Weber [1908], with simplifications made possible by more modest objectives. Item 3 is a consequence of the class field theory initiated by Hilbert in his [1897] report on number theory. This provides an overview of the totality of commutative extensions of $\mathbf{K}$. It falls outside the scope of the present account, but see Borel et al. [1966], Cox [1989], and Vladut [1991] for more information, and also Reid [1970], who tells the story behind this report and some details of Hilbert's life. Siegel [1949] found a remarkable supplement to item 1: If ω is an algebraic number but not a quadratic irrationality, then $j(\omega)$ is a transcendental number; see Baker [1975] for details.

Aside 1. Just as the class field of $\mathbb{Q}(\sqrt{D})$ is produced by the adjunction of **singular values** of the absolute invariant $j(\mathbf{k})$ taken at suitable quadratic irrationalities, so also the cyclotomic fields of the Kronecker–Weber theorem are produced by the adjunction to $\mathbb{Q}$ of singular values of the invariant $j_\infty(\omega) = \exp(2\pi\sqrt{-1}\omega)$ taken at rational values of ω.

Aside 2. $h(D) = 1$ if and only if $j(\mathbf{k}) \in \mathbb{Z}$ for every class $\mathbf{k}$; for example, $j(\sqrt{-3}) = 54000$, by ex. 4.12.3, so $h(-3) = 1$, as in example 2.6.

[3] $\mathbb{SZ}[x]$ is the class of polynomials $\mathbb{Z}[x]$ with the added restriction that the top coefficient be 1.

6.4 Application of the Modular Equation

The proof that the class invariants are algebraic integers employs the modular equations $F_n(x, j(\omega)) = 0$ of levels $n \geq 2$ discussed in Section 4.13.

Step 1. The class invariant $j(\mathbf{k})$ is a root of the diagonal polynomial $G_n(x) = F_n(x, x)$ for suitable n.

Proof. Fix an ideal $\mathfrak{i}$ of class $\mathbf{k}$ and an integer $m \in \mathcal{O}(\mathbf{K})$ with square-free norm $n = N_{\mathbf{K}/\mathbb{Q}}(m) \geq 2$: $m = 1+\sqrt{-1}$ will do if $D = -1$ and $m = \sqrt{D}$ if $D \leq -2$, so that $n = -D$, with the exception of $n = 2$ for $D = -1$. Then m is a complex multiplier of $\mathfrak{i}$ viewed as a lattice $\mathbb{Z}\omega_1 \oplus \mathbb{Z}\omega_2$, that is, $m\omega_2 = a\omega_2 + b\omega_1$ and $m\omega_1 = c\omega_2 + d\omega_1$ with $a, b, c, d \in \mathbb{Z}$, and n appears as the index $|ad - bc|$ of $m\mathfrak{i}$ in $\mathfrak{i}$, by comparison of areas of fundamental cells: area$(m\mathfrak{i}) = (m\bar{m}) \times$ area $\mathfrak{i}$. Well and good. Now n is square-free, so the numbers a, b, c, d have highest common divisor 1, and the substitution $[ab/cd]$ is a modular correspondence of level n in the language of Section 4.13, and since the period ratio $\omega = \omega_2/\omega_1$ is a fixed point of that substitution, the class invariant $x = j(\mathbf{k}) = j(\omega)$ is a root of the modular equation $F_n(x, y) = 0$ with $y = j(\mathbf{k})$.

Step 2. The last statement is *not void*; in fact, the diagonal polynomial is of positive degree equal to $2\times$ the sum of the divisors d of n with $d > \sqrt{n}$.

Proof. $F_n(x, j(\omega))$ is the product of $x - j(d^{-1}(a\omega + b))$ for $[ab/0d]$ running over the full family of reduced correspondences with a, b, d coprime, $ad = n$, and $0 \leq b < d$. The square-free character of n simplifies life: If d divides n, then it is already prime to $a = n/d$ and every $0 \leq b < d$ is permitted, so

$$F_n(x, j(\omega)) = \prod_{\substack{0 \leq b < d \\ d|n}} \left[x - j\left(\frac{a\omega + b}{d}\right)\right].$$

The degree of G_n is now read off with the aid of the expansion $j(\omega) = q^{-1} + \cdots$ at the cusp $\omega = \sqrt{-1}\infty$:[4] To leading order,

$$G_n(j(\omega)) = \prod_{d|n, 0 \leq b < d} [q^{-1} - q^{-a/d} e^{-2\pi\sqrt{-1}b/d}],$$

[4] Here $q = e^{2\pi\sqrt{-1}\omega}$.

and no factor vanishes since $a/d = 1$ violates the square-free character of n. It follows that G_n is of degree

$$\sum_{\substack{0 \le b < d \\ a/d < 1}} 1 + \sum_{\substack{0 \le b < d \\ a/d > 1}} \frac{a}{d} = \sum_{d^2 > n} d + \sum_{d^2 < n} n/d = 2 \times \sum_{d > \sqrt{n}} d.$$

Exercise 1. The degree of F_n is the sum of the divisors of n, by ex. 4.13.2 and a little further thought, taking account of the square-free character of n; for example, $\deg F_{10} = 1+2+5+10 = 18$ while $\deg G_{10} = 2 \times (5+10) = 30$.

Step 3. The proof that $j(\mathbf{k})$ is an algebraic integer is finished by confirming that $G_n \in S\mathbb{Z}[x]$. The top coefficient is easy: The expansion at the cusp confirms that its value is

$$\prod_{\substack{0 \le b < d \\ d^2 < n}} (-1) \times e^{-2\pi\sqrt{-1}b/d} = \prod_{d^2 < n} (-1) \times e^{-\pi\sqrt{-1}(d-1)} = \pm 1.$$

The rest follows from $G_n \in \mathbb{Z}[x]$, but it is just as easy to prove a little more: that *the full polynomial $F_n(x, y)$ has rational integral coefficients.*

Proof. Let $y = j(\omega)$ as usual. The elementary symmetric functions of the reduced invariants $x = j(d^{-1}(a\omega+b))$ are polynomials in y. It is the coefficients of *these* polynomials that are in question. The expansion $j = q^{-1} + \cdots$ has rational integral coefficients, as noted in sample 5 of Section 3.5, so

$$j\left(\frac{a\omega + b}{d}\right) = e^{-2\pi\sqrt{-1}b/d} q^{-a/d}$$

$$\times \left[\begin{matrix} 1 + \text{ a sum of whole powers of} \\ e^{2\pi\sqrt{-1}b/d} q^{a/d} \text{ with rational coefficients} \end{matrix}\right].$$

Let $\epsilon = e^{2\pi\sqrt{-1}/n}$, observe that $e^{2\pi\sqrt{-1}b/d} = \epsilon^{ab}$ is an integer of the field $\mathbb{Q}(\epsilon)$, and let σ be a Galois substitution of $\mathbb{Q}(\epsilon)$ over the ground field $\mathbb{Q}$. Then $\sigma(\epsilon)$ is a primitive nth root of unity $= \epsilon^f$, say, with f coprime to n, and σ maps $e^{2\pi\sqrt{-1}b/d} = \epsilon^{ab}$ to ϵ^{fab}. This effects a permutation of the numbers $e^{2\pi\sqrt{-1}b/d}$, f being coprime to d, and so fixes the coefficients of the expansion at the cusp of any symmetric function of the reduced invariants. It follows that these coefficients could only be rational integers. The rest will be plain from the self-evident principle that if $P \in \mathbb{C}[x]$ and if the expansion $P(j(\omega))$ in powers of q has coefficients in $\mathbb{Z}$, then so does $P(x)$.

Exercise 2. Check it, from the top power of P on down.

6.5 The Class Polynomial

The class invariants $j(\mathbf{k})$ are now known to be roots of the diagonal polynomial $G_n(x)$ for $n = 2$ if $D = -1$ and for $n = -D$ if $D \le -2$. The next task is to extract from G_n the **class polynomial** $H(x) = \prod[x - j(\mathbf{k})]$ corresponding to the totality of class invariants of $\mathbf{K}$ and to prove that it has rational integral coefficients. The proof of its irreducibility is postponed. The cases $D = -1, -2$, and -3 may be excluded since the class numbers are 1 and the class invariants belong to $\mathbb{Z}$ by explicit evaluation:

$$D = -1\colon \mathcal{O}(\mathbf{K}) = [1, \sqrt{-1}], \qquad j(\mathbf{k}) = j(\sqrt{-1}) = 1728 = 2^6 \cdot 3^3;$$

$$D = -2\colon \mathcal{O}(\mathbf{K}) = [1, \sqrt{-2}], \qquad j(\mathbf{k}) = j(\sqrt{-2}) = 2^6 \cdot 5^3;$$

$$D = -3\colon \mathcal{O}(\mathbf{K}) = \left[1, \left(\frac{1}{2}(-1 + \sqrt{-3})\right)\right] \; j(\mathbf{k}) = j\left(\frac{1}{2}(-1 + \sqrt{-3})\right) = 0.$$

The proof for $D \le -5$ relies on the following facts.

Fact 1. If $D \not\equiv 1 \bmod 4$, then the class invariants are the simple roots of $G_n(x) = 0$.

Fact 2. If $D \equiv 1 \bmod 4$, then the class invariants are still simple roots of $G_n(x) = 0$, but not the only ones. Now $(n + 1)/4$ is a whole number and the class invariants can be distinguished from the extraneous simple roots of $G_n(x) = 0$ by the fact that they are *also* roots of $G_{(n+1)/4}(x) = 0$ whereas the extraneous roots are *not*.

Unlike $n = 2$ or $-D$, the number $(n + 1)/4$ need not be square-free for $D \equiv 1 \bmod 4$; for example, $n = -D = 15$ produces $(15 + 1)/4 = 2^2$. Luckily this is irrelevant. The proof is postponed in favor of the moral, which is stated in two parts according to the value of $D \bmod 4$.

Case 1. $D \not\equiv 1 \bmod 4$. $G_n \in \mathbb{SZ}[x]$ and the class invariants are its simple roots. Move up to the splitting field of $G_n(x) = 0$ over $\mathbb{Q}$ and note that the equality $G'_n(x) = 0$ is invariant under its Galois group. It follows that the simple roots of $G_n(x) = 0$ are permuted by the group, so the class polynomial (= the product of the simple factors of G_n) is fixed thereby and so belongs to $\mathbb{SQ}[x]$.[5] The rest is plain from Gauss's lemma: *If a polynomial of class* $\mathbb{SZ}[x]$ *factors in* $\mathbb{SQ}[x]$, *then each factor belongs to* $\mathbb{SZ}[x]$ *already*. Pollard [1950: 27–9] gives the standard proof.

[5] $\mathbb{SQ}[x]$ means elements of $\mathbb{Q}[x]$ with top coefficient ± 1.

Case 2. $D \equiv 1 \bmod 4$. The product of the simple factors of G_n belongs to $\mathbb{SZ}[x]$ as before. Now the class polynomial is the highest common divisor of this product and $G_{(n+1)/4} \in \mathbb{SQ}[x]$. Gauss's lemma does the rest.

Now comes the proof of facts 1 and 2 concerning the class invariants and the roots of G_n and $G_{(n+1)/4}$.

Step 1. The root $x = j(\omega)$ of $G_n(x) = 0$ is simple or not according as $j(\omega)$ coincides with one or more than one of the roots $j(d^{-1}(a\omega + b))$, $ad = n$, $0 \le b < d$, of $F_n(x, j(\omega)) = 0$, with the proviso, in the case of one coincidence, that ω does not belong to $\mathbb{Q}(\sqrt{-1})$ or $\mathbb{Q}(\sqrt{-3})$.

Proof. A coincidence is necessary to have a root at all. Let the fixed root $x_0 = j(\omega)$ have f such coincidences and let $x' = j(\omega')$ be variable. Then $G_n(x')$ contains an equal number of vanishing factors $j(\omega') - j((a\omega' + b)/(c\omega' + d))$ in which $[ab/cd]$ is an unreduced modular correspondence of level n *fixing* ω. The other factors of $G_n(x')$ do not vanish at $x' = x$. Now, with the fields $\mathbb{Q}(\sqrt{-1})$ and $\mathbb{Q}(\sqrt{-3})$ excluded by the proviso, $j(\omega')$ takes the value $j(\omega)$ simply to leading order, so

$$j(\omega') - j\left(\frac{a\omega' + b}{c\omega' + d}\right) = [j(\omega') - j(\omega)] \times \left[1 - \frac{ad - bc}{(c\omega + d)^2}\right],$$

and $G_n(x')$ vanishes simply at $x' = x$ only if $f = 1$ and $(c\omega + d)^2 \ne ad - bc = n$. The proof is finished by noting that $c\omega + d = \pm\sqrt{n}$ requires $c = 0$ and $d^2 = n$, violating the square-free character of n.

Step 2. If $D \not\equiv 1 \bmod 4$, the class invariants of $\mathbb{Q}(\sqrt{D})$ are the simple roots of $G_n(x) = 0$.

Proof. This is more complicated but be patient. The number $x = j(\omega)$ is a root of $G_n(x) = 0$ if and only if ω is fixed by a modular correspondence $[ab/cd]$ of level n, which is to say, $c\omega^2 + (d - a)\omega - b = 0$. Let A, B, C be the quotients of $c, d - a, -b$ by their highest common divisor $l \ge 1$, so that $A\omega^2 + B\omega + C = 0$ and $\omega = (2A)^{-1}(-B + \sqrt{D'})$ with $D' = B^2 - 4AC < 0$. A, B, C are specified by ω up to a common sign; in particular, D' is unique. What is complicated is that the *correspondence* is not unique: If $a + d = m$, then

$$a = \frac{m - lB}{2}, \quad b = -lC, \quad c = lA, \quad d = \frac{m + lB}{2}$$

and

$$4n = 4(ad - bc) = (a + d)^2 - (a - d)^2 - 4bc = m^2 - l^2 D',$$

so the number of distinct classes of correspondences fixing ω may, and in fact will, depend upon the number of representations of $4n$ by the quadratic form $m^2 - l^2 D'$. There are two cases.

Case 1. If $4n$ can be represented as $m^2 - l^2 D'$ with $m \neq 0$, then $x = j(\omega)$ is a multiple root of $G_n(x) = 0$.

Proof. The reason is that if $M = [ab/cd]$ is a correspondence of level n fixing ω, then the same is true of $nM^{-1} = [d-b/-ca]$, and these cannot be equivalent under the left action of the modular group; indeed, equivalence takes place only if $n^{-2}M^2$ is a modular substitution, and this requires n to divide the trace

$$\begin{aligned}
\operatorname{tr}(M^2) = c^2 + 2bc + d^2 &= \left(\frac{m - lB}{2}\right)^2 - 2l^2AC + \left(\frac{m + lB}{2}\right)^2 \\
&= \frac{1}{2}m^2 + \frac{1}{2}l^2B^2 - 2l^2AC \\
&= \frac{1}{2}(m^2 + l^2D') \\
&= m^2 - 2n.
\end{aligned}$$

But then n divides m^2, and this cannot happen unless $n = 1, 2$, or 3, which was expressly excluded; in fact, if n divides m^2, it must also divide m because it is square-free, and that is contradictory if $n \geq 5$, in view of $4n = n^2 - l^2D' \geq n^2$ which entails $4 \geq m^2/n \geq m \geq n \geq 5$.

Case 2. If $4n$ can be represented by $m^2 - l^2 D'$ only with $m = 0$, then the correspondence $[ab/cd]$ fixing ω is fully specified and $x = j(\omega)$ is a simple root of $G_n[x] = 0$ *provided* ω does not belong to $\mathbb{Q}(\sqrt{-1})$ or $\mathbb{Q}(\sqrt{-3})$, by step 1. But what, in fact, are these simple roots and do they satisfy the proviso? The answer is the content of the present step: *They are the class invariants and the proviso is met.* The proof is broken into two parts.

Part 1. The class invariant $j(\mathbf{k})$ is a root satisfying the proviso in view of the exclusion of $D = -1$ and $D = -3$; also, the expression $\omega = (2A)^{-1}(-B + \sqrt{D'})$ implies $D' = f^2 D$ with $f \in \mathbb{Z}$, D being square-free. Now $n = -D$, so

$$4n = m^2 - l^2 D' = m^2 - l^2 f^2 D = m^2 + l^2 f^2 n,$$

$lf = 0, 1$, or 2, and either $lf = 0$ and $m^2 = 4n^2$, violating the fact that n is square-free; or $lf = 1$, $3n = m^2$, and $n = 3$, which is excluded; or else $lf = 2$

and $m = 0$, which is what is wanted. In short, case 2 prevails, the proviso is satisfied, and $x = j(\mathbf{k})$ is a simple root.

Part 2. It remains to prove that every root $x = j(\omega)$ that falls under case 2 is a class invariant $j(\mathbf{k})$. Now you begin from $4n = -l^2 D'$ and you notice that either $l = 1$ and $D' = 4D$, or $l = 2$ and $D = D'$, or else the square-free character of n is violated by the fact that l is divisible by either an odd prime or 4. If $l = 1$ and $D' = 4D$, then $B = 2a \equiv 0 \bmod 2$, $B^2/4 \equiv D \bmod A$ in view of $4D = D' = B^2 - 4AC$, and ex. 2.1 guarantees that the lattice $[1, \omega] = [1, A^{-1}(-B/2 + \sqrt{D})]$ is closed under multiplication by $\mathcal{O}(\mathbf{K})$; in particular, ω is the period ratio of an ideal of $\mathcal{O}(\mathbf{K})$ and $j(\omega)$ is a class invariant. If $l = 2$ and $D' = D$, the same conclusion is reached since $B^2 = D + 4AC \equiv D \bmod 4A$, as required for the application of ex. 2.1 to $[1, \omega] = [1, (2A)^{-1}(-B + \sqrt{D})]$. The proof is finished.

Step 3. If $D \equiv 1 \bmod 4$, then the class invariants $j(\mathbf{k})$ of $\mathbb{Q}(\sqrt{D})$ are still simple roots of $G_n(x) = 0$, *but there are more.* The class invariants arise, as in part 2 of the preceding step, from $l = 2$, $D' = D$, whereas the *extraneous* simple roots arise from $l = 1$, $D' = 4D$.

Proof. $D \not\equiv 1 \bmod 4$ was not used in step 2 until part 2, so the conclusions are the same up to that point. The possibilities are as before: Either $l = 1$ and $D' = 4D$ or $l = 2$ and $D' = D$; in both cases; $\omega = (2A)^{-1}(-B + \sqrt{D'})$ does not belong to $\mathbb{Q}(\sqrt{-1})$ or to $\mathbb{Q}(\sqrt{-3})$. Now if $l = 2$ and $D' = D$, then $\omega = (2A)^{-1}(-B + \sqrt{D})$, $B \equiv 1 \bmod 2$, and $B^2 \equiv D \bmod 4A$ in view of $B^2 - 4AC = D$ and the indivisibility of D by 4, whereupon ex. 2.1 may be used as before to check that $x = j(\omega)$ is a class invariant of $\mathbb{Q}(\sqrt{D})$. It remains to show that $x = j(\omega)$ *is not* a class invariant if $l = 1$ and $D' = 4D$. But in this case ex. 2.1 states that $\omega = A^{-1}(-B/2 + \sqrt{D})$ is the period ratio of an ideal of $\mathcal{O}(\mathbf{K})$ only if $B/2 \equiv 1 \bmod 2$, $A \equiv 0 \bmod 2$, and $B^2/4 \equiv D \bmod 2A$, in which case $4AC = B^2 - 4D' = B^2 - D \equiv 0 \bmod 8A$ and 2 divides C, violating the coprime character of A, B, C. The proof is finished.

Step 4. Check that for $D \equiv 1 \bmod 4$, the class invariants are roots of $G_{(n+1)/4}(x) = 0$.

Proof. $n' = (n + 1)/4 \in \mathbb{Z}$ since $n = -D \equiv 3 \bmod 4$. Now the class invariant $x = j(\omega)$ is a simple root of $G_n(x) = 0$, so, in the parlance of step 3, $l = 2$, $D' = D$, $B \equiv 1 \bmod 2$, and $4n' = 1 + n = 1^2 - 1^2 D'$. It is immediate that the substitution $[ab/cd]$ with $a = (1 - B)/2$, $b = -C$, $c = A$, and $d = (AB)/2$ is a correspondence of level n' fixing ω, which is to say $G_{n'}(x) = 0$.

Step 5. Verify that for $D \equiv 1 \bmod 4$, the extraneous simple roots of $G_n(x) = 0$ arising from $l = 1$ and $D' = 4D$ *do not solve* $G_{(n+1)/4}(x) = 0$.

Proof. The point is that $4n' = n + 1 = (m')^2 - (l')^2 D' = (m')^2 + 4(l')^2 n$ is not solvable in whole numbers: l' has to vanish, so any correspondence $[ab/cd]$ of level n' fixing ω must have $a = m'/2$, $b = 0$, $c = 0$, $d = m'/2$. This is not even a correspondence unless $m' = 2$ since a, b, c, d have to be coprime. But if $m' = 2$ then $n + 1 = (m')^2 = 4$ and n has the excluded value 3. The discussion is finished.

6.6 Class Invariants at a Prime Level

The modular equation $F_p(x, j(\mathbf{k})) = 0$ at a prime level p provides new information about the class invariants $j(\mathbf{k})$, preparatory to the proof that the class polynomial is irreducible.

Ambiguous Primes. The prime ideals $\mathfrak{p} \subset \mathcal{O}(\mathbf{K})$ are detected by factoring the rational primes $p \in \mathbb{Z}$. The main facts are recapitulated from Section 2: $(p) = \mathfrak{p}$ is prime as it stands if $\left(\frac{D}{p}\right) = -1$ or if $p = 2$ and $\left(\frac{D}{p}\right) = +1$; otherwise, $(p) = \mathfrak{p}\mathfrak{p}^{-1}$, $\mathfrak{p}^{-1}$ being the ideal $\bar{\mathfrak{p}}$ conjugate to $\mathfrak{p}$. The new notation emphasizes the fact that $\mathfrak{p}^{-1}$ is reciprocal to $\mathfrak{p}$ in the class group. The ideal $\mathfrak{p}$ is said to be **ambiguous** if $\mathfrak{p}^{-1} = \mathfrak{p}$ in the class group. The nomenclature derives from the fact that $\mathfrak{p}$ is ambiguous if and only if its period ratio, reduced by the action of Γ_1, falls on the boundary of the fundamental cell. $\left(\frac{D}{p}\right) = 0$ produces only ambiguous primes, while $\left(\frac{D}{p}\right) = +1$ can produce both kinds, as you may easily check.

Exercise 1. Do that.

Exercise 2. Check that $\mathfrak{p} = [3, 1 + \sqrt{-53}]$ is an unambiguous prime ideal of $\mathbb{Q}(\sqrt{-53})$, whereas $\mathfrak{q} = [31, 3 + \sqrt{-53}]$ is an ambiguous one. *Hint* for $\mathfrak{q}$: The period ratio is equivalent to $1 - [(1/31)(3 + \sqrt{-53})]^{-1} = (-1/2)(1 + \sqrt{-53})$ under the action of the modular group.

Class Invariants. The principal facts to be elicited can now be stated. Fix an odd rational prime p with $\left(\frac{D}{p}\right) = 0$ or $+1$, so that $(p) = \mathfrak{p} \times \mathfrak{p}^{-1}$, and note that the class invariants $j(\mathfrak{p}\mathbf{k})$ and $j(\mathfrak{p}^{-1}\mathbf{k})$ are distinct from $j(\mathbf{k})$ and coincide precisely when $\mathfrak{p}$ is ambiguous.

Fact 1. $j(\mathfrak{p}\mathbf{k})$ and $j(\mathfrak{p}^{-1}\mathbf{k})$ are the only class invariants that satisfy $F_p(x, j(\mathbf{k})) = 0$.

Fact 2. If $\mathfrak{p}$ is ambiguous, then $j(\mathfrak{p}\mathbf{k}) = j(\mathfrak{p}^{-1}\mathbf{k})$ belongs to the field $\mathbb{Q}_\mathbf{k}$ obtained from $\mathbb{Q}$ by adjunction of $j(\mathbf{k})$.

Fact 3. If $\mathfrak{p}$ is unambiguous, then either the distinct invariants $j(\mathfrak{p}\mathbf{k})$ and $j(\mathfrak{p}^{-1}\mathbf{k})$ belong to $\mathbb{Q}_\mathbf{k}$ already or else they are conjugate over $\mathbb{Q}_\mathbf{k}$ and their common splitting field over the latter is $\mathbb{Q}_\mathbf{k}(\sqrt{D}) = \mathbb{Q}(\sqrt{D})$ with $j(\mathbf{k})$ adjoined.

Proof of fact 1. Fix an ideal $\mathfrak{i}$ of class $\mathbf{k}$. Then $\mathfrak{p}\mathfrak{i} \subset \mathfrak{i}$ is of index[6] $N_{\mathbf{K}/\mathbb{Q}}(\mathfrak{p}) = p$ in the latter. Now $\mathfrak{i} = [\omega_1, \omega_2]$ with integral ω_1 and ω_2 from $\mathcal{O}(\mathbf{K})$, and $\mathfrak{p}\mathfrak{i} = [c\omega_2 + d\omega_1, a\omega_2 + b\omega_1]$ with $a, b, c, d \in \mathbb{Z}$ and $ad - bc =$ the stated index p. This means that $[ab/cd]$ is a modular correspondence of level p mapping the period ratio $\omega = \omega_2/\omega_1$ of $\mathfrak{i}$ to the period ratio ω' of $\mathfrak{p}\mathfrak{i}$, and it follows from $j(\mathbf{k}) = j(\omega)$ that $x = j(\mathfrak{p}\mathfrak{i}) = j(\omega')$ is a root of $F_p(x, j(\mathbf{k})) = 0$; naturally, the same applies to $j(\mathfrak{p}^{-1}\mathbf{k})$. The rest of the proof is similar. The class invariant $x = j(\mathbf{k}') = j(\omega')$ is a root of $F_p(x, j(\mathbf{k})) = 0$ only if ω' is the image of ω under a correspondence $[ab/cd]$ of level p. This means that the class $\mathbf{k}'$ contains an ideal $\mathfrak{i}' = [c\omega_2 + d\omega_1, a\omega_2 + b\omega_1]$ of index $p = ad - bc$ in the ideal $\mathfrak{i} = [\omega_1, \omega_2] \in \mathbf{k}$ employed before. Now inclusion of ideals is the same as divisibility, so $\mathfrak{i}'$ is the product of $\mathfrak{i}$ by some ideal $\mathfrak{q} \subset \mathcal{O}(\mathbf{K})$, of norm $N_{\mathbf{K}/\mathbb{Q}}(\mathfrak{q}) = p$. But then $\mathfrak{q}$ must be $\mathfrak{p}$ or $\mathfrak{p}^{-1}$, as you can check by splitting $\mathfrak{q}$ into prime ideals and using the multiplicative property of the norm; in short, $j(\mathbf{k}')$ can only be $j(\mathfrak{p}\mathbf{k})$ or $j(\mathfrak{p}^{-1}\mathbf{k})$.

Proof of fact 2. This is easy: $\mathfrak{p}$ is ambiguous, so $\mathfrak{p} = \mathfrak{p}^{-1}$ in the class group and $x = j(\mathfrak{p}\mathbf{k}) = j(\mathfrak{p}^{-1}\mathbf{k})$ counted once is the *only* common root of the class polynomial and $F_p(x, j(\mathbf{k}))$. Their highest common divisor belongs to $\mathbb{Q}_\mathbf{k}[x]$, and it is of degree 1, so its only root $x = j(\mathfrak{p}\mathbf{k})$ lies in $\mathbb{Q}_\mathbf{k}$.

Proof of fact 3. At first, the argument proceeds just as for fact 2: $\mathfrak{p}$ is now unambiguous, the distinct invariants $j(\mathfrak{p}\mathbf{k})$ and $j(\mathfrak{p}^{-1}\mathbf{k})$ are the only common roots of the class polynomial and $F_p(x, j(\mathbf{k}))$, and either they belong to $\mathbb{Q}_\mathbf{k}$ already, or else they are the roots of an irreducible quadratic polynomial $[x -$

[6] $\mathbf{K} = \mathbb{Q}(\sqrt{D})$ as usual.

$j(\mathfrak{p}\mathbf{k})][x - j(\mathfrak{p}^{-1}\mathbf{k})]$ of class $\mathbb{Q}_{\mathbf{k}}[x]$, which is to say that their common splitting field is of degree 2 over $\mathbb{Q}_{\mathbf{k}}$. The rest of the proof deals with the second case. Its goal is to check that $j(\mathfrak{p}\mathbf{k})$ and $j(\mathfrak{p}^{-1}\mathbf{k})$ split precisely in $\mathbb{Q}_{\mathbf{k}}(\sqrt{D})$, which is to say that $j(\mathfrak{p}\mathbf{k}) - j(\mathfrak{p}^{-1}\mathbf{k})$ belongs to that field.

Step 1. The correspondences of level p produce, from $\mathbb{L} = [\omega_1, \omega_2]$ any lattice, $p + 1$ distinct lattices

$$\mathbb{L}_\infty = [\omega_1, p\omega_2], \quad \mathbb{L}_q = [p\omega_1, q\omega_1 + p\omega_2]\ (0 \le q < p)$$

with relative discriminants

$$k_\infty = \frac{\Delta(\mathbb{L}_\infty)}{\Delta(\mathbb{L})}, \quad k_q = \frac{\Delta(\mathbb{L}_q)}{\Delta(\mathbb{L})}\ (0 \le q < p).$$

These depend upon only the period ratio $\omega = \omega_2/\omega_1$, and the action of the modular group on it permutes them in the same way as it permutes the roots

$$j_\infty = j(\mathbb{L}_\infty), \quad j_q = j(\mathbb{L}_q)\ (0 \le q < p)$$

of the modular equation $F_p(x, j(\omega)) = 0$; in particular,

$$\frac{k_\infty}{x - j_\infty} + \sum_{0 \le q < p} \frac{k_q}{x - j_q} \equiv \frac{N(x, j(\omega))}{F_p(x, j(\omega))}$$

is invariant under the modular group. Here, $N(x, j(\omega)) \in \mathbb{Q}[x, j(\omega)]$, as the notation is meant to convey and as you may check from the expansions $j = q^{-1} + \cdots$ and $\Delta = (2\pi)^{12} q + \cdots$ at the cusp $\sqrt{-1}\infty$, which reveal that, for fixed x, $N(x, j(\omega))$ has no poles except, perhaps, at the cusp. Now take the residue of the display at $x = j_\infty$ to produce

$$\frac{\Delta(\mathbb{L}_\infty)}{\Delta(\mathbb{L})} = k_\infty = \frac{N(j_\infty, j(\omega))}{F'_p(j_\infty, j(\omega))}.$$

This was the goal of step 1.

Step 2. In step 1, j is construed as a *transcendental* element, so $F'_p(j_\infty, j(\omega))$, as an element of $\mathbb{Q}(j_\infty, j(\omega))$, does not vanish. But step 3 requires the *numerical* use of the final formula in case $\mathbb{L} = \mathfrak{i}$ is an ideal of $\mathcal{O}(\mathbf{K})$, so you need to know that $F'_p(j_\infty, j)$ does not vanish when ω is the period ratio of such an ideal.

Proof. $F'_p(j_\infty(\omega), j(\omega))$ vanishes only if $j_\infty(\omega) = j(p\omega)$ coincides with $j(p^{-1}(\omega + q))$ for some $0 \le q < p$, which is to say that the lattices $\mathfrak{i}_\infty = [\omega_1, p\omega_2]$ and $\mathfrak{i}_q = [p\omega_1, q\omega_1 + \omega_2]$ are related by integers α and β of $\mathcal{O}(\mathbf{K})$: $\alpha\mathfrak{i}_\infty = \beta\mathfrak{i}_q$. Notice that α and β have equal norm since $\mathfrak{i}_\infty$ and $\mathfrak{i}_q$ have equal index p in

$\mathfrak{i}$. Now $\alpha p\mathfrak{i} \subset \alpha\mathfrak{i}_\infty = \beta\mathfrak{i}_q \subset \beta\mathfrak{i}$, and as inclusion of ideals is the same as divisibility, so $\alpha p\mathfrak{i} = \beta\mathfrak{q}\mathfrak{i}$ for some ideal $\mathfrak{q}$ of norm $N_{\mathbf{K}/\mathbb{Q}}\mathfrak{q} = p^2$. The prime factorization of $\mathfrak{q}$ now reveals that there are only three possibilities: $\mathfrak{q} = \mathfrak{p}^2$, or $\mathfrak{q} = \mathfrak{p}^{-2}$, or else $\mathfrak{q} = \mathfrak{p}\mathfrak{p}^{-1}$. These are tested against $\alpha(p) = \beta\mathfrak{q}$ and each of them fails: The unambiguous character of $\mathfrak{p}$ spoils the first two, and the third implies that α/β is a unit (± 1) of $\mathbf{K}$ so $\mathfrak{i}_\infty = \mathfrak{i}_q$, and this is not possible either since $q\omega_1 + \omega_2$ does not belong to $\mathfrak{i}_\infty$.

Step 3. The formula of step 1 applies not just to $\mathfrak{i}_\infty$ and the corresponding invariant j_∞ but to $\mathfrak{i}_q$ and j_q for every $0 \le q < p$, as well. Now $\mathfrak{p}\mathfrak{i}$ is related to $\mathfrak{p}$ by a correspondence of level p, as in the proof of fact 1, so

$$\frac{\Delta(\mathfrak{p}\mathfrak{i})}{\Delta(\mathfrak{i})} = \frac{N(j(\mathfrak{p}\mathbf{k}), j(\mathbf{k}))}{F_p'(j(\mathfrak{p}\mathbf{k}), j(\mathbf{k}))},$$

in which the denominator does not vanish, by step 2. Let e be the order of $\mathfrak{p}$ in the class group, so $\mathfrak{p}^e$ is a principal ideal (m) with $m \in \mathcal{O}(\mathbf{K})$, and apply the formula with $\mathfrak{p}^f\mathfrak{i}$ in place of $\mathfrak{i}$ for $0 \le f < e$ to verify that

$$m^{-12} = \frac{\Delta(m\mathfrak{i})}{\Delta(\mathfrak{i})} = \frac{\Delta(\mathfrak{p}^e\mathfrak{i})}{\Delta(\mathfrak{i})} = \frac{\Delta(\mathfrak{p}^e\mathfrak{i})}{\Delta(\mathfrak{p}^{e-1}\mathfrak{i})} \cdots \frac{\Delta(\mathfrak{p}^2\mathfrak{i})}{\Delta(\mathfrak{p}\mathfrak{i})}\frac{\Delta(\mathfrak{p}\mathfrak{i})}{\Delta(\mathfrak{i})}$$

is a rational function of the class invariants $j(\mathfrak{p}^f\mathbf{k})$, $0 \le f < e$. This can be improved: $j(\mathfrak{p}^{f+1}\mathbf{k}) + j(\mathfrak{p}^{f-1}\mathbf{k})$ belongs to $\mathbb{Q}_{\mathbf{k}'}$ with $\mathbf{k}' = \mathfrak{p}^f\mathbf{k}$, so $j(\mathfrak{p}^{f+1}\mathbf{k})$ is rationally expressible in terms of $j(\mathfrak{p}^f\mathbf{k})$ and $j(\mathfrak{p}^{f-1}\mathbf{k})$, with the result that m^{-12} is a rational function of $j(\mathfrak{p}\mathbf{k})$ and $j(k)$ with coefficients from $\mathbb{Q}$:

$$m^{-12} = L(j(\mathfrak{p}\mathbf{k}), j(\mathbf{k})).$$

The same procedure applies with $\mathfrak{p}^{-1} = \bar{\mathfrak{p}}$ in place of $\mathfrak{p}$: The order e is the same, m is replaced by its conjugate $\bar{m}$, and $\bar{m}^{-12}$ is the same rational function of $j(\mathfrak{p}^{-1}\mathbf{k})$ and $j(\mathbf{k})$, so

$$m^{-12} - \bar{m}^{-12} = L(j(\mathfrak{p}\mathbf{k}), j(\mathbf{k})) - L(j(\mathfrak{p}^{-1}\mathbf{k}), j(\mathbf{k})) = c[j(\mathfrak{p}\mathbf{k}) - j(\mathfrak{p}^{-1}\mathbf{k})]$$

with $c \in \mathbb{Q}_{\mathbf{k}}$, the only nontrivial Galois automorphism of the common splitting field of $j(\mathfrak{p}\mathbf{k})$ and $j(\mathfrak{p}^{-1}\mathbf{k})$ over $\mathbb{Q}_{\mathbf{k}}$ being the exchange of these two class invariants. This was the goal of step 3.

Step 4. It remains to prove that the left-hand side of the last display is a non-vanishing rational multiple of $\sqrt{D}$. Then $j(\mathfrak{p}\mathbf{k}) - j(\mathfrak{p}^{-1}\mathbf{k})$ is itself a rational multiple of $\sqrt{D}$, which serves to identify the splitting field as $\mathbb{Q}_{\mathbf{k}}(\sqrt{D})$. The point is that the unambiguous character of $\mathfrak{p}$ requires that common powers of

$\mathfrak{p}$ and of $\mathfrak{p}^{-1} = \bar{\mathfrak{p}}$ are different, so $(m)^{12} = \mathfrak{p}^{12e}$ and $\bar{m}^{12} = \bar{\mathfrak{p}}^{12e}$ cannot agree. The proof of fact 3 is finished.

The final piece of information needed from level p is embodied in

Weber's Congruences. $F_p(x, y) \equiv (x^p - y)(x - y^p) \bmod p\mathbb{Z}[x, y]$.

Proof. $j(\omega)$ is expanded at the cusp in the familiar way: $j(\omega) = q^{-1} + \cdots \in \mathbb{Z}[q]$, with the temporary abuse of notation that $p\mathbb{Z}[q]$ denotes the ring of sums $\sum c_n q^n$ having rational integral coefficients that vanish for large negative n. Now the elementary congruence $n^p \equiv n \bmod p$ $(n \in \mathbb{Z})$ implies[7]

$$j_\infty = j(p\omega) = q^{-p} + \cdots \equiv [j(\omega)]^p \bmod p\mathbb{Z}[q],$$

and the same idea applies to

$$j_0^p = [j(\omega/p)]^p = [q^{-1}/p + \cdots]^p \equiv j(\omega) \bmod p\mathbb{Z}[q^{1/p}].$$

The other roots of $F_p(x, j(\omega)) = 0$ are

$$j_n = j\left(\frac{\omega + n}{p}\right) = q^{-1/p} e^{2\pi\sqrt{-1}n/p} + \cdots \quad (1 \le n < p),$$

and these are congruent to j_0 modulo $(\epsilon - 1)\mathbb{Z}_\epsilon[q^{1/p}]$, in which $\mathbb{Z}_\epsilon$ is the ring of integers of the cyclotomic field $\mathbb{Q}_\epsilon = \mathbb{Q}$ with $\epsilon = e^{2\pi\sqrt{-1}/p}$ adjoined. The upshot is

$$\begin{aligned} F_p(x, j(\omega)) &= (x - j_\infty) \prod_{0 \le n < p} (x - j_n) \\ &\equiv (x - j^p)(x - j_0)^p \\ &\equiv (x - j^p)(x^p - j_0^p) \\ &\equiv (x - j^p)(x^p - j) \end{aligned}$$

modulo $(\epsilon - 1)\mathbb{Z}_\epsilon[q^{1/p}]$, account being taken in line 3 of $(x + y)^p = x^p + y^p \bmod p\mathbb{Z}[x, y]$ and of the divisibility of p by $\epsilon - 1$ in $\mathbb{Z}_\epsilon$:

$$0 = \sum_{0 \le n < p} \epsilon^n = \sum_{0 \le n < p} (1 + \epsilon - 1)^n = \sum_{0 \le n < p} 1 \bmod (\epsilon - 1),$$

so $(\epsilon - 1)$ divides $p = N_{\mathbb{Q}_\epsilon/\mathbb{Q}}(\epsilon - 1)$. Now, in reality, the discrepancy between $F_p(x, j)$ and $(x^p - j)(x - j^p)$ is a polynomial in x and j with rational integral coefficients m. Obviously these lie in $(\epsilon - 1)\mathbb{Z}_\epsilon$ in view of the last display, so $m^{p-1} = N_{\mathbb{Q}_\epsilon/\mathbb{Q}}(m)$ is divisible by $p = N_{\mathbb{Q}_\epsilon/\mathbb{Q}}(\epsilon - 1)$. The proof is finished.

[7] $(x + y)^p = x^p + y^p \bmod p\mathbb{Z}[x, y]$.

6.7 Irreducibility of the Class Polynomial

It is to be proved that the class polynomial $H(x)$ is irreducible not only over $\mathbb{Q}$ but over $\mathbf{K} = \mathbb{Q}(\sqrt{D})$ as well.

Step 1. Let $H_0(x) = \prod[x - j(\mathbf{k}')]$ be some factor of $H(x)$ that is irreducible over $\mathbf{K}$, let $y = j(\mathbf{k})$ be one of its roots, and fix an unambiguous prime ideal $\mathfrak{p} \subset \mathcal{O}(\mathbf{K})$, *not dividing* the discriminant $d \in \mathbb{Z}$ of $H(x)$, if any such ideal exists. Then $x = j(\mathfrak{p}\mathbf{k})$ *is also a root* of $H_0(x) = 0$.

Proof. Let $\mathfrak{q}$ be a prime ideal of the larger field $\mathbb{Q}_{\mathbf{k}}(\sqrt{D})$, dividing $\mathfrak{p}$ but not d. This must exist, for if each $\mathfrak{q} \supset \mathfrak{p}$ contains d as well, then $\mathfrak{p}$ contains a power of d and, being prime, must contain d itself. Weber's congruence for the rational prime p associated to $\mathfrak{p}$, is now applied *numerically* to $x = j(\mathfrak{p}\mathbf{k})$ and $y = j(\mathbf{k})$ with the result that

$$0 = F_p(j(\mathfrak{p}\mathbf{k}), j(\mathbf{k})) = [j^p(\mathfrak{p}\mathbf{k}) - j(\mathbf{k})][j(\mathfrak{p}\mathbf{k}) - j^p(\mathbf{k})] \bmod \mathfrak{q} \supseteq \mathfrak{p},$$

since $p \in \mathfrak{q}$ and $j(\mathfrak{p}\mathbf{k})$ and $j(\mathbf{k})$ are integers of $\mathbb{Q}_{\mathbf{k}}(\sqrt{D})$. The conclusion is that either $[j(\mathfrak{p}\mathbf{k})]^p \equiv j(\mathbf{k}) \bmod \mathfrak{q}$ or the other way around: $j(\mathfrak{p}\mathbf{k}) \equiv [j(\mathbf{k})]^p \bmod \mathfrak{q}$. Now H_0 is a factor of $H \in S\mathbb{Z}[x]$ in the ring $\mathbf{K}[x]$, so its coefficients c lie in $\mathcal{O}(\mathbf{K})$, by a simple extension of Gauss's lemma, and these satisfy $c^p \equiv c \bmod \mathfrak{q}$ in $\mathbb{Q}_{\mathbf{k}}(\sqrt{D})$. For example, if $D \not\equiv 1 \bmod 4$, then $c = a + b\sqrt{D}$ with $a, b \in \mathbb{Z}$, $D \equiv x^2 \bmod p$ with $1 \le x < p$, and $x^{p-1} \equiv 1 \bmod p$, so $(\sqrt{D})^{p-1} \equiv x^{p-1} \equiv 1 \bmod p$, p being odd, and $c^p \equiv a^p + b^p(\sqrt{D})^p \equiv a + b\sqrt{D} \bmod \mathfrak{q}$. The situation is now as follows: *Either* $[j(\mathfrak{p}\mathbf{k})]^p \equiv j(\mathbf{k}) \bmod \mathfrak{q}$ and

$$H_0^p(j(\mathfrak{p}\mathbf{k})) \equiv H_0(j^p(\mathfrak{p}\mathbf{k})) \equiv H_0(j(\mathbf{k})) = 0 \bmod \mathfrak{q}$$

or else $j(\mathfrak{p}\mathbf{k}) \equiv [j(\mathbf{k})]^p \bmod \mathfrak{q}$ and

$$H_0(j(\mathfrak{p}\mathbf{k})) \equiv H_0(j^p(\mathbf{k})) \equiv H_0^p(j(\mathbf{k})) = 0 \bmod \mathfrak{q}.$$

The upshot is that some factor $j(\mathfrak{p}\mathbf{k}) - j(\mathbf{k}')$ of $H_0(j(\mathfrak{p}\mathbf{k}))$ lies in the ideal $\mathfrak{q}$. But $j(\mathfrak{p}\mathbf{k})$ is distinct from any other class invariant in the quotient $\mathbb{Q}_{\mathbf{k}}(\sqrt{D})/\mathfrak{q}$ since the discriminant d does not belong to $\mathfrak{q}$, so you will have to agree that $j(\mathfrak{p}\mathbf{k})$ is not just congruent but equal to some $j(\mathbf{k}')$. The proof is finished.

Step 2. The next step is to understand just how much has been achieved: The discussion of step 1 excludes $p = 2$ and also primes that divide the discriminant d or else are ambiguous, which is to say that they divide D, so the situation can be summed up as follows: *If* $j(\mathbf{k})$ *is a root of* $H_0(x) = 0$, *then so is* $j(\mathfrak{p}\mathbf{k})$ *provided* p *does not divide* $2dD$. Now the fact is that you may pass from any fixed class $\mathbf{k}$ to any other class $\mathbf{k}'$ by successive multiplications of $\mathbf{k}$ by a series of *nonexceptional* prime ideals $\mathfrak{p}$ not dividing $2dD$, so every class invariant $x = j(\mathbf{k}')$ is a root of $H_0(x) = 0$; in short, *the class polynomial is irreducible*, and this not only over $\mathbb{Q}$ but over $\mathbf{K} = \mathbb{Q}(\sqrt{D})$ as well.

Step 3. Verify that such a passage from $\mathbf{k}$ to $\mathbf{k}'$ is possible. This can be done in a perfectly elementary but tiresome way. It is nicer at this juncture to use a big machine. Dirichlet [1840] proved that the prime ideals are equidistributed over the classes $\mathbf{k}$ in a suitable technical sense; in particular, every class contains an infinite number of primes, and since the proviso $2dD \notin \mathfrak{p}$ excludes only a finite number of primes, every class may be represented by a nonexceptional prime. The possibility of passing from $\mathbf{k}$ to $\mathbf{k}'$ in the manner described will now be plain.

Exercise 1. Look up the proof of Dirichlet's theorem in, for example, Weyl [1940].

Exercise 2. Give the kind of elementary proof, hinted at before, that the passage from $\mathbf{k}$ to $\mathbf{k}'$, via nonexceptional primes $\mathfrak{p}$, is possible.

6.8 Class Field and Galois Group

The invariants $j(\mathfrak{p}\mathbf{k})$ and $j(\mathfrak{p}^{-1}\mathbf{k})$ belong to $\mathbb{Q}_{\mathbf{k}}(\sqrt{D}) = \mathbb{Q}(\sqrt{D})[j(\mathbf{k})]$ and repetitions with nonexceptional primes show that $\mathbf{K}_1 = \mathbb{Q}_{\mathbf{k}}(\sqrt{D})$ is independent of the class $\mathbf{k}$ and contains every one of the class invariants. This was the content of Sections 4–7, serving to identify $\mathbf{K}_1$ as the splitting field of the (irreducible) class polynomial over the ground field $\mathbf{K} = \mathbb{Q}(\sqrt{D})$. Now the elements of $\mathbf{K}_1$ are uniquely expressible as

$$c_0 + c_1 j(\mathbf{k}') + \cdots + c_{h-1} j^{h-1}(\mathbf{k}')$$

with $c_0, \ldots, c_{h-1} \in \mathbb{Q}(\sqrt{D})$ for any fixed class $\mathbf{k}'$, so each element $\mathbf{k}$ of the class group $C(\mathbf{K})$ induces an automorphism σ of $\mathbf{K}_1$ over $\mathbf{K}$ mapping the general invariant $j(\mathbf{k}')$ to $\sigma j(\mathbf{k}') = j(\mathbf{k}^{-1}\mathbf{k}')$. These are distinct and so provide a faithful copy of $C(\mathbf{K})$. Besides, they fix only the ground field $\mathbf{K} = \mathbb{Q}(\sqrt{D})$, with the implication that *the Galois group* $\mathfrak{G}$ *of* $\mathbf{K}_1$ *over* $\mathbf{K}$ *is a copy of the class group.* This finishes the program outlined in Section 3. The final identification of $\mathbf{K}_1$ as Hilbert's class field, this being the biggest, unramified, commutative extension of $\mathbf{K}$, lies outside the scope of the present discussion, but see Vladut [1991] for a nice account.

Table 6.9.1. *Values of the absolute invariant*

D	Class number	Class invariants	Class field
-1	1	$j(\sqrt{-1}) = 2^6 \cdot 3^3$	$\mathbb{Q}(\sqrt{-1})$
-2	1	$j(\sqrt{-2}) = 2^6 \cdot 5^3$	$\mathbb{Q}(\sqrt{-1})$
-3	1	$j\left(\frac{1}{2}(-1+\sqrt{-3})\right) = 0$	$\mathbb{Q}(\sqrt{-3})$
-5	2	$j(\sqrt{-5}) = 2^3 \cdot 5\sqrt{5} \cdot (1+\sqrt{5})^3(3+2\sqrt{5})^3$	$\mathbb{Q}(\sqrt{-1}, \sqrt{5})$
		$j\left(\frac{1}{2}(-1+\sqrt{-5})\right) = 2^6 \cdot 5\sqrt{5} \cdot (1-\sqrt{5})^3(3-2\sqrt{5})^3$	
-7	1	$j\left(\frac{1}{2}(-1+\sqrt{-7})\right) = -3^3 \cdot 5^3$	$\mathbb{Q}(\sqrt{-7})$
-15	2	$j\left(\frac{1}{2}(-1+\sqrt{-15})\right) = -3^3 \cdot 5\sqrt{5} \cdot \left(\frac{1}{2}(1+\sqrt{5})\right)^2 (4+\sqrt{5})^3$	$\mathbb{Q}(\sqrt{-3}, \sqrt{5})$
		$j\left(\frac{1}{4}(-1-\sqrt{-15})\right) = 3^3 \cdot 5\sqrt{5} \cdot \left(\frac{1}{2}(1-\sqrt{5})\right)^2 (4-\sqrt{5})^3$	
-51	2	$j\left(\frac{1}{2}(-1+\sqrt{-51})\right) = -2^{11} \cdot 3^3 \cdot (4+\sqrt{17})$ $\cdot (1+\sqrt{17})^2 \cdot (3+\sqrt{17})$	$\mathbb{Q}(\sqrt{-3}, \sqrt{17})$
		$j\left(\frac{1}{6}(-3+\sqrt{-51})\right) = -2^{11} \cdot 3^3 \cdot (4-\sqrt{17})$ $\cdot (1-\sqrt{17})^2 \cdot (3+\sqrt{17})$	
-67	1	$j\left(\frac{1}{2}(-1+\sqrt{-67})\right) = -2^{15} \cdot 3^3 \cdot 5^3 \cdot 11^3$	$\mathbb{Q}(\sqrt{-67})$

6.9 Computation of the Class Invariants

The commutativity of the Galois group of the class polynomial over $\mathbf{K} = \mathbb{Q}(\sqrt{D})$ has the arithmetic consequence that the class invariants can be expressed by radicals. A few samples are tabulated in Table 6.9.1; for more, see Fricke [1928: 390–502], Weber [1908: 457–99], Cohen [1992], and Husemoller [1987].

Sample Computations.
$D = -1$: $j(\sqrt{-1}) = 1728 = 2^6 \cdot 3^3$; see Section 4.7, aside 1.
$D = -2$: $j(\sqrt{-2}) = 2^6 \cdot 5^3$; see ex. 4.12.6.
$D = -3$: $j\left(\frac{1}{2}(-1+\sqrt{-3})\right) = 0$; see Section 4.7, aside 1.
$D = -5$: $[1, \sqrt{-5}]$ and $[2, -1+\sqrt{-5}]$ are ideals of $\mathbb{Q}(\sqrt{-5})$ belonging to different classes: If their period ratios $\sqrt{-5}$ and $(1/2)(-1+\sqrt{-5})$ were

related by a modular substitution $[ab/cd]$, as in $(1/2)(-1+\sqrt{-5}) = (a\sqrt{-5}+b)/(c\sqrt{-5}+d)$, you would have $5c^2 + d^2 = 2$, and that cannot be solved in whole numbers. For the rest, it suffices to compute $j(\sqrt{-5}) = 844+395\sqrt{5}$, as will be done shortly; its value lies in $\mathbb{Q}(\sqrt{5})$ so the class field is $\mathbb{Q}(\sqrt{-5})$ with $\sqrt{5}$ adjoined, the class number is 2, and the other class invariant $j\left(\frac{1}{2}(-1+\sqrt{-5})\right)$ can only be the conjugate of $j(\sqrt{-5})$ over $\mathbb{Q}(\sqrt{-5})$. Now the moduli $y^4 = k(-1/\sqrt{-5}) = k'(\sqrt{-5})$ and $x^4 = k(5 \times -1/\sqrt{-5}) = k(\sqrt{-5})$ satisfy $x^8 + y^8 = 1$ as well as Jacobi's modular equation of level 5 described in ex. 4.14.5:

$$F_5(x, y) = x^6 + 4x^5y^5 + 5x^4y^2 - 5x^2y^4 - 4xy - y^6 = 0;$$

in particular, using ex. 4.14.6,

$$\begin{aligned} 0 = F_5(x, y)F_5(x, -y) &= (x^2 - y^2)^6 - 16x^2y^2(1 - x^8)(1 - y^8) \\ &= (x^2 - y^2)^6 - 16x^{10}y^{10}, \end{aligned}$$

so $(x^2 - y^2)^3 + 4x^5y^5 = 0$ in view of $0 < x < y$. This serves to eliminate $(x^2-y^2)^3$ from $F_5(x, y) = 0$, as in ex. 4.14.6, with the result that $c = (4x^8y^8)^{1/6}$ is a root of

$$c^3 + 2c^2 - 1 = (c + 1)(c^2 + c - 1) = 0.$$

This yields, in turn, $c = (1/2)(-1 + \sqrt{5})$, $(xy)^8 = 9/4 - \sqrt{5}$, and $x^8 = (1/2) - \sqrt{\sqrt{5} - 2} = k^2(\sqrt{-5})$. The class invariant is now obtained from

$$j(\sqrt{-5}) = 2^8 \frac{(k^4 - k^2 + 1)^3}{k^4(1 - k^2)^2} = 2^6 \cdot \frac{(4\sqrt{5} - 5)^3}{(9 - 4\sqrt{5})^2} = 844 + 395\sqrt{5}.$$

Exercise 1. Check it and reconcile this value with the table.

Exercise 2. What is the class polynomial of $\mathbb{Q}(\sqrt{-5})$? *Answer*: $H(x) = x^2 - 2^7 \cdot 5^3 \cdot 331x - 2^{12} \cdot 5^3 \cdot 11^3$.

Naturally, the complexity of the computation of class invariants increases as you proceed. Fricke [1928] deals directly with the absolute invariant in the manner indicated in Section 4.16. Weber [1908] invokes modular functions of higher level such as the eighth root of the Jacobi modulus k^2 employed here. This is more efficient in special cases, though not so systematic as Fricke's recipe.

7

Arithmetic of Elliptic Curves

The arithmetic of elliptic curves concerns itself with the points of a nonsingular cubic $y^2 = x^3 + ax + b$ with coordinates from the field $\mathbb{Q}$ of rational numbers or, more generally, from a fixed algebraic number field. The subject is reviewed in Cassels [1966] and Silverman [1986,1994] from a sophisticated point of view; for something accessible to the beginner see Nagell [1964: 188–265], Skolem [1938: 73–83], and especially the splendid book of Silverman and Tate [1992].

7.1 Arithmetic of the Projective Line

The question of arithmetic on curves goes back to antiquity. Take the circle $\mathbf{X}: x^2 + y^2 = 1$. It is uniformized by the projective line $\mathbb{P}^1$ via the substitution $w \mapsto (x, y) = (w^2 + 1)^{-1}(w^2 - 1, 2w)$; plainly, w belongs to $\mathbb{Q}$ if and only if x and y do. Now write $w = n/m$ with coprime $n > m \geq 1$. Then $x^2 + y^2 = 1$ is equivalent to $a^2 + b^2 = c^2$ with coprime whole numbers $a = n^2 - m^2$, $b = 2nm$, $c = n^2 + m^2$; in short, the rational points of the circle represent the triples of Pythagoras (520 BC), epitomized by $3^2 + 4^2 = 5^2$ as every school child knows.

Exercise 1. Let a and b be coprime whole numbers. Then $a^2 + b^2$ is a square in $\mathbb{Q}$ only if $a + \sqrt{-1}b$ is a square in $\mathbb{Q}(\sqrt{-1})$. Check this and use it to confirm the recipe for Pythagorean triples. *Hint*: $\mathbb{Q}(\sqrt{-1})$ has class number 1 so its integers may be factored into prime integers in just one way; also, $a + \sqrt{-1}b$ and $a - \sqrt{-1}b$ are mostly coprime in $\mathbb{Q}(\sqrt{-1})$.

The arithmetic of the general ellipse $\mathbf{X}: Ax^2 + By^2 = C$ $(0 < A, B, C \in \mathbb{Z})$ is more complicated. If it has at least five rational points, then $-D = B/A$ and $n = C/A$ may be taken to be square-free integers, and the question takes the form: Is n the norm $x^2 - Dy^2$ of a number $x + y\sqrt{D}$ of the imaginary quadratic field $\mathbb{Q}(\sqrt{-D})$? This may or may not happen, as the next exercises show.

Exercise 2. Check that the number 3 is not a norm for $\mathbb{Q}(\sqrt{-1})$ so $x^2 + y^2 = 3$ cannot be solved in $\mathbb{Q}$.

If n *is* the norm of a number $\omega \in \mathbb{Q}(\sqrt{D})$, then every other such number of $\mathbb{Q}(\sqrt{D})$ is the product of ω by $(w^2 - D)^{-1}(w^2 + D + 2w\sqrt{D})$ with $w \in \mathbb{Q}$, the upshot being that the rational points of $\mathbf{X}$ can be naturally identified with $\mathbb{Q}$ itself, if any such points are present.

Exercise 3. Check it.

Exercise 4. What about hyperbolas?

7.2 Cubics: The Mordell–Weil Theorem

Let $\mathbf{X}$ be a nonsingular cubic $y^2 = (x - e_1)(x - e_2)(x - e_3) = x^3 + ax + b$ with points $\mathfrak{p} = (x, y)$. The numbers a and b belong to $\mathbb{Q}$ as soon as $\mathbf{X}$ has two rational points with different abscissas and it is no loss of generality to let them be whole numbers. The customary factor 4 is missing so the addition theorem of Section 2.14 is a little modified:

$$x(\mathfrak{p}_1 + \mathfrak{p}_2) = -x(\mathfrak{p}_1) - x(\mathfrak{p}_2) + \left[\frac{y(\mathfrak{p}_2) - y(\mathfrak{p}_1)}{x(\mathfrak{p}_2) - x(\mathfrak{p}_1)}\right]^2,$$

$$y(\mathfrak{p}_1 + \mathfrak{p}_2) = \frac{y(\mathfrak{p}_1)[x(\mathfrak{p}_1 + \mathfrak{p}_2) - x(\mathfrak{p}_2)] - y(\mathfrak{p}_2)[x(\mathfrak{p}_1 + \mathfrak{p}_2)] - x(\mathfrak{p}_1)}{x(\mathfrak{p}_2) - x(\mathfrak{p}_1)}.$$

Rational Points. The formulas make it plain that the rational points $\mathbf{X}(\mathbb{Q})$ of $\mathbf{X}$ are closed under the addition of the curve, so they form a module over $\mathbb{Z}$ after adjunction of the identity element $\mathfrak{o} = (\infty, \infty)$; see Fig. 2.21 for the geometric interpretation of the addition. Now for the projective lines of Section 1.10 either $\mathbf{X}(\mathbb{Q})$ is void as in ex. 2 or else it can be naturally identified with $\mathbb{Q}$ as in ex. 3. The facts are quite different for cubics. Poincaré [1901] conjectured and Mordell [1922] proved that $\mathbf{X}(\mathbb{Q})$ *is of finite rank over* $\mathbb{Z}$. This means that $\mathbf{X}(\mathbb{Q})$ contains a finite number of points $\mathfrak{p}_1, \mathfrak{p}_2, \ldots, \mathfrak{p}_d$ from which all others may be obtained by addition[1] $\mathfrak{p} = n_1\mathfrak{p}_1 + \cdots + n_d\mathfrak{p}_d$. In more detail $\mathbf{X}(\mathbb{Q})$ is of the form $\mathbb{Z}^r \oplus$ *a finite commutative group.* The latter possibility is really present. For example, the half-periods of $\mathbf{X}$ are of order 2 under the addition

[1] $n\mathfrak{p}$ signifies the sum of n copies of $\mathfrak{p}$.

and may produce rational points as for $y^2 = x(x^2 - 1)$; indeed, they are the only rational points of this curve, as will be seen in example 3.1. Mazur [1976] has a complete description of the torsion. The number r is the **rank** of $\mathbf{X}(\mathbb{Q})$, and it is conjectured that r can be arbitrarily large: 21 is the highest rank known to date; see Ireland and Rosen [1990], Mestre [1992], and Nagao and Kouya [1994].

Reduction to Standard Form. The same applies to any nonsingular curve $\mathbf{X}: y^2 = c_0x^4 + \cdots + c_4$, genuinely of degree 4 or 3, with $c_0, \ldots, c_4 \in \mathbb{Z}$. The fact is that if $\mathbf{X}$ has a rational point, then it can be reduced to standard form $\mathbf{X}': y^2 = x^3 + ax + b$ so as to make $\mathbf{X}(\mathbb{Q})$ and $\mathbf{X}'(\mathbb{Q})$ correspond. This is what is meant by saying that $\mathbf{X}$ and $\mathbf{X}'$ are *equivalent over* $\mathbb{Q}$. The idea is illustrated by two examples.

Example 1. $\mathbf{X}: y^2 = x^4 - 1$ is equivalent over $\mathbb{Q}$ to $y^2 = x^3 + 4x$ via the substitutions

$$x \mapsto 1 + 4(x-2)^{-1}, \quad y \mapsto -4y(x-2)^{-2},$$

as in ex. 2.11.6.

Example 2. $\mathbf{X}: y^2 = 2x^4 - 1$ is not so simple. The analogous substitutions

$$x \mapsto 2^{-1/4} + 8 \cdot 2^{-3/4}(x - 4 \cdot 2^{-1/2})^{-1}, \quad y \mapsto -8 \cdot 2^{-3/4} y (x - 4 \cdot 2^{-1/2})^{-2}$$

reduce $\mathbf{X}$ to the standard form $y^2 = x^3 + 8x$, but do not match up the rational points. The new coordinates are expressed in terms of the old as

$$x' = 4c^2(x+c)^{-1}(x-c), \quad y' = -8c^3(x-c)^{-3}y \quad \text{with} \quad c = 2^{-1/4},$$

of which the first term represents $4\times$ the $\wp$-function of $\mathbf{X}$ with pole at $(c, 0)$. The mistake is that the pole is placed at an *irrational* point. $\mathbf{X}$ has the self-evident rational point $(1, -1)$ and if you will only move the pole to *that* place, the difficulty will be overcome. The shift is produced by addition of the point $(1, 1) \in \mathbf{X}$, alias $(4c^2(1+c)(1-c)^{-1}, -8c^3(1-c)^{-3}) \in \mathbf{X}'$, the new coordinates of $\mathbf{X}'$ being

$$x' = 2(x-1)^{-2}[2x^2 - 1 - y], \quad y' = 4(x-1)^{-3}[2x^3 - 1 - (2x-1)y].$$

Exercise 1. Check the details of example 2, including the matching up of the rational points by the second recipe.

7.3 Examples

The splendid proof of Mordell's theorem sketched in Section 4 is due to Weil [1930] with improvements by Silverman [1986] and Knapp [1993], but first here are some examples. These illustrate the method of infinite descent of Fermat (1645) which is also the key to Weil's proof.

Example 1. The curve $\mathbf{X}$: $y^2 = x^3 - x$ comes from the square lattice; see Section 2.9, example 1. It has the four obvious rational points (∞, ∞), $(1, 0)$, $(0, 0)$, and $(-1, 0)$ produced by the half-periods, and no others.

Proof.[2] Let $x = a/b$, $y = c/d$ be a rational point of $\mathbf{X}$ expressed in lowest terms ($a, b \in \mathbb{Z}$ coprime, etc.) and let it be distinct from the half-periods noted before. Take $b, c, d \geq 1$ and use the involution $(x, y) \mapsto (-1/x, y/x^2)$ to make $a \geq 1$, too, if it is not so already. Now $b^3c^2 = d^2a(a-b)(a+b)$ implies $b^3 = d^2$ (b is coprime to $a(a-b)(a+b)$ and d is coprime to c), so b is the square of a whole number $B \geq 1$. Next, a prime of c cannot divide two factors of $a(a-b)(a+b)$ unless it is the prime 2 and divides both $a - b$ and $a + b$, and now it follows, from $c^2 = a(a-b)(a+b)$, that a is the square of a whole number $A \geq 1$ and that $a^2 - b^2 = A^4 - B^4 = c^2/a$ is the square of a whole number $C \geq 1$. The *descent* consists in producing, from such a pair A, B of coprime whole numbers with $A^4 - B^4 = C^2$, a second pair of properly smaller **height** $H = A$, so that repeated descent is contradictory ($H \geq 1$).

Case 1. $B \equiv 0 \bmod 2$. Then $(A^2 - C)(A^2 + C) = B^4$ is divisible by 16 while the left-hand factors have highest common divisor 2, so one of them is divisible by 2 and the other one by 8. The division makes them coprime, permitting you to write

$$A^2 \pm C = 2a^4, \;\; A^2 - (\pm C) = 8b^4, \;\; B = 2ab,$$

with coprime $a, b \geq 1$. Now $A^2 = a^4 + 4b^4$, by addition; also, $(A + a^2)(A - a^2) = 4b^4$, and as the highest common divisor of the left-hand factors is 2, so

$$A + a^2 = 2c^2, \;\; A - a^2 = 2d^4, b = cd,$$

in which $c, d \geq 1$ may be taken coprime. The descent is accomplished: In fact, $c^4 - d^4 = a^2$ and *the height is diminished*, from $H = A$ to

$$c < 2acd = 2ab = B < A.$$

[2] Nagell [1964: 227–9].

Case 2. $B \equiv 1 \bmod 2$. Then $(A^2 + B^2)(A^2 - B^2) = C^2$ is even since $B^2 \equiv 1 \bmod 4$ and $A^2 \equiv 0$ or $1 \bmod 4$, so $C^2 \equiv A^4 - 1$ is congruent to 0 or to -1 mod 4, and $C^2 \equiv -1 \bmod 4$ is not possible. As before, the left-hand factors have highest common divisor 2, so

$$A^2 + B^2 = 2a^2, \ A^2 - B^2 = 2b^2, \ 2C = ab$$

with coprime $a, b \geq 1$, and from $A^2 = a^2 + b^2$ and $B^2 = a^2 - b^2$, you obtain a new solution $a^4 - b^4 = (AB)^2$ of *smaller height*

$$a < \sqrt{a^2 + b^2} = A = H.$$

The descent is over.

Exercise 1. $\mathbf{X}$: $y^2 = x^4 - 1$ is equivalent over $\mathbb{Q}$ to $\mathbf{X}'$: $y^2 = x^3 + 4x$, as in example 2.1. Check that it has just the two rational points $(\pm 1, 0)$. *Hint*: Use the recipe for Pythagorean triples of Section 1 and the descent of example 1.

Example 2. Euler (1770) nearly proved that the curve $\mathbf{X}$: $y^2 = x^3 + 1$ has only the six rational points

$$(\infty, \infty), \ (2, 3), \ (0, 1), \ (-1, 0), \ (0, -1), (2, -3)$$

produced by the fractional periods $k\omega/6$ $(0 \leq k \leq 5)$, ω being the real primitive period $2\int_{-1}^{\infty}(x^3 + 1)^{-1/2}dx$ of the triangular lattice; compare ex. 1 and, for the proof, see Nagell [1964: 244–6].

Exercise 2. Prove that $\mathfrak{p} = (2, 3)$ generates the five rational points different from $\mathfrak{o} = (\infty, \infty)$, to wit, $(0, 1) = 2\mathfrak{p}$, $(-1, 0) = 3\mathfrak{p}$, and so on, and identify the period.

Example 3. The curve $\mathbf{X}$: $y^2 = 2x^4 - 1$ is equivalent over $\mathbb{Q}$ to $\mathbf{X}'$: $y^2 = x^3 + 8x$, as in example 2 of Section 2. Its rational points form a copy of $\mathbb{Z} \oplus \mathbb{Z}_2$; in particular, it is of rank 1.

Proof.[3] As before, the proof is by descent, but with a twist. Let $x = a/b$, $y = c/d$ be a rational point in lowest terms, with $a, b, c, d \geq 1$. Then $(2a^4 - b^4)d^2 = b^4c^2$, and it is easy to see that $d = b^2$, so $2a^4 - b^4 = c^2$. Now b and c are of like parity, $a^4 = [(b^2 + c)/2]^2 + [(b^2 - c)/2]^2$, and either $a = 1 = b = c = d$ and $\mathfrak{p} = (x, y)$ is the self-evident rational point $(1, 1)$, or else $a > 1$ and

$$a^2 = A^2 + B^2, \ b^2 = A^2 - B^2 + 2AB, \ \pm c = A^2 - B^2 - 2AB$$

[3] Nagell [1964: 232–5], with a geometric amplification.

with coprime $A, B \geq 1$, by the recipe for Pythagorean triples of Section 1. Here $A = -B$ is not possible, so $A + B \geq 1$ can be assumed. Also, A and B are of opposite parity, a and b being coprime: in fact, A is odd and B is even since, in the opposite case, $b^2 \equiv -1 \bmod 4$ and -1 is not a quadratic residue of 4. The recipe for Pythagorean triples is now applied to $a^2 = A^2 + B^2$ to produce

$$a = C^2 + D^2, \ A = C^2 - D^2, \ B = 2CD$$

with $C, D \geq 1$. The identity $b^2 = A^2 - B^2 + 2AB$ produces

$$(C^2 + 2CD - D^2)^2 - b^2 = 8C^2D^2$$

with $C^2 + 2CD - D^2 = A + B \geq 1$, and the left-hand side of the display factors in a self-evident way into pieces with highest common divisor 2, so one piece is twice a square, the other is four times a square, and half their sum is $C^2 + 2CD - D^2 = M^2 + 2N^2$ with $M, N \geq 1$ and $MN = CD$. This new identity is recast as

$$(2 + D^2/N^2)\left(\frac{N}{C}\right)^2 - 2\frac{D}{N}\frac{N}{C} = 1 - N^2/C^2 = 1 - D^2/N^2$$

and solved for N/C:

$$\frac{N}{C} = \frac{DN \pm \sqrt{2N^4 - D^4}}{2N^2 + D^2} = \frac{x \pm \sqrt{2x^4 - 1}}{2x^2 + 1} \quad \text{with} \quad x = \frac{N}{D},$$

showing that $2x^4 - 1$ is a square in $\mathbb{Q}$. In fact,

$$x = \frac{N}{D}, \ y = \pm\left[\frac{N}{C}(2x^2 + 1) - x\right] > 0$$

is a new rational point of $\mathbf{X}$ *descended* from the old: If $x = N/D$ is expressed in lowest terms, then its numerator cannot exceed

$$N \leq CD = B/2 < B < \sqrt{A^2 + B^2} = a.$$

The reasoning is invalid if $a = 1$: Then repeated descent ends at the point $(1, 1)$. The recipe can be reversed to express the higher point in terms of the lower: The higher abscissa is

$$\frac{a}{b} = \frac{C^2 + D^2}{M^2 - 2N^2} = \frac{x^2(2x^2 + 1)^2 + (x \pm y)^2}{(2x^2 + 1)^2 - 2x^2(x \pm y)^2},$$

$(x, y = \sqrt{2x^4 - 1})$ being the lower point. The descent shows that all rational points of $\mathbf{X}$ are obtained from the single point $(1, 1)$ by application of this formula and the involutions $(x, y) \mapsto (\pm x, \pm y)$.

Exercise 3. One application of the recipe produces the point (13, 239). Check it. What does the next produce? *Answer*: (1525/1343, 2750257/1343).

It remains to understand what all this means geometrically. To do this, use the addition theorem in its Jacobian form expounded in Section 2.16. The modulus is $k^2 = -1$, x is identified with $2^{-1/4} \times$ sin amp with its argument multiplied by $\sqrt{-1} \times 2^{1/4}$, and the addition theorem states that

$$x_3 = \frac{1}{\sqrt{-1}} \frac{x_1 y_2 + x_2 y_1}{1 + 2x_1^2 x_2^2},$$

$$y_3 = \frac{1}{\sqrt{-1}} \frac{y_1 y_2 (1 - 2x_1^2 x_2^2) + 4x_1 x_2 (x_1^2 + x_2^2)}{(1 + 2x_1^2 x_2^2)^2}$$

with a self-evident notation. Now a short computation shows that the abscissa of the sum of $2\mathfrak{p} + \mathfrak{q}$ with $\mathfrak{p} = (x, y)$ and $\mathfrak{q} = (1, \pm 1)$ is precisely the higher abscissa a/b already displayed, with an extra minus sign, so the geometric content of the descent is that any rational point $\mathfrak{p}_0$ can be expressed as $2\mathfrak{p}_1 + \mathfrak{q}$ with $\mathfrak{q} = (1, \pm 1)$ and a new rational point $\mathfrak{p}_1$.

Exercise 4. Check the computation of $2\mathfrak{p} + \mathfrak{q}$.

The descent can be repeated ($\mathfrak{p}_1 = 2\mathfrak{p}_2 \pm \mathfrak{q}_2$) unless $\mathfrak{p}_1$ has abscissa ± 1; it ends at one of the four points $(\pm 1, \pm 1)$, after a finite number of steps. Now $\mathfrak{o} = (0, \sqrt{-1})$ is the origin for the addition; also, $-(1, 1) = (-1, 1)$, $-(1, -1) = (-1, -1)$, and $2(1, 1) + 2(1, -1) = \mathfrak{o}$; and now it appears that $\mathbf{X}(\mathbb{Q})$ is included in

$$2\mathbb{Z}(1, 1) \oplus \pm[(1, 1) + (1, -1)] \oplus (1, -1),$$

in which the summand $(1, -1)$ corresponds to the shift of origin employed in example 2.2. The fact is that this is a faithful representation of $\mathbf{X}(\mathbb{Q})$, that is, the latter is a copy of $\mathbb{Z} \oplus \mathbb{Z}_2$, as advertised.

Exercise 5. Check the final statement. *Hint*: $n\mathfrak{p} = (0, \pm\sqrt{-1})$ only if n is even and $(n/2)\mathfrak{p} = (0, \pm\sqrt{-1})$. Descend by this route.

The descent from $\mathfrak{p}_0$ to $\mathfrak{p}_1$ epitomizes the main idea of Weil's proof, which will now be sketched.

7.4 Proof of the Mordell–Weil Theorem

The curve is $\mathbf{X}$: $y^2 = x^3 + ax + b = (x - e_1)(x - e_2)(x - e_3)$ with $a, b \in \mathbb{Z}$. Let also $e_1, e_2, e_3 \in \mathbb{Z}$ to simplify life. The proof is made in four steps: Step 1 explains what it takes for $\mathfrak{p}_2 \in \mathbf{X}(\mathbb{Q})$ to be the double $2\mathfrak{p}_1$ of some other point $\mathfrak{p}_1 \in \mathbf{X}(\mathbb{Q})$. Step 2 introduces the proper **height**, first in a **naive** version and then in an improved **absolute** version. Step 3 confirms that the quotient of $\mathbf{X}(\mathbb{Q})$ by $2\mathbf{X}(\mathbb{Q})$ is finite. Step 4 finishes the proof by application of heights to $\mathbf{X}(\mathbb{Q})/2\mathbf{X}(\mathbb{Q})$.

Step 1. $\mathfrak{p} = (x, y) \in 2\mathbf{X}(\mathbb{Q})$ if and only if each of the numbers $x - e$ is a rational square.

Proof. If $\mathfrak{p} = \mathfrak{p}_2 = (x_2, y_2)$ is the double of $\mathfrak{p}_1 = (x_1, y_1)$, and $y = mx + n$ is the tangent line to $\mathbf{X}$ at $\mathfrak{p}_1$, then

$$(x - e_1)(x - e_2)(x - e_3) - (mx + n)^2 = (x - x_1)^2(x - x_2),$$

and at each root $x = e$ this reduces to the stated condition: $x_2 - e = (me + n)^2(e - x_1)^{-2}$ is a rational square. Now let $\mathfrak{p}_2 = (x_2, y_2) \in \mathbf{X}(\mathbb{Q})$ satisfy the condition that each of the numbers $x - e$ is a rational square. It is convenient, for the derivation of this step only, to take $x_2 = 0$ at the price of violating the standard normalization $e_1 + e_2 + e_3 = 0$. Then $-e = f^2$ with $f \in \mathbb{Q}$ for each of the three numbers e. It is desired to find $\mathfrak{p}_1 = (x_1, y_1)$ so that

$$\begin{aligned}[x^3 + ax^2 + bx + c = (x - e_1)(x - e_2)(x - e_3)] - (mx + n)^2 \\ = x(x - x_1)^2.\end{aligned}$$

The left-hand side vanishes at $x = 0$ $(c = n^2)$ and division by x produces

$$x^2 + ax + b - m^2x - 2mn = (x - x_1)^2,$$

which requires the left-hand discriminant to vanish: $(m^2 - a)^2 = 4(b - 2mn)$. Now introduce an artificial variable e, via

$$(m^2 - a - 2e)^2 = -4em^2 - 8mn + 4(e^2 + ae + b),$$

chosen so that the right-hand side has a double root in the variable m. This requires the vanishing of *its* discriminant $64n^2 + 64e(e^2 + ae + b)$, which is to say $e^3 + ae^2 + be + (c = n^2) = 0$, the upshot being $e = e_1, e_2$, or e_3. Take $e = e_1$, say, and recall $-e = f^2$. Then from $-a = (e_1+e_2+e_3)$, $b = e_1e_2+e_2e_3+e_3e_1$, and $c = -e_1e_2e_3 = n^2$ you find

$$\begin{aligned}(m^2 - e_1 + e_2 + e_3)^2 &= 4e_1m^2 - 8\sqrt{-e_1e_2e_3}m \\ &\quad +4\left[e_1^2 - e_1(e_1+e_2+e_3) + e_1e_2 + e_2e_3 + e_3e_1\right] \\ &= 4(f_1m - f_2 + f_3)^2,\end{aligned}$$

which is to say, $(m^2 \mp f_1)^2 = (f_2 \mp f_3)^2$, that is, $m = \pm f_1 \pm (f_2 \mp f_3)$, in which the third sign conforms to the first. Let m_0 be any of these roots. Then $x_1 = (1/2)(m_0^2 - a)$ is the (double) root of $x^2 + ax + b - m^2x - 2mn = 0$, and $\mathfrak{p}_2 = (x_2, y_2)$ is the double of $\mathfrak{p}_1 = (x_1, m_0x_1 + n) \in \mathbf{X}(\mathbb{Q})$.

Step 2. For $\mathfrak{p} = (x, y) \in \mathbf{X}(\mathbb{Q})$ and $x = p/q$ expressed in lowest terms, the naive height of p is declared to be $H_0(\mathfrak{p}) = \log m$, m being the larger of $|p|$ and $|q|$, with the understanding that the origin $\mathfrak{o} = (\infty, \infty)$ is the ground level, that is, $H_0(\mathfrak{o}) = 0$. Plainly, there are only a finite number of points of $\mathbf{X}(\mathbb{Q})$ of limited height $H_0(\mathfrak{p}) \le h < \infty$. The chief feature of the height is its behavior under doubling: $H_0(2\mathfrak{p}) - 4H_0(\mathfrak{p})$ is bounded above and below, by constants depending not upon $\mathfrak{p}$, but upon $\mathbf{X}$ alone.

Proof of the upper bound. The condition $e_1 + e_2 + e_3 = 0$ is back in force. The duplication rule

$$x(2\mathfrak{p}) = \frac{x^4 - 2ax^2 - 8bx + a^2}{4(x^3 + ax + b)} \quad \text{with} \quad x = x(\mathfrak{p})$$

shows that, if $x = p/q$, then $x(2\mathfrak{p}) = p_2/q_2$ with

$$p_2 = p^4 - 2ap^2q^2 - 8bpq^3 + a^2q^4 \le c_1(\mathbf{X})m^4$$

and

$$q_2 = 4q(p^3 + apq^2 + bq^3) \le c_2(\mathbf{X})m^4.$$

Unlike p/q, the new fraction p_2/q_2 need not be in lowest terms, but it does not matter: $H_0(2\mathfrak{p})$ is overestimated by $4\log m = 4H_0(\mathfrak{p})$ *plus* $c_3(\mathbf{X}) =$ the larger of $\log c_1(\mathbf{X})$ and $\log c_2(\mathbf{X})$.

Proof of the lower bound. This is trickier. The fact is, as you may verify with tears, that if d is the discriminant $-4a^3 - 27b^2$ of the cubic, then

$$-4dp^7 = Ap_2 + Bq_2$$

and

$$-4dq^7 = Cp_2 + Dq_2$$

with

$$A = 4 \times \left[-dp^3 - a^2bp^2q + a(3a^3 + 22b^2)pq^2 + 3b(a^3 + 8b^2)q^3\right],$$
$$B = a^2bp^3 + a(5a^3 + 32b^2)p^2q + 2b(3a^3 + 96b^2)pq^2 - 3a^2(a^3 + 8b^2)q^3,$$
$$C = 4 \times (3p^2q + 4aq^2), \quad \text{and} \quad D = 3p^3 - 5apq^2 - 27q^3,$$

not the numbers that would spring to mind! Anyhow, p and q being coprime, the common divisor of p_2 and q_2 is seen to divide d and so to be capable of only a limited number of values. Besides, $m^7 \le c_4(\mathbf{X})m^3m_2$, and writing $x_2 = p_2/q_2$ in lowest terms p_3/q_3 reveals that

$$\begin{aligned} H_0(2\mathfrak{p}) = \log m_3 &\ge \log m_2 - c_5(d) \\ &\ge 4\log m - c_5(d) - \log c_4(\mathbf{X}) \\ &= 4H_0(\mathfrak{p}) - c_6(\mathbf{X}), \end{aligned}$$

as advertised.

Now it is plain that $4^{-n}H_0(2^n\mathfrak{p})$ tends to a limit $H(\mathfrak{p})$ as $n \to \infty$. This is the **absolute height** of $\mathfrak{p} \in \mathbf{X}(\mathbb{Q})$, obeying the neater duplication rule $H(2\mathfrak{p}) = 4H(\mathfrak{p})$. The rest of the present step elicits the more general rule

$$H(\mathfrak{p}_1 + \mathfrak{p}_2) + H(\mathfrak{p}_1 - \mathfrak{p}_2) = 2H(\mathfrak{p}_1) + 2H(\mathfrak{p}_2).$$

Proof. It suffices to check that the left side is smaller than the right: Just apply *that* inequality with $\mathfrak{p}_1 \pm \mathfrak{p}_2$ in place of $\mathfrak{p}_1$ and $\mathfrak{p}_2$, use the duplication rule, and divide by 4. In fact, you need only the inequality with an additive constant $c_0(\mathbf{X})$ on the right, and *this* is proved as follows. Let $x_1 = x(\mathfrak{p}_1) = p_1/q_1$ in lowest terms, let m_1 be the larger of $|p_1|$ and $|q_1|$, and similarly construct m_2, m_+, and m_- for $\mathfrak{p}_2, \mathfrak{p}_+ = \mathfrak{p}_1 + \mathfrak{p}_2$, and $\mathfrak{p}_- = \mathfrak{p}_1 - \mathfrak{p}_2$. Then

$$x_+ + x_- = 2 \times \frac{(p_1p_2 + aq_1q_2)(p_2q_1 + p_1q_2) + 2bq_1^2q_2^2}{(p_2q_1 - p_1q_2)^2} \equiv \frac{A}{C},$$

$$x_-x_+ = \frac{(p_1p_2 - aq_1q_2)^2 - 4q_1q_2b(p_2q_1 + p_1q_2)}{(p_2q_1 - p_1q_2)^2} \equiv \frac{B}{C}$$

with $A, B, C \in \mathbb{Z}$ and $|A| + |B| + |C| \le c_0(\mathbf{X})m_1^2m_2^2$, and now from $x_\pm = (2C)^{-1}[-A \pm \sqrt{A^2 - 4BC}]$ it appears that both q_- and q_+ divide $2C$, whence

$$(1) \qquad 4C^2 = n^2 q_- q_+ \quad \text{with} \quad n \in \mathbb{Z},$$

$$(2) \qquad 4AC = 4C^2 \frac{p_- q_+ + q_- p_+}{q_- q_+} = n(p_- q_+ + q_- p_+),$$

$$(3) \qquad BC = 4C^2 \frac{p_- p_+}{q_- q_+} = n p_- p_+.$$

But $q_- q_+$, $p_- q_+ + q_- p_+$, and $p_- p_+$ are coprime, so C divides n, $|q_- q_+| \le 4|C|$ by (1), $|p_- q_+ + q_- p_+| \le 4|A|$ by (2), $|p_- p_+| \le 4|B|$ by (3), and $|p_- q_+ - q_- p_+| \le |q_- q_+||x_- - x_+| \le 4|\sqrt{A^2 - 4BC}|$. Now the end of the proof is here: $m_- m_+$ cannot exceed

$$4 \times \left(|A| + |B| + |C| + |\sqrt{A^2 - 4BC}|\right) \le c_0(\mathbf{X}) m_1^2 m_2^2,$$

so

$$H_0(\mathfrak{p}_+) + H_0(\mathfrak{p}_-) \le 2H_0(\mathfrak{p}_1) + 2H_0(\mathfrak{p}_2) + \log c_0(\mathbf{X}).$$

Step 3. This step shows that the quotient $\mathbf{X}(\mathbb{Q})$ is finite. Let $\mathbb{Q}^\times$ be the *multiplicative* group of nonvanishing rational numbers. Any number $p/q \in \mathbb{Q}$ can also be expressed as P/Q^2, whence the quotient

$$G = \mathbb{Q}^\times/(\mathbb{Q}^\times)^2 = \left\{(\pm 1) \times 2^{n_1} \cdot 3^{n_2} \cdot 5^{n_3} \cdots \text{with powers } n_i = 0 \text{ or } 1\right\}$$

is the direct sum of copies of $\mathbb{Z}_2$, one for the sign ± 1 and one for each prime p. The discussion is divided into three steps.

Item 1. The map $\varphi\colon \mathbf{X}(\mathbb{Q}) \to \mathbb{Q}^\times/(\mathbb{Q}^\times)^2$ defined by

$$\varphi(\mathfrak{p}) = \begin{cases} (\mathbb{Q}^\times)^2 & \text{if } \mathfrak{p} = (\infty, \infty) = \mathfrak{o}, \\ (e_1 - e_2)(e_1 - e_3)(\mathbb{Q}^\times)^2 & \text{if } \mathfrak{p} = (e_1, 0), \\ (x(\mathfrak{p}) - e_1)(\mathbb{Q}^\times)^2 & \text{otherwise} \end{cases}$$

is a homomorphism.

Proof. It is to be shown that $\varphi(\mathfrak{p}_1)\varphi(\mathfrak{p}_2) \subset \varphi(\mathfrak{p}_1 + \mathfrak{p}_2)(\mathbb{Q}^\times)^2$. Now $x(\mathfrak{p}) = x(-\mathfrak{p})$ so $\varphi(\mathfrak{p}) = \varphi(-\mathfrak{p}) = [\varphi(\mathfrak{p})]^{-1} \bmod (\mathbb{Q}^\times)^2$, so what you must prove is that $\varphi(\mathfrak{p}_1)\varphi(\mathfrak{p}_2)\varphi(\mathfrak{p}_3) = (\mathbb{Q}^\times)^2$ if $\mathfrak{p}_1 + \mathfrak{p}_2 + \mathfrak{p}_3 = 0$.

Case 0. $\mathfrak{p}_1 = \mathfrak{o}$. Then $\mathfrak{p}_3 = -\mathfrak{p}_2$ and $\varphi(\mathfrak{p}_1)\varphi(\mathfrak{p}_2)\varphi(\mathfrak{p}_3) = [\varphi(\mathfrak{p}_2)]^2 \in (\mathbb{Q}^\times)^2$.

Case 1. $\mathfrak{p}_1, \mathfrak{p}_2, \mathfrak{p}_3$ are all different from $\mathfrak{o}$ and from $(e_1, 0)$. Let $y = mx + n$ be the line passing through these points. Then

$$(x - e_1)(x - e_2)(x - e_3) - (mx + n)^2 = (x - x_1)(x - x_2)(x - x_3)$$

read backward with $x = e_1$ states that

$$\begin{aligned}\varphi(\mathfrak{p}_1)\varphi(\mathfrak{p}_2)\varphi(\mathfrak{p}_3) &= (x_1 - e_1)(x_2 - e_1)(x_3 - e_1)(\mathbb{Q}^\times)^2 \\ &= (me_1 + n)^2(\mathbb{Q}^\times)^2 \subset (\mathbb{Q}^\times)^2.\end{aligned}$$

Case 2. $\mathfrak{p}_1 = (e_1, 0)$ and $\mathfrak{p}_2, \mathfrak{p}_3 \neq \mathfrak{o}$. Then $\mathfrak{p}_2$ and $\mathfrak{p}_3$ are different from $\mathfrak{p}_1$ since $2(e_1, 0) = \mathfrak{o}$. As before,

$$(x - e_1)(x - e_2)(x - e_3) - (mx + n)^2 = (x - x_1)(x - x_2)(x - x_3)$$

so $x - e_1$ divides $mx + n = m(x - e_1)$, and canceling this factor produces $(x - e_2)(x - e_3) - m^2(x - e_1) = (x - x_2)(x - x_3)$. Now evaluate at $x = e_1$ to obtain

$$\begin{aligned}[\varphi(\mathfrak{p}_1)]^{-1} &= (e_1 - e_2)(e_1 - e_3)(\mathbb{Q}^\times)^2 = (e_1 - x_2)(e_1 - x_3)(\mathbb{Q}^\times)^2 \\ &= \varphi(\mathfrak{p}_2)\varphi(\mathfrak{p}_3).\end{aligned}$$

The proof is finished.

The duplication rule $\varphi(2\mathfrak{p}) = [\varphi(\mathfrak{p})]^2 \subset (\mathbb{Q}^\times)^2$ permits you to pass to the quotient, legitimizing the induced homomorphism $\varphi\colon \mathbf{X}(\mathbb{Q})/2\mathbf{X}(\mathbb{Q}) \to \mathbb{Q}^\times/(\mathbb{Q}^\times)^2$.

Item 2. Now write $\varphi = \varphi_1$ and form the allied map φ_2 with e_2 in place of e_1. Then $\psi = \varphi_1 \bigotimes \varphi_2\colon \mathbf{X}(\mathbb{Q})/\,2\mathbf{X}(\mathbb{Q}) \to \mathbb{Q}^\times/(\mathbb{Q}^\times)^2 \bigotimes \mathbb{Q}^\times/(\mathbb{Q}^\times)^2$ is 1:1.

Proof. If $\mathfrak{p} \in \mathbf{X}(\mathbb{Q})$ is different from $\mathfrak{o}$, $(e_1, 0)$, and $(e_2, 0)$, then $\psi(\mathfrak{p}) = \text{id}$ precisely when $x(\mathfrak{p}) - e_1$ and $x(\mathfrak{p}) - e_2$ are squares in $\mathbb{Q}$, and $(x - e_1)(x - e_2)(x - e_3) = y^2$ implies that $x(\mathfrak{p}) - e_3$, too, is a square. Now use step 1 to confirm that $\mathfrak{p} \in 2\mathbf{X}(\mathbb{Q})$. The same things works if, for example, $\mathfrak{p} = (e_1, 0)$. Then $\varphi_1(\mathfrak{p}) = (e_1 - e_2)(e_1 - e_3)(\mathbb{Q}^\times)^2 \subset (\mathbb{Q}^\times)^2$, and likewise $\varphi_2(\mathfrak{p}) = (e_1 - e_2)(\mathbb{Q}^\times)^2$, so $e_1 - e_1 = 0$, $e_1 - e_2$, and $e_1 - e_3$ are squares, with the same result as before.

Item 3. The image of $\mathbf{X}(\mathbb{Q})/2\mathbf{X}(\mathbb{Q})$ under ψ is contained in a finite group of the form $\bigoplus \mathbb{Z}_2 \subset \mathbb{Q}^\times/(\mathbb{Q}^\times)^2$, with one copy of $\mathbb{Z}_2$ for the sign ± 1 and one copy for each of the prime factors of the discriminant $d = (e_1 - e_2)^2(e_1 - e_3)^2(e_3 - e_1)^2$.

Proof. Let $\mathfrak{p} = (x, y) \in \mathbf{X}(\mathbb{Q})$ be distinct from $\mathfrak{o} = (\infty, \infty)$ and distinguish cases according as $\mathfrak{p}$ is a half-period or not.

Case 1. $\mathfrak{p}$ is not a half-period. Let $n = n_1, n_2, n_3 \in \mathbb{Z}$, positive or negative, be the powers of the fixed prime p appearing in the numbers $x(\mathfrak{p}) - e$, these being expressed in lowest terms, and note that $(x - e_1)(x - e_2)(x - e_3) \in (\mathbb{Q}^\times)^2$ forces $n_1 + n_2 + n_3 \equiv 0 \bmod 2$. Now $x(\mathfrak{p}) \in \mathbb{Q}$ is fixed, independently of $e \in \mathbb{Z}$, with the result that $n_1 = n_2 = n_3 \equiv 0 \bmod 2$ if any one of them is negative. Then

$$\psi(\mathfrak{p}) = [x(\mathfrak{p}) - e_1](\mathbb{Q}^\times)^2 \bigotimes [x(\mathfrak{p}) - e_2](\mathbb{Q}^\times)^2 = (\mathbb{Q}^\times)^2 \bigotimes (\mathbb{Q}^\times)^2.$$

If now $n_1 > 0$ and p *does not divide* the discriminant d, then it cannot appear in the numerator of $x(\mathfrak{p}) - e_2 = x(\mathfrak{p}) - e_1 + e_1 - e_2$, that is, $n_2 = 0$ and likewise $n_3 = 0$, with the implication that $n_1 \equiv 0 \bmod 2$ and $\psi(\mathfrak{p}) = (\mathbb{Q}^\times)^2 \otimes (\mathbb{Q}^\times)^2$, if no prime p dividing d appears.

Case 2. $x(\mathfrak{p}) = e_1, e_2$, or e_3. Then $\varphi_1(\mathfrak{p})$ and $\varphi_2(\mathfrak{p})$ are products of $e_1 - e_2$, $e_2 - e_3$, and $e_3 - e_1$, and these are prime to p unless the latter divides d. The proof of item 3 is finished.

The finite character of $\mathbf{X}(\mathbb{Q})/2\mathbf{X}(\mathbb{Q})$ is plain from items 2 and 3.

Summary to date.(1) $\mathbf{X}(\mathbb{Q})/2\mathbf{X}(\mathbb{Q})$ is finite, (2)$H(2\mathfrak{p}) = 4H(\mathfrak{p})$, (3)$H(\mathfrak{p}_1 + \mathfrak{p}_2) + H(\mathfrak{p}_1 - \mathfrak{p}_2) = 2H(\mathfrak{p}_1) + 2H(\mathfrak{p}_2)$, (4)$\mathbf{X}(\mathbb{Q}) \bigcap [\mathfrak{p}\colon H(\mathfrak{p}) \le h < \infty]$ is finite.

Step 4. The finite generation of $\mathbf{X}(\mathbb{Q})$ now follows from (1), (2), (3), and (4).

Proof. Choose $h < \infty$ so large that $\mathbf{X}(\mathbb{Q}) \bigcap \{\mathfrak{p}\colon H(\mathfrak{p}) \le h < \infty\}$ contains some member of each of the finite number of classes of $\mathbf{X}(\mathbb{Q})/2\mathbf{X}(\mathbb{Q})$. It is a fact that these special points of $\mathbf{X}(\mathbb{Q})$ generate the whole. If not, you could pick a point $\mathfrak{p} \in \mathbf{X}(\mathbb{Q})$ that cannot be so produced, and of smallest possible height $H(\mathfrak{p})$. Let $\mathfrak{p}_0$ be the special point in the class of $\mathfrak{p}$ and write $\mathfrak{p} = \mathfrak{p}_0 + 2\mathfrak{q}_0$. Then either $H(\mathfrak{p} + \mathfrak{p}_0) \le H(\mathfrak{p}) + H(\mathfrak{p}_0)$ or else $H(\mathfrak{p} - \mathfrak{p}_0) \le H(\mathfrak{p}) + H(\mathfrak{p}_0)$ in accordance with (3). Take, for instance, the positive sign. Then $\mathfrak{p} + \mathfrak{p}_0 = 2\mathfrak{p}_0 + 2\mathfrak{q}_0 = 2\mathfrak{q}$, and

$$\begin{aligned} 4H(\mathfrak{q}) = H(2\mathfrak{q}) = H(\mathfrak{p} + \mathfrak{p}_0) &\le H(\mathfrak{p}) + H(\mathfrak{p}_0) \\ &\le H(\mathfrak{p}) + h \quad \text{(since } \mathfrak{p}_0 \text{ is special)} \\ &< 2H(\mathfrak{p}) \quad \text{(since } \mathfrak{p} \text{ is not).} \end{aligned}$$

But now $\mathfrak{q}$ has to be special, $H(\mathfrak{p})$ being the smallest height of any nonspecial point, and that is contradictory, $\mathfrak{p}$ being the sum $\mathfrak{q} + \mathfrak{q} - \mathfrak{p}_0$ of three special points. The proof is finished and isn't it remarkable!

References

Abel, N. H. 1826. Beweis der Unmöglichkeit algebraische Gleichungen von höheren Graden als dem vierten allgemeinen aufzulösen. *J. reine angew. Math.* 1: 67–84. Reprinted in *Oeuvres Complètes*, vol. 1, 66–94. Grondahl, Christiania, Sweden, 1881.

Abel, N. H. 1827. Recherches sur les fonctions elliptiques. *J. reine angew. Math.* 2: 101–81. Reprinted in *Oeuvres Complètes*, vol. 1, 263–388. Grondahl, Christiania, Sweden, 1881.

Ablowitz, M. J., and Clarkson, P. A. 1991. *Solitons, Nonlinear Evolution Equations, and Inverse Scattering.* London Math. Soc. Lecture Notes Series, no. 149. London Math. Soc., London.

Ahlfors, L. V. 1973. *Conformal Invariants: Topics in Geometric Function Theory.* McGraw-Hill, New York.

Ahlfors, L. V. 1979. *Complex Analysis.* 3d ed. McGraw-Hill, New York.

Airault, H.; McKean, H.; and Moser, J. 1977. Rational and elliptic solutions of the Korteweg–de Vries equation and a related many body problem. *Comm. Pure Applied Math.* 30: 95–148.

Akhiezer, N. I. 1970. *Elements of the Theory of Elliptic Functions.* Nauka, Moscow. English translation, AMS, Providence, RI, 1990.

Almkvist, G., and Berndt, B. 1988. Gauss, Landen, Ramanujan, the arithmetic–geometric mean, ellipses, π, and the Ladies Diary. *Am. Math. Mon.* 95: 585–608.

Andrews, G. 1976. *The Theory of Partitions.* Encyclopedia of Mathematics and Its Applications, vol. 2. Addison-Wesley, Reading, MA.

Andrews, G. A. 1989. On the proofs of the Rogers–Ramanujan identities. In *q-Series and Partitions*, ed. D. Stanton, 1–14. Springer-Verlag, New York.

Arnold, V. I. 1978. *Mathematical Methods of Classical Mechanics.* Trans. K. Vogtmann and A. Weinstein. Springer-Verlag, New York.

Artin, E. 1953. *Galois Theory.* Notre Dame Math. Lect. Notes, no. 2. Notre Dame, South Bend, IN.

Artin, E. 1964. *The Gamma Function.* Holt, Rinehart and Winston, New York.

Atkin, A. O. L., and O'Brien, J. N. 1967. Some properties of $p(n)$ and $c(n)$ modulo powers of 13. *Trans. AMS* 126: 442–59.

Ayoub, R. 1982. On the nonsolvability of the general polynomial. *Am. Math. Mon.* 89: 397–401.

Ayoub, R. 1984. The lemniscate and Fagnano's contributions to elliptic integrals. *Arch. Hist. Exact Sci.* 29: 131–49.

Baker, A. 1971a. Imaginary quadratic fields with class number two. *Ann. of Math.* (2) 94: 139–52.
Baker, A. 1971b. On the class number of imaginary quadratic fields. *Bull. AMS* 77: 678–84.
Baker, A. 1975. *Transcendental Number Theory.* Cambridge University Press, Cambridge.
Baker, H. F. 1897. *Abel's Theorem and the Allied Theory Including the Theory of Theta Functions.* Cambridge Univeristy Press, Cambridge. Reprinted 1995.
Baker, H. F. 1911. Function. *Encyclopedia Britannica* 11: 301–29.
Bashmakova, I. G. 1981. Arithmetic of algebraic curves from Diophantus to Poincaré. *Hist. Math.* 8: 393–416.
Bateman, H. 1953. *Higher Transcendental Functions.* McGraw-Hill, New York.
Beardon, A. F. 1983. *The Geometry of Discrete Groups.* GTM 91. Springer-Verlag, New York.
Bell, E. T. 1937. *Men of Mathematics.* Simon and Schuster, New York.
Belokolos, E. D.; Bobenko, A. I.; Enol'skii, V.; Its, A.; and Matveev, V. 1994. *Algebraic–Geometric Approach to Nonlinear Integrable Systems.* Springer Series in Nonlinear Dynamics. Springer-Verlag, New York.
Belokolos, E. D., and Enol'skii, V. 1994. Reduction of theta functions and elliptic finite-gap potentials. *Acta Appl. Math.* 36: 87–117.
Berndt, B. 1991. *Ramanujan's Notebooks.* Part 3. Springer-Verlag, New York.
Bernoulli, J. 1713. *Ars Conjectandi.* Reprinted by Culture et Civilization, Bruxelles, 1968.
Betti, E. 1851. Sopra la risolubilita per radicali delle equazioni algebriche irredutibiliti di grado primo. *Ann. Sci. mat. fis.* 2: 5–19. Reprinted in *Opere Matematiche*, vol. 1, 17–27. Ulrico Hoepli, Milano, 1903.
Betti, E. 1852. Sulla risoluzione delle equazioni algebriche. *Ann. Sci. mat. fis.* 3: 49–115. Reprinted in *Opere Matematiche*, vol. 1, 31–80. Ulrico Hoepli, Milano, 1903.
Bezout, E. 1779. *Théorie générale des équations algébriques.* Impr. de PhD Pierres, Paris.
Birch, B. 1973. Some calculations of modular relations. In *Modular Functions of One Variable*, ed. W. Kuyk, Springer Lecture Notes in Math., no. 320, 175–86. Springer, Berlin.
Birkhoff, G. 1973. *A Source Book in Classical Analysis.* Harvard University Press, Cambridge, MA.
Birkhoff, G., and Maclane, S. 1948. *Modern Algebra.* Macmillan, New York.
Bliss, G. A. 1933. *Algebraic Functions.* AMS Colloquium Publications, vol. 16. AMS, Providence, RI.
Borel, A.; Chowla, S.; Herz, C. S.; Iwasawa, K.; and Serre, J. P. 1966. *Seminar on Complex Multiplication.* Springer Lecture Notes in Mathematics, no. 21. Springer-Verlag, New York.
Borevich, Z. I., and Shafarevich, I. R. 1966. *Number Theory.* Academic Press, New York.
Borwein, J. M., and Borwein, P. B. 1987. *Pi and the AGM.* Wiley, New York.
Bos, H.; Kers, C.; Oort, F.; and Raven, D. 1987. Poncelet's closure theorem. *Expositiones Mathematicae* 5: 289–364.
Brioschi, F. 1858. Sulla risoluzione delle equazioni del quinto grado. *Annali Math.* 1: 326.
Brioschi, F. 1878. Über die Auflösung der Gleichungen von fünften Grade. *Math. Ann.* 13: 109–60.

Briot, C., and Bouquet, J. C. 1875. *Théorie des fonctions elliptiques.* Gauthier-Villars, Paris.
Buhler, W. K. 1981. *Gauss: A Biographical Study.* Springer-Verlag, New York.
Byrd, P., and Friedman, M. 1954. *Handbook of Elliptic Integrals for Engineers and Physicists.* Springer, Berlin.
Cardano, G. 1545. *Artis Magnae sive de regvlis algebraicis.* English translation: *The Great Art, or the Rules of Algebra.* Translated and edited by T. R. Witmer. MIT Press, Cambridge, MA, 1968.
Carlitz, L., and Subbarao, M. 1972. A simple proof of the quintuple product identity. *Proc. AMS* 32: 42–4.
Cassels, J. W. S. 1966. Diophantine equations with special reference to elliptic curves. *J. London Math. Soc.* 41, 193–291.
Cayley, A. 1853. Note on the porism of the in- and circumscribed polygon. *Philosophical Magazine (4th ser.)*: 99–103, 376–7. Reprinted in *Collected Papers*, vol. 2, 138–44.
Cayley, A. 1859. Sixth memoir upon quantics. *Phil. Trans. Roy. Soc.* 149: 61–91. Reprinted in *Collected Papers*, vol. 2, 561–91.
Cayley, A. 1861. On the porism of the in- and circumscribed polygon. *Phil. Trans. Roy. Soc.* 149: 225–39. Reprinted in *Collected Papers*, vol. 4, 292–308.
Cayley, A. 1895. *An Elementary Treatise on Elliptic Functions.* Constable, London. Reprinted by Dover, New York, 1961.
Cayley, A. 1889–97. *The Collected Mathematical Papers.* 13 vols. Cambridge University Press, Cambridge.
Chalal, J. S. 1988. *Topics in Number Theory.* Plenum, New York.
Chazy, J. 1911. Sur les équations différentielles du troisième et d'ordre supérieur dont l'intégrale générale a ses points critiques fixes. *Acta Math.* 34: 317–85.
Clemens, C. H. 1980. *A Scrapbook of Complex Function Theory.* Plenum, New York.
Cohen, H. 1992. Elliptic curves. In *From Number Theory to Physics*, ed. M. Waldschmidt et al., 212–37. Springer-Verlag, New York.
Cohen, P. 1984. On the coefficients of the transformation polynomials for the elliptic modular function. *Math. Proc. Cambridge Phil. Soc.* 95: 389–402.
Cohn, H. 1985. *Introduction to the Construction of Class Fields.* Cambridge University Press, Cambridge.
Cole, F. N. 1887. Klein's Ikosaeder. *Amer. Journal of Math.* 9: 45–61.
Conway, J. H., and Norton, S. P. 1979. Monstrous moonshine. *Bull. London Math. Soc.* 11, 308–39.
Copson, E. T. 1935. *Theory of Functions of a Complex Variable.* Oxford University Press, Oxford.
Cox, D. A. 1984. The arithmetic–geometric mean of Gauss. *L'Enseig. Math.* 30: 275–330.
Cox, D. A. 1989. *Primes of the Form $x^2 + ny^2$: Fermat, Class Field Theory, and Complex Multiplication.* Wiley, New York.
Coxeter, H. S. M. 1963. *Regular Polytopes.* Dover, New York.
Coxeter, H. S. M. 1980. *Introduction to Geometry.* 2d ed. Wiley, New York.
Coxeter, H. S. M., and Moser, W. O. J. 1980. *Generators and Relations for Discrete Groups.* 4th ed. Springer-Verlag, New York.
Dedekind, R. 1877. Schreiben an Herrn Borchardt über die Theorie der elliptischen Modulfunktionen. *J. reine angew. Math.* 83: 265–92.
Deuring, M. 1958. Die Klassenkörper der komplexen Multiplikation. *Enzy. Math. Wiss. Algebra und Zahlentheorie* 10 (11): 1–60.

Dieudonne, J. 1978. *Abrégé d'histoire des mathématiques 1700–1900.* Hermann et Cie., Paris.
Dieudonne, J. 1985. *History of Algebraic Geometry.* Wadsworth, Belmont, CA.
Dirichlet, P. G. L. 1840. Auszug aus einer Akademie der Wissenschaften zu Berlin am 5ten März 1840 vorgelesenen Abhandlung. *J. reine angew. Math.* 21: 98–100.
Dirichlet, P. G. L. 1889. *Werke.* 2 vols. Reimer, Berlin. Reprinted by Chelsea, New York, 1969.
Dirichlet, P. G. L. 1894. Vorlesungen über Zahlentheorie. 4th ed. Vieweg, Braunschweig.
Drazin, P. G., and Johnson, R. S. 1989. *Solitons: An Introduction.* Cambridge University Press, New York.
Dym, H., and McKean, H. 1972. *Fourier Series and Integrals.* Academic Press, New York.
Edwards, H. 1984. *Galois Theory.* Springer-Verlag, New York.
Ehrenpreis, L. 1994. Singularities, functional equations, and the circle method. In *The Rademacher Legacy to Mathematics*, ed. G. Andrews, D. Bressoud, and L. A. Parsen, Contemporary Math., vol. 166, 35–80. AMS, Providence, RI.
Eichler, M., and Zagier, D. 1982. On the zeros of the Weierstrass $\wp$-function. *Math. Ann.* 258: 399–407.
Eisenstein, G. 1847. Beitrage zur Theorie der elliptischen Funktionen. *J. reine angew. Math.* 35: 135–274.
Eisenstein, G. 1975. *Mathematische Werke.* 2 vols. Chelsea, New York.
Euler, L. 1743. De summis serierum reciprocarum ex .... *Miscellania Berolinensia* 7: 172–92. Reprinted in *Opera Omnia*, ser. 1, vol. 14, 138–55.
Euler, L. 1748. *Introductio in analysin infinitorum.* Marcum-Michaelem Bousquet, Lausanne. English translation: *Introduction to Analysis of the Infinite.* Trans. J. D. Blantan. Springer-Verlag, New York, 1988.
Euler, L. 1752. Elementa doctrinae solidorum. *Novi comm. acad. sci. Petrop.* 4: 109–40. Reprinted in *Opera Omnia*, ser. 1, vol. 26, 71–93.
Euler, L. 1763. Consideratio formularum quarum integratio per arcus sectionum conicarum absolvi potest. *Novi comm. acad. sci. Petrop.* 8: 129–49. Reprinted in *Opera Omnia*, ser. 1, vol. 20, 235–55.
Euler, L. 1782. De miris proprietatibus curvae elasticae sub aequatione $y = \int xx\,dx/\sqrt{(1-x^4)}$ contentae. *Acta academiae scientiarum Petrop.* 2: 34–61. Reprinted in *Opera Omnia*, ser. 1, vol. 21, 91–118.
Euler, L. 1911–76. *Opera Omnia.* Teubner, Leipzig.
Fagnano, G. 1750. Metodo per misurare la lemniscata. *Giorn. lett. d'Italia* 29: 258. Reprinted in *Opere Matematiche*, vol. 2, papers 32, 33, and 34. Allerighi e Segati, Milano, 1911.
Farkas, H. M., and Kopeliovich, Y. 1993. New theta constant identities. *Israel J. Math.* 82: 133–41.
Farkas, H. M., and Kopeliovich, Y. 1994. New theta constant identities II. *Proc. AMS* 123: 1009–20.
Farkas, H. M., and Kra, I. 1992. *Riemann Surfaces.* 2d ed. Springer-Verlag, New York.
Farkas, H. M., and Kra, I. 1993. Automorphic forms for subgroups of the modular group. *Israel J. Math.* 82: 87–131.
Fermat, P. 1894. Letter to Carcavi, communicated to Huygens, Aug. 14, 1659. In *Oeuvres*, vol. 2, 431–6. Gauthier-Villars, Paris, 1894.
Ford, J. 1972. *Automorphic Functions.* 2d ed. Chelsea, New York.
Forster, O. 1981. *Lectures on Riemann Surfaces.* Trans. B. Gilligan. GTM 81. Springer-Verlag, New York.

Fricke, R. 1913. Elliptische Funktionen. *Encyclopedia der Math. Wiss.* 2 (2): 177–348.
Fricke, R. 1916, 1922. *Die elliptischen Funktionen und ihre Anwendungen.* 2 vols. Teubner, Leipzig.
Fricke, R. 1924–8. *Lehrbuch der Algebra.* 3 vols. Vieweg, Braunschweig.
Fricke, R., and Klein, F. 1926. *Vorlesungen über die Theorie der Automorphen Functionen.* Teubner, Leipzig.
Gaal, L. 1973. *Classical Galois Theory, with Examples.* Chelsea, New York.
Garvan, F. 1995. A combinatorial proof of the Farkas–Kra theta function identities and their generalizations. *Journal Math. Anal. Appl.* 195: 354–75.
Gauss, C. F. 1797. Elegantiores integralis $\int(1-x^4)^{-1/2}dx$ proprietates et de curva lemniscata. In *Werke*, vol. 3, 404–32.
Gauss, C. F. 1799. Arithmetisch Geometrisches Mittel. In *Werke*, vol. 3, 361–432.
Gauss, C. F. 1801. *Disquisitiones arithmeticae.* Fleischer, Leipzig. Reprinted in *Werke*, vol. 1. English translation: Yale University Press, New Haven, 1966. Reprinted by Springer-Verlag, New York, 1986.
Gauss, C. F. 1827. Disquisitiones generales circa superficies curvas. In *Werke*, vol. 4, 217–58.
Gauss, C. F. 1870–1929. *Werke.* 11 vols. Königliche Gesellschaft der Wissenschaft, Göttingen. Reprinted by Olms, Hildescheim, 1981.
Goldfeld, D. 1985. Gauss' class number problem for imaginary quadratic fields. *Bull. AMS* 13: 23–37.
Gradshteyn, I. S., and Ryzhik, I. M. 1994. *Tables of Integrals, Sums, and Products.* 5th ed. Ed. A. Jeffrey. Academic Press, New York.
Gray, J. 1986. *Linear Differential Equations and Group Theory from Riemann to Poincaré.* Birkhäuser, Boston.
Green, M. L. 1978. On the analytic solution of the equation of fifth degree. *Compos. Math.* 37: 233–41.
Greenberg, M. J. 1974. An elementary proof of the Kronecker–Weber theorem. *Am. Math. Mon.* 81: 601–7.
Greenhill, A. G. 1892. *The Applications of Elliptic Functions.* Macmillan Press, London.
Griffiths, P. 1976. Variations on a theme of Abel. *Invent. Math.* 35: 321–90.
Griffiths, P., and Harris, J. 1978a. On Cayley's explicit solution of Poncelet's porism. *L'Enseignement Math.* 24: 31–40.
Griffiths, P., and Harris, J. 1978b. *Principles of Algebraic Geometry.* Wiley, New York.
Grosswald, E. 1984a. *Representations of Integers as Sums of Squares.* Springer-Verlag, New York.
Grosswald, E. 1984b. *Topics from the Theory of Numbers.* 2d ed. Birkhäuser, Boston.
Halphen, G. H. 1886–91. *Traité des fonctions elliptiques et de leurs applications.* 3 vols. Gauthier-Villars, Paris.
Hancock, H. 1958. *Lectures on the Theory of Elliptic Functions.* Dover, New York.
Hannah, M. 1928. The modular equations. *Proc. Lon. Math. Soc.* 28: 46–52.
Hardy, G. H. 1937. The Indian mathematician Ramanujan. *Am. Math. Mon.* 44: 137–55. Reprinted in *Ramanujan*, 1–22.
Hardy, G. H. 1940. *Ramanujan.* Cambridge University Press, Cambridge. Reprinted by Chelsea, New York, 1960.
Hardy, G. H., and Wright, E. M. 1979. *An Introduction to the Theory of Numbers.* 5th ed. Oxford University Press, Oxford.
Hartshorne, R. 1977. *Algebraic Geometry.* GTM 52. Springer-Verlag, New York.
Hasse, H. 1967. *Vorlesungen über Klassenkörpertheorie.* Physica-Verlag, Würzberg.

Hasse, H. 1980. *Number Theory.* Grunderlehren der mathematischen Wissenchaften 229. Springer-Verlag, New York.

Heegner, K. 1952. Diophantische Analysis und Modulfunktionen. *Math. Z.* 56: 227–53.

Hermite, C. 1859. Considérations sur la résolution algébrique de l'équation du cinquième degré. *C.R. Acad. Sci. Paris* 46: 508–15. Reprinted in *Oeuvres*, vol. 1, 3–9. Gauthier-Villars, Paris, 1905–17.

Hermite, C. 1862. Note sur la théorie des fonctions elliptiques. In *Calcul différentiel et calcul integral de Lacroix.* Mallet-Bachelier, Paris. Reprinted in *Oeuvres*, vol. 2, 125–238. Gauthier-Villars, Paris, 1908.

Hermite, C. 1865–6. Sur l' équation du cinquième degré. *C.R. Acad. Sci. Paris.* Reprinted in *Oeuvres*, vol. 2, 347–424. Gauthier-Villars, Paris, 1908.

Hermite, C. 1877–82. Sur quelques applications des fonctions elliptiques. *C. R. Acad. Paris.* Reprinted in *Oeuvres*, vol. 3, 266–418. Gauthier-Villars, Paris, 1912.

Hilbert, D. 1897. Die Theorie der algebraischen Zahlkörper. *Jahresbericht D. Math. Ver.* 4: 175–546.

Hirschhorn, M. D. 1985. A simple proof of Jacobi's two-square theorem. *Am. Math. Mon.* 92: 579–80.

Hirschhorn, M. D. 1987. A simple proof of Jacobi's four-square theorem. *Proc. AMS* 101: 436–8.

Hodgkin, A. 1971. *The Conduction of the Nervous Impulse.* Liverpool University Press, Liverpool.

Houzel, C. 1978. Fonctions elliptiques et intégrales abeliennes. In *Abrégé d'histoire des mathématiques, 1700–1900*, ed. J. Dieudonne, 1–112. Hermann et Cie., Paris.

Hurwitz, A., and Courant, R. 1964. *Vorlesungen über allgemeine Funktionentheorie und elliptische Funktionen.* Springer, Berlin.

Husemoller, D. 1987. *Elliptic Curves.* GTM 111. Springer-Verlag, New York.

Imayoshi, Y., and Taniguchi, M. 1992. *An Introduction to Teichmüller Spaces.* Springer-Verlag, New York.

Ireland, K., and Rosen, M. 1990. *A Classical Introduction to Modern Number Theory.* 2d ed. GTM 84. Springer-Verlag, New York.

Ivory, J. 1796. A new series for the rectification of the ellipsis; together with some observations on the evolution of the formula $(a^2 + b^2 - 2ab\cos\varphi)^n$. *Trans. Royal Soc. Edinburgh* 4: 177–90.

Jacobi, C. G. J. 1827. Extraits de deux lettres de M. Jacobi de l'Université de Königsberg à M. Schumacher. *Astronomische Nachrichten*, vol. 6, no. 123. Reprinted in *Gesammelte Werke*, vol. 1, 29–36. Reprinted by Chelsea, New York, 1969.

Jacobi, C. G. J. 1828. Note sur la décomposition d'un nombre donné en quatre carrés. *J. reine angew. Math.* 3: 191. Reprinted in *Gesammelte Werke*, vol. 1, 247.

Jacobi, C. G. J. 1829. *Fundamenta nova theoriae functionarum ellipticarum.* Bornträger, Königsberg. Reprinted in *Gesammelte Werke*, vol. 1, 49–239.

Jacobi, C. G. J. 1832. Anzeige von Legendre *Theorie des fonctions elliptiques. J. reine angew. Math.* 8: 413–17. Reprinted in *Gesammelte Werke*, vol. 1, 373–82.

Jacobi, C. G. J. 1838. Theorie der elliptischen Funktionen aus der Eigenschaften der Thetareihen abgeleitet. In *Gesammelte Werke*, vol. 1, 497–538.

Jacobi, C. G. J. 1847. Über die Differentialgleichung welcher die reihen $1 \pm 2q + 2q^4 \pm 2q^9 +$ etc., $2q^{1/4} + 2q^{9/4} + 2q^{25/4} +$ etc. genüge leisten. *J. reine angew. Math.* 36: 97–112. Reprinted in *Gesammelte Werke*, vol. 2, 171–90.

Jacobi, C. G. J. 1866. *Vorlesungen über Dynamik.* Reprinted as *Gesammelte Werke*, supplementary vol. Reimer, Berlin, 1884.

Jacobi, C. G. J. 1881–91. *Gesammelte Werke.* 7 vols. Reimer, Berlin. Reprinted by Chelsea, New York, 1969.

Janusz, G. 1973. *Algebraic Number Fields.* Academic Press, New York.

Jones, G. A., and Singerman, D. 1987. *Complex Functions.* Cambridge University Press, Cambridge.

Jordan, C. 1870. *Traité de substitutions et des équations algébriques.* Gauthier-Villars, Paris.

Jordan, C. 1894. *Cours d'analyse de l'École Polytechnique.* Vol. 2, *Calcul Intégral,* Gauthier-Villars, Paris.

Kaltofen, E., and Yui, N. 1984. On the modular equation of order 11. In *Third MACSYMA User's Conference, Proceedings,* ed. V. E. Golden, 472–85. General Electric, Schenectady, NY.

Kazdan, J. L. 1985. *Prescribing the Curvature of a Riemannian Manifold.* CBMS Regional Conf. Series in Mathematics. AMS, Providence, RI.

Kepler, J. 1596. *Mysterium Cosmographicum.*

Kirwan, F. 1992. *Complex Algebraic Curves.* London Mathematical Society Students Texts, no. 23. Cambridge University Press, Cambridge.

Klein, F. 1882. Über eindeutige Funktionen mit linearen Transformationen in sich. *Math. Ann.* 19: 565–8; 20: 49–51.

Klein, F. 1884. *Vorlesungen über das Ikosaeder und die Auflösung der Gleichungen vom fünften Grade.* Teubner, Leipzig. English translation: *The Ikosahedron and the Solution of Equations of the Fifth Degree.* Trans. G. G. Morrice. Dover, New York, 1956. Reprinted with comments by Peter Slodowy, 1993.

Klein, F. 1893. *On Riemann's Theory of Algebraic Functions and Their Integrals.* Macmillan and Bowes, Cambridge.

Klein, F. 1908. *Elementarmathematik vom höheren Standpunkt aus.* Springer, Berlin. English translation: *Elementary Mathematics from an Advanced Standpoint.* Dover, New York, 1939.

Klein, F., and Fricke, R. 1890, 1892. *Vorlesungen über die Theorie der elliptischen Modulfunctionen.* Teubner, Leipzig.

Knapp, A. 1993. *Elliptic Curves.* Mathematical Notes 40. Princeton Univeristy Press, Princeton, NJ.

Knopp, M. 1970. *Modular Forms in Analytic Number Theory.* Markham, Chicago.

Koblitz, N. 1993. *Introduction to Elliptic Curves and Modular Forms.* 2d. ed. GTM 97. Springer-Verlag, New York.

Koebe, P. 1909–14. Über die Uniformizierung der algebraischen Kurven. *Math. Ann.* 67 (1909): 145–224; 69 (1910): 1–81; 72 (1912): 437–88; 75 (1914): 42–129.

Koike, M. 1994. Moonshine: A mysterious relationship between simple groups and automorphic forms. *AMS Translations* (2) 160: 33–45.

Kollros, L. 1949. *Evariste Galois.* Birkhäuser, Basel.

Korteweg, D. J., and de Vries, G. 1895. On the change of form of long waves advancing in a rectangular canal, and on a new type of long stationary waves. *Phil. Mag.* 39: 422–43.

Kovalevskaya, S. 1889. Sur le problème de la rotation d'un corps solide autour d'un point fixe. *Acta Math.* 12: 177–232.

Kra, I. 1972. *Automorphic Forms and Kleinian Groups.* Benjamin, Reading, MA.

Krazer, A. 1903. *Lehrbuch der Thetafunktionen.* Teubner, Leipzig. Reprinted by Chelsea, New York, 1970.

Krichever, I. M. 1980. Elliptic solutions of the KP equation and integrable systems of particles. *Funct. Anal.* 14: 45–54.

Kronecker, L. 1858. Sur la résolution de l'équation du cinquième degré. *C. R. Acad. Paris* 46: 1150–2. In *Werke*, vol. 4, 43–7.
Kronecker, L. 1895–1931. *Werke.* 5 vols. Teubner, Leipzig. Reprinted by Chelsea, New York, 1968.
Kronecker, L. 1901. *Vorlesungen über Zahlentheorie.* Teubner, Leipzig. Reprinted by Springer-Verlag, New York, 1978.
Lagrange, J. L. 1868. *Oeuvres.* Vol. 2. Gauthier-Villars, Paris.
Landen, J. 1771. A disquisition concerning certain fluents, which are assignable by the arcs of the conic sections; wherein are investigated some new and useful theorems for computing such fluents. *Philos. Trans. Royal Soc. London* 61: 298–309.
Landen, J. 1775. An investigation of a general theorem for finding the length of any arc of any conic hyperbola, by means of two elliptic arcs, with some other new and useful theorems deduced therefrom. *Philos. Trans. Royal Soc. London* 65: 283–9.
Landsberg, M. 1893. Zur Theorie der Gauss'schen Summen und der linearen Transformationen der Thetafunktionen. *J. reine. angew. Math.* 111: 234–53.
Lang, S. 1984. *Algebra.* 2d ed. Addison-Wesely, Reading, MA.
Lang, S. 1987. *Elliptic Functions.* 2d. ed. Springer, Berlin.
Lawden, D. F. 1989. *Elliptic Functions and Applications.* Applied Mathematical Sciences, vol. 80. Springer-Verlag, New York.
Lefschetz, S. 1921. On certain numerical invariants of algebraic varieties. *Trans. AMS* 22: 327–482.
Legendre, A. M. 1811. *Exercices de calcul intégral sur diverses ordres de transcendants.* Paris.
Legendre, A. M. 1825. *Traité des fonctiones elliptiques et des intégrales Euleriennes.* Paris.
Lehner, J. 1964. *Discontinuous Groups and Automorphic Functions.* Mathematical Surveys, no. 8. AMS, Providence, RI.
Leibowitz, A., and Rauch, H. E. 1973. *Elliptic Functions, Theta Functions, and Riemann Surfaces.* Williams and Wilkins, Baltimore, MD.
Lindeman, F. 1884,1892. Über die Auflösung algebraischer Gleichungen durch transzendente Functionen. Parts 1 and 2. *Göttingen Nach.* (1884): 245–8; (1892): 292–8.
Liouville, J. 1880. Leçons sur les fonctions doublement périodiques faites en 1847 par M. J. Liouville. *J. reine angew. Math.* 88: 277–310.
Luroth, J. 1876. Beweis eines Satzes über rationalen Curven. *Math. Ann.* 9: 163–5.
Magnus, W. 1974. *Noneuclidean Tessellations and Their Groups.* Academic Press, New York.
Massey, W. S. 1991. *A Basic Course in Algebraic Topology.* GTM 127. Springer-Verlag, New York.
Matthews, G. B. 1911. Number. *Encyclopedia Britannica* 29: 847–63.
Mazur, B. 1976. Modular curves and the Eisenstein ideal. *IHES Publ. Math.* 47: 33–186.
Mazur, B. 1986. Arithmetic on curves. *Bull. AMS* 14: 207–59.
McKean, H. P. 1970. Nagumo's equation. *Adv. Math.* 4: 209–23.
McKean, H. P. 1978. Integrable systems and algebraic curves. In *Global Analysis*, ed. M. Grmela and J. E. Marsden, Springer Lecture Notes in Mathematics, no. 755, 83–200. Springer, Berlin.
Mestre, J. F. 1992. Un exemple de courbes elliptiques sur $\mathbb{Q}$ de rang ≥ 15. *C. R. Acad. Sci. Paris* 314: 453–5.
Milnor, J. 1982. Hyperbolic geometry: The first 150 years. *Bull. AMS* 6: 9–24.

Mordell, L. J. 1917. On Mr. Ramanujan's empirical expansion of modular functions. *Proc. Camb. Phil. Soc.* 19: 117–24.
Mordell, L. J. 1922. On the rational solutions of the indeterminate equations of the third and fourth degree. *Proc. Camb. Phil. Soc.* 21: 179–92.
Mordell, L. J. 1969. *Diophantine Equations.* Academic Press, New York.
Moser, J. 1980. Various aspects of integrable Hamiltonian systems. *Progress in Math.* 8: 233–89.
Mostow, G. D. 1966. Quasi-conformal mappings in n-space and the rigidity of hyperbolic space-forms. *IHES Publ. Math.* 34: 203–14.
Mumford, D. 1975. *Curves and Their Jacobians.* University of Michigan Press, Ann Arbor, MI.
Mumford, D. 1976. *Algebraic Geometry I: Complex Projective Varieties.* Springer, Berlin.
Mumford, D. 1983. *Tata Lectures on Theta I.* Birkhäuser, Boston.
Mumford, D. 1984. *Tata Lectures on Theta II.* Birkhäuser, Boston.
Nagao, K., and Kouya, T. 1994. An example of an elliptic curve over $\mathbb{Q}$ with rank ≥ 21. *Proc. Japan Acad.* 70: 104–5.
Nagell, J. 1964. *Introduction to Number Theory.* Chelsea, New York. Originally printed by Almqvist and Wiskell, Stockholm, 1951.
Nagumo, J. S.; Arimoto, S.; and Yoshizawa, S. 1962. An active pulse transmission line simulating nerve axon. *Proc. IRE* 50: 2061–70.
Narasimhan, R. 1992. *Compact Riemann Surfaces.* Lectures in Mathematics ETH Zürich. Birkhäuser, Basel.
Neugebauer, O. 1957. *The Exact Sciences in Antiquity.* Brown University Press, Providence, RI. Reprinted by Dover, New York, 1969.
Neumann, C. 1859. De problemate quodam mechanica, quod ad primam integralium ultra-ellipticorum classem revocatur. *J. reine angew. Math.* 56: 54–66.
Neumann, C. 1865. *Vorlesungen über Riemanns Theorie der Abel'schen Integral.* Teubner, Leipzig.
Nevanlinna, R. 1970. *Analytic Functions.* Springer Grundlehren 162. Springer, Berlin.
Newman, D. J. 1985. A simplified version of the fast algorithms of Brent and Salamin. *Math. Comp.* 44: 207–10.
Novikov, S.; Manakov, S.; Pitaevskii, L.; and Zakharov, V. 1984. *Theory of Solitons: The Inverse Scattering Method.* Plenum, New York.
Oesterle, J. 1985. Nombre de classes des corps quadratiques imaginaires. In *Séminaire N. Bourbaki, 1983–1984*, Exp. 631, 309–23. *Astérisque*, vol. 121–2. CNRS, Paris.
Ore, O. 1965. *Cardano, the Gambling Scholar.* Dover, New York. Originally printed by Princeton University Press, Princeton, NJ, 1953.
Ore, O. 1957. *Niels Henrik Abel.* Chelsea, New York.
Peterson, H. 1932. Über die Entwicklungskoeffizienten der automorphen Formen. *Acta Math.* 58: 169–215.
Picard, E. 1879. Sur une propriété des fonctions entiers. *C. R. Acad. Sci. Paris* 88: 1024–7. Reprinted in *Oeuvres*, vol. 1, 19–22. CNRS, Paris, 1978.
Picard, E. 1882. Sur la réduction de intégrales abeliennes aux intégrales elliptiques. *C. R. Acad. Sci. Paris* 94: 1704–7. Reprinted in *Oeuvres*, vol. 3, 11–14. CNRS, Paris, 1978.
Pogorelov, A. V. 1967. *Differential Geometry.* Nordhoff, Groningen.
Poincaré, H. 1882. Theorie des groupes fuchsiens. *Acta Math.* 1: 1–62. Reprinted in *Oeuvres*, vol. 2, 108–68. Gauthier-Villars, Paris, 1946.

Poincaré, H. 1898. Les fonctions fuchsiennes et l'équation $\Delta u = e^u$. *J. de Math.* 4: 137–230. Reprinted in *Oeuvres*, vol. 2, 512–91. Gauthier-Villars, Paris, 1946.

Poincaré, H. 1901. Sur les propriétés arithmétiques des courbes algébriques. *J. de Math.* 7: 161–233. Reprinted in *Oeuvres*, vol. 5, 483–550. Gauthier-Villars, Paris, 1950.

Poincaré, H. 1907. Sur l'uniformisation des fonctions analytiques. *Acta Math.* 31: 1–63. Reprinted in *Oeuvres*, vol. 4, 70–139. Gauthier-Villars, Paris, 1950.

Poincaré, H. 1985. *Papers on Fuchsian Groups.* Trans. J. Stillwell. Springer-Verlag, New York.

Pollard, H. 1950. *The Theory of Algebraic Numbers.* Carus Math. Mono., no. 9. Wiley, New York.

Pólya, G. 1921. Über eine Aufgabe der Wahrscheinlichkeitsrechnung betreffend die Irrfahrt im Strassennetz. *Math. Ann.* 84, 149–60. Reprinted in *Collected Papers*, vol. 4, 69–80. MIT Press, Cambridge, MA, 1974–84.

Pólya, G. 1927. Elementaren Beweis einer Thetaformel. In *Sitz. der Phys. Math. Klasse*, 158–61. Preuss. Akad. der Wiss., Berlin. Reprinted in *Collected Papers*, vol. 1, 303–6. MIT Press, Cambridge, MA, 1974–84.

Poncelet, J. V. 1822. *Traité des propriétés projectives des figures.* Paris, 1822.

Rademacher, H. 1973. *Topics in Analytic Number Theory.* Springer-Verlag, New York.

Ramanujan, S. 1916. On certain arithmetical functions. *Trans. Camb. Phil. Soc.* 22: 159–84. Reprinted in *Collected Papers*, 136–62.

Ramanujan, S. 1927. *Collected Papers of Srinivasa Ramanujan.* Cambridge University Press, Cambridge.

Reid, C. 1970. *Hilbert.* Springer-Verlag, New York.

Reyssat, E. 1989. *Quelques aspects des surfaces de Riemann.* Birkhäuser, Boston.

Riemann, B. 1851. Grundlagen für eine allgemeine Theorie der Functionen einer veränderlichen complexen Grösse. Inaugural dissertation, Göttingen, 1851; reprinted, Göttingen, 1867. Reprinted in *Gesammelte mathematische Werke*, 35–80.

Riemann, B. 1857. Theorie der Abel'schen Functionen. *J. reine angew. Math.* 54: 88–142.

Riemann, B. 1858. *Elliptische Funktionen.* Teubner, Leipzig.

Riemann, B. 1990. *Gesammelte mathematische Werke and wissenschaftlicher Nachlass.* Springer, Berlin.

Ritt, J. F. 1923. Permutable rational functions. *Trans. AMS* 25: 399–448.

Robert, A. 1973. *Elliptic Curves.* Springer-Verlag, New York.

Roberts, J. 1977. *Elementary Number Theory: A Problem Oriented Approach.* MIT Press, Cambridge, MA.

Roch, G. 1865. Über die Anzahl der willkürlischen Constanten in algebraischen Functionen. *J. reine angew. Math.* 64: 372–6.

Rogers, L. J. 1894. Second memoir on the expansion of certain infinite products. *Proc. Lon. Math. Soc.* 25: 318–43.

Rosen, M. I. 1981. Abel's theorem on the lemniscate. *Am. Math. Mon.* 88: 387–93.

Rosen, M. I. 1995. Niels Hendrik Abel and equations of the fifth degree. *Am. Math. Mon.* 102: 495–505.

Ruffini, P. 1799. *Teoria generale delle equazioni in cui si dimostra impossibile la soluzione algebraica delle equazioni generali di grado superiore al quarto.* Tommaso d'Aquino, Bologna. Reprinted in *Collected Works*, vol. 1, 159–86.

Ruffini, P. 1950. *Collected Works.* Vol. 1. Math. Circle of Palermo. Tipografia Matematica, Palermo, Italy.

Sansone, G., and Gerretsen, G. 1969. *Lectures on the Theory of Functions of a Complex Variable.* 2 vols. Wolters-Noordhoff, Groningen.
Sarnak, P. 1990. *Some Applications of Modular Forms.* Cambridge University Press, Cambridge.
Schoenberg, I. J. 1981. A direct derivation of a Jacobian identity from elliptic functions. *Am. Math. Mon.* 88: 616–18.
Schwarz, H. A. 1872. Über diejenigen Fälle, in welchen die Gaussische hypergeometrische Reihe eine algebraische Function ihres vierten Elementes darstellt. *J. reine angew. Math.* 75: 292–335.
Schwarz, H. A. 1893. *Formeln und Lehrsätze zum Gebräuche der elliptischen Funktionen, nach Vorlesungen und Aufzeichnungen des Herrn Prof. K. Weierstrass.* Springer, Berlin.
Segal, S. 1981. *Nine Introductions in Complex Analysis.* North-Holland Math. Studies, no. 53. North-Holland, Amsterdam.
Serre, J. P. 1973. *A Course in Arithmetic.* GTM 7. Springer-Verlag, Berlin.
Shafarevich, I. R. 1977. *Basic Algebraic Geometry.* Trans. K. A. Hirsch. Grundlehren 213. Springer, Berlin.
Shanks, D. 1951. A short proof of an identity of Euler. *Proc. AMS* 2: 747–9.
Shanks, D. 1958. Two theorems of Gauss. *Pacific Journal of Math.* 8, 609.
Shimura, G. 1971. *Introduction to the Arithmetic Theory of Automorphic Functions.* Princeton University Press, Princeton, NJ.
Siegel, C. L. 1936. Über die Classenzahl quadratischer Zahlkörper. *Acta Arith.* 1: 83–6.
Siegel, C. L. 1949. *Transcendental Numbers.* Ann. Math. Studies, no. 16. Princeton University Press, Princeton, NJ.
Siegel, C. L. 1969,1971,1973. *Topics in Complex Function Theory.* 3 vols. Wiley, New York.
Silverman, J. 1986. *The Arithmetic of Elliptic Curves.* GTM 106. Springer-Verlag, New York.
Silverman, J. 1994. *Advanced Topics in the Arithmetic of Elliptic Curves.* GTM 151. Springer-Verlag, New York.
Silverman, J., and Tate, J. 1992. *Rational Points on Elliptic Curves.* Springer-Verlag, New York.
Skolem, T. 1938. *Diophantische Gleichungen.* Ergeb. der Math. und ihrer Grenzgeb., vol. 5. Springer, Berlin. Reprinted by Chelsea, New York, 1950.
Sohnke, L. A. 1836. Equationumes modulares pro transformatione functionum ellipticarum. *J. reine angew. Math.* 16: 97–130.
Springer, G. 1981. *Introduction to Riemann Surfaces.* 2d ed. Chelsea, New York.
Stark, H. M. 1967. A complete determination of the complex number fields of class number one. *Michigan Math. J.* 14: 1–27.
Stark, H. M. 1969. On the "gap" in a theorem of Heegner. *Journal of Number Theory* 1: 16–27.
Stark, H. M. 1970. *An Introduction to Number Theory.* Markham, Chicago.
Stark, H. M. 1971. A transcendence theorem for class number problems. *Ann. of Math.* (2) 94: 153–73.
Stark, H. M. 1973. Class-numbers of complex quadratic fields. In *Modular Functions of One Variable*, ed. W. Kuyk, Springer Lect. Notes, no. 320, 153–74. Springer, Berlin.
Stewart, I. 1973. *Galois Theory.* Chapman Hall, London.
Stillwell, J. 1989. *Mathematics and Its History.* Springer-Verlag, New York.

Stillwell, J. 1992. *Geometry of Surfaces.* Springer-Verlag, New York.

Stillwell. J. 1994. *Elements of Algebra.* Springer-Verlag, New York.

Swinnerton-Dyer, H. P. F. 1988. Congruence properties of $\tau(n)$. In *Ramanujan Revisited: Proceedings of the Centenary Conference, University of Illinois at Urbana-Champain, June 1–5, 1987*, ed. G. E. Andrews et al., 289–311. Academic Press, Boston.

Tannery, J., and Molk, J. 1893–1902. *Elements de la théorie des fonctions elliptiques.* 4 vols. Gauthier-Villars, Paris. Reprinted by Chelsea. New York, 1972.

Tartaglia, N. 1546. *Quesiti et inventioni diverse.* Fascimile of 1554 edition. Ed. A. Masotti. Ateneo di Brescia, Brescia.

Terras, A. 1985. *Harmonic Analysis on Symmetric Spaces and Applications.* Vol. 1. Springer-Verlag, New York.

Thompson, J. G. 1979. Some numerology between the Fischer–Griess monster and the elliptic modular function. *Bull. Lon. Math. Soc.* 11: 352–3.

Treibich, A. 1994. New elliptic potentials. *Acta Appl. Math.* 36: 27–48.

Treibich, A., and Verdier, J. 1990. Solitons elliptiques. In *Progress in Math.*, ed. P. Cartier et al., vol. 88, 437–80. Birkhäuser, Boston.

Van der Pol, B. 1951. On a non-linear partial differential equation satisfied by the logarithm of the Jacobian theta-functions, with arithmetical applications. Parts 1 and 2. *Indagationes Math.* 13: 261–84.

Vladut, S. G. 1991. *Kronecker's Jugendtraum and Modular Functions.* Studies and Developments of Modern Mathematics, vol. 2. Gordon and Breach, New York.

Waerden, B. L. van der. 1970. *Modern Algebra.* 2 vols. Trans. F. Blum and J. R. Schulenberger. 5th ed. Ungar, New York.

Walker, R. 1978. *Algebraic Curves.* Springer-Verlag, New York.

Watson, G. N. 1929. Theorems stated by Ramanujan (7): Theorems on continued fractions. *J. London Math. Soc.* 4: 39–48.

Watson, G. N. 1933. The marquis and the land-agent; a tale of the eighteenth century. *Mathematical Gazette* 17: 5–17.

Watson, G. N. 1938. Ramanujans Vermutung über Zerfallungsanzahlen. *J. reine angew. Math.* 179: 97–128.

Watson, G. N. 1939. Three triple integrals. *Oxford Quart. Journal Math.* 10: 266–76.

Weber, H. 1886. Theorie der Abel'schen Körper. *Acta Math.* 8: 193–263.

Weber, H. 1891. *Elliptische Funktionen und algebraische Zahlen.* Vieweg, Braunschweig.

Weber, H. 1908. *Lehrbuch der Algebra.* Vol. 3. 2d ed. Vieweg, Braunschweig. Reprinted by Chelsea, New York, 1961.

Weierstrass, K. 1915. *Vorlesungen über die Theorie der elliptischen Funktionen.* Vol. 5 of *Math. Werke*. Mayer and Müller, Berlin.

Weil, A. 1928. L'arithmetique sur les courbes algébriques. *Acta Mathematica* 52: 281–315. Reprinted in *Oeuvres Scientifiques, Collected Papers*, vol. 1, 11–45. Springer-Verlag, New York, 1980.

Weil, A. 1930. Sur un théorème de Mordell. *Bull. Sci. Math.* 54: 182–91. Reprinted in *Oeuvres Scientifiques, Collected Papers*, vol. 1, 47–56. Springer-Verlag, New York, 1980.

Weil, A. 1974. Two lectures on number theory, past and present. *L'Enseign. Math.* 20: 87–110. Reprinted in *Oeuvres Scientifiques, Collected Papers*, vol. 3, 279–302. Springer-Verlag, New York, 1980.

Weil, A. 1975. *Elliptic functions according to Eisenstein and Kronecker.* Springer, Berlin.

Weil, A. 1983. *Number Theory: An Approach Through History.* Birkhäuser, Boston.

Weyl, H. 1940. *Algebraic Theory of Numbers.* Ann. Math. Studies, no. 1. Princeton University Press, Princeton, NJ.
Weyl, H. 1955. *The Concept of a Riemann Surface.* 3d ed. Addison-Wesley, Reading, MA. Translation of *Die Idee der Riemannschen Fläche*. Teubner, Berlin, 1913.
Whittaker, E. T. 1937. *A Treatise on the Analytical Dynamics of Particles and Rigid Bodies.* Cambridge University Press, Cambridge.
Whittaker, E. T., and Watson, G. N. 1963. *A Course in Modern Analysis.* 4th ed. Cambridge University Press, Cambridge.
Wijngaarden, A. van 1953. On the coefficients of the modular invariant $J(\tau)$. *Indag. Math.* 15: 389–400.
Winquist, L. 1969. An elementary proof of $p(11m + 6) \equiv 0 \bmod 11$. *J. Combin. Theory* 6: 56–9.
Wyman, B. F. 1972. What is a reciprocity law? *Am. Math. Mon.* 79: 571–86.
Yui, N. 1978. Explicit form of the modular equation. *J. reine angew. Math.* 299: 185–200.
Zagier, D. 1984. L-series of elliptic curves, the Birch–Swinnerton-Dyer conjecture, and the class number problem of Gauss. *Notices AMS* 31: 739–43.
Zagier, D. 1992. Introduction to modular forms. In *From Number Theory to Physics*, ed. M. Waldschmidt et al., 238–91. Springer-Verlag, New York.
Zuckerman, H. S. 1939. Computation of the smaller coefficients of $J(\tau)$. *Bull. AMS* 45: 917–19.

Index